FUEL INJECTION

Installation, Performance, Tuning & Modification

FUEL INJECTION
Installation, Performance, Tuning & Modification

Nagma Ali

2025
Agri Horti Press
New Delhi-110 084

ISBN: 978-93-9296-961-4 (Hardbound)

Published By

Agri Horti Press

New Delhi

E-mail: agrihortipress@gmail.com

Preface

In the realm of automotive performance, fuel injection stands as a cornerstone technology, revolutionizing the way engines deliver power, efficiency, and reliability. As enthusiasts, engineers, and tuners seek to unlock the full potential of their vehicles, mastering the intricacies of fuel injection installation, performance tuning, and modification becomes essential to achieving optimal engine performance and unleashing the thrill of the drive.

This handbook, aptly titled "Fuel Injection: Installation, Performance, Tuning, Modification," serves as a comprehensive guide to navigating the dynamic landscape of fuel injection systems in automotive applications. Drawing upon the collective expertise of automotive engineers, tuners, and enthusiasts, this handbook offers a wealth of practical insights, technical know-how, and best practices to empower enthusiasts at every level of experience.

From the fundamentals of fuel injection technology to advanced tuning techniques and performance modifications, "Fuel Injection: Installation, Performance, Tuning, Modification" provides readers with a systematic approach to optimizing engine performance and achieving desired outcomes. Through detailed explanations, step-by-step instructions, and case studies, we explore the intricacies of fuel system design, injector selection, engine calibration, and after market upgrades tailored to specific vehicle platforms and driving preferences.

Moreover, this handbook delves into the nuances of fuel injection installation, maintenance, and troubleshooting, addressing common challenges and pitfalls encountered by enthusiasts and professionals alike. By demystifying complex concepts and providing practical solutions, we aim to equip readers with the knowledge and skills to confidently tackle fuel injection projects and unleash the full potential of their vehicles.

As we embark on this journey of exploration and discovery, it is our aspiration that "Fuel Injection: Installation, Performance, Tuning, Modification" will serve as an indispensable resource for automotive enthusiasts, tuners, mechanics, and performance shops alike. Whether you are seeking to enhance engine power, improve fuel efficiency, or fine-tune vehicle performance for the track or street, this handbook invites you to delve into the exciting world of fuel injection and experience the thrill of unlocking peak performance.

Nagma Ali

Contents

1

Gasoline direct injection

1. Introduction

The basic goals of the automotive industry; a high power, low specific fuel consumption, low emissions, low noise and better drive comfort. With increasing the vehicle number, the role of the vehicles in air pollution has been increasing significantly day by day. The environment protection agencies have drawn down the emission limits annually. Furthermore, continuously increasing price of the fuel necessitates improving the engine efficiency. Since the engines with carburetor do not hold the air fuel ratio close to the stoichiometric at different working conditions, catalytic converter cannot be used in these engines. Therefore these engines have high emission values and low efficiency. Electronic controlled Port Fuel Injection (PFI) systems instead of fuel system with carburetor have been used since 1980's. In fuel injection systems, induced air can be metered precisely and the fuel is injected in the manifold to air amount. By using the lambda sensor in exhaust system, air/fuel ratio is held of stable value. Fuel systems without electronic controlled it is impossible to comply with the increasingly emissions legislation.

If port fuel injection system is compared with carburetor system, it is seen that has some advantages. These are;

1. Lower exhaust emissions.
2. Increased volumetric efficiency and therefore increased output power and torque. The carburetor venturi prevents air and, in turn, volumetric efficiency decrease.
3. Low specific fuel consumption. In the engine with carburetor, fuel cannot be delivered the same amount and the same air/fuel ratio per cycle, for each cylinder.
4. The more rapid engine response to changes in throttle position. This increases the drive comfort.
5. For less rotation components in fuel injection system, the noise decreases (Heywood, 2000; Ferguson, 1986).

Though the port fuel injection system has some advantages, it cannot be meet continuously increased the demands about performance, emission legislation and fuel economy, at the present day (Stone, 1999). The electronic controlled gasoline direct injection systems were started to be used instead of port fuel injection system since 1990's.

The Gasoline Direct Injection (GDI) engines give a number of features, which could not be realized with port injected engines: avoiding fuel wall film in the manifold, improved accuracy of air/fuel ratio during dynamics, reducing throttling losses of the gas exchange by stratified and homogeneous lean operation, higher thermal efficiency by stratified operation and increased compression ratio, decreasing the fuel consumption and CO_2 emissions, lower heat losses, fast heating of the catalyst by injection during the gas expansion phase, increased performance and volumetric efficiency due to cooling of air charge, better cold-start performance and better the drive comfort (Zhao et al., 1999; Karamangil, 2004; Smith et al., 2006).

2. The Performance and Exhaust Emissions of The Gasoline Direct Injection (GDI) Engine

2.1 Performance of the GDI Engine

The parameters that have the greatest influence on engine efficiency are compression ratio and air/fuel ratio. The effect of raising compression ratio is to increase the power output and to reduce the fuel consumption. The maximum efficiency (or minimum specific fuel consumption) occurs with a mixture that is weaker than stoichiometric (Çelik, 2007). Because the port fuel injection engines work at stoichiometric air/fuel ratio, it is impossible to see more improvement in the fuel economy. In these engines, the compression ratio is about 9/1-10/1. To prevent the knock, the compression ratio cannot be increased more. For the same engine volume, the increasing volumetric efficiency also raises the engine power output.

GDI engine operate with lean mixture and unthrottled at part loads, this operation provide significantly improvements in fuel economy. At full load, as the GDI engine operates with homogeneous charge and stoichiometric or slightly rich mixture, this engine gives a better power output (Spicher et al., 2000). In GDI engine, fuel is injected into cylinder before spark plug ignites at low and medium loads. At this condition, Air/Fuel (A/F) ratio in cylinder vary, that is, mixture in front of spark plug is rich, in other places is lean. In all cylinder A/F ratio is lean and A/F ratio can access until 40/1. In homogeneous operation, fuel starts injecting into cylinder at intake stroke at full loads (Alger et al., 2000; Çınar, 2001). The fuel, which is injected in the intake stoke, evaporates in the cylinder. The evaporation of the fuel cools the intake charge. The cooling effect permits higher compression ratios and increasing of the volumetric efficiency and thus higher torque is obtained (Muñoz et al., 2005). In the GDI engines, compression ratio can gain until 12/1 (Kume, 1996). The knock does not occur because only air is compressed at low and medium loads. At full load, since fuel is injected into cylinder, the charge air cool and this, in turn, decreases knock tendency.

Since the vehicles are used usually in urban traffic, studies on improving the urban driving fuel economy have increased. Engines have run usually at part loads (low and medium loads) in urban driving. Volumetric efficiency is lower at part loads, so engine effective compression ratio decreases (e.g. from 8/1 to 3/1-4/1), engine efficiency decreases and fuel consumption increases. The urban driving fuel economy of the vehicles is very high (Çelik, 1999). Distinction between the highway fuel economies of vehicles is very little. As majority of the life time of the vehicles pass in the urban driving, the owners of the vehicles prefer the vehicles of which the urban driving fuel economy is low.

At full load, as the GDI engine operate with throttle, only a small reduction of fuel consumption can be obtained to the PFI engine. There is the more fuel economy potential at part load. At compression stroke, since air is given the cylinders without throttle for stratified charge mode, pumping losses of the GDI engine is minimum at part loads, Fig.1 (Baumgarten, 2006). The improvements in thermal efficiency have been obtained as a result of reduced pumping losses, higher compression ratios and further extension of the lean operating limit under stratified combustion conditions at low engine loads. In the DI gasoline engines, fuel consumption can be decreased by up to 20%, and a 10% power output improvement can be achieved over traditional PFI engines (Fan et al., 1999).

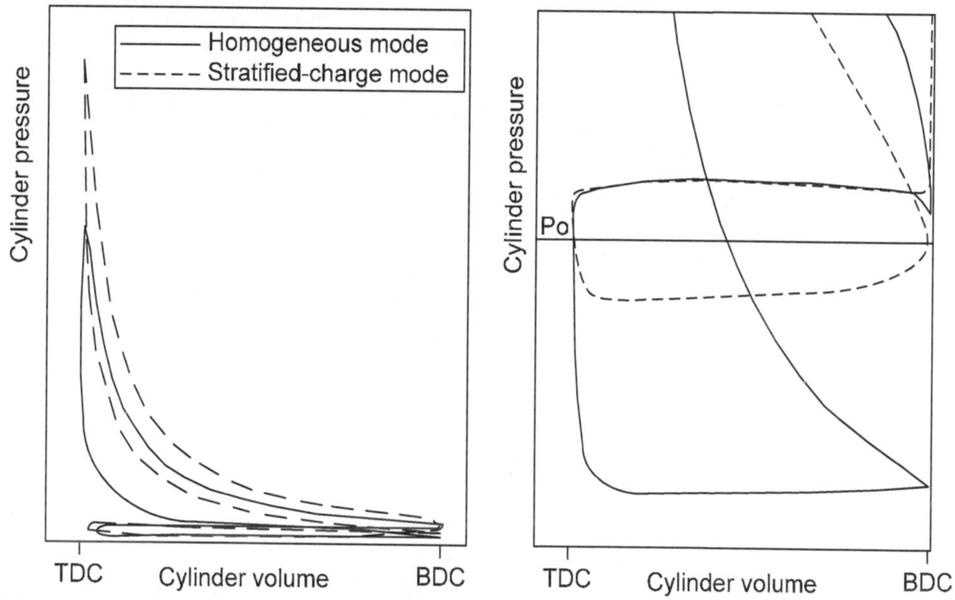

Fig. 1. Reduction of throttle losses in the stratified-charge combustion (Baumgarten, 2006).

The CO_2 emissions, which are one of the gases, bring about the global warming. To decrease CO_2 emitted from vehicles, it is required to decrease fuel consumption. Downsizing (reduction of the engine size) is seen as a major way of improving fuel consumption and reducing greenhouse emissions of spark ignited engines. In the same weight and size, significant decreases in CO_2 emissions, more power and higher break mean effective pressure can be obtained. GDI engines are very suitable for turbocharger applications. The use of GDI engine with turbocharger provides also high engine knock resistance especially at high load and low engine speed where PFI turbocharged engines are still limited (Lecointe & Monnier, 2003; Stoffels, 2005). Turbocharged GDI engines have showed great potential to meet the contradictory targets of lower fuel consumption as well as high torque and power output (Kleeberg, 2006).

In GDI engine, by using twin charging system drive comfort, engine torque and power can be increased for the same engine size. For example, Volkswagen (VW) has used the dual charging system in TSI (twin charged stratified injection) engine. The system includes a roots-type supercharger as well as a turbocharger. The supercharger is basically an air compressor. A mechanical device driven off the engine's crankshaft, it employs rotating vanes which spin in opposite directions to compress air in the engine's intake system. The high and constant torque is obtained at wide range speed by activate supercharger at low speeds and turbo charger at high speeds (Anon, 2006).

In Table 1, it is given specifications of the two different engines belonging to the 2009 model VW Passat vehicle, for example. TSI engine urban driving fuel economy is 18% lower than that of PFI engine. CO_2 emission is 12% lower than that of PFI engine. Although TSI engine swept volume is lower than PFI engine, power and torque is higher by 20% and 35%, respectively (Table 1). As engine torque is maximum at interval 1500-4000 1/min, shifting is not necessary at the acceleration and thus drive comfort increase (Anon, 2009).

Engine Type	Swept volume	Max. Power	Max. Torque	Mixture formation system	Fuel economy (urban driving) L/100km	Fuel economy (highway driving) L/100km	CO_2 emission g/km
Gasoline engine	1.6 L	75 kW 5600 1/min	148 Nm 3800 1/min	PFI (port fuel injection)	10,5	6,0	179
TSI gasoline engine	1.4 L	90 kW 5000 - 5500 1/min	200 Nm 1500 - 4000 1/min	GDI (Gasoline direct injection)	8,6	5,5	157

Table 1. Comparison of the GDI and PFI engines (Anon, 2009).

2.2 Exhaust Emissions of the GDI Engine

CO emission is very low in GDI engine. CO varies depending on air /fuel ratio. CO is high at rich mixtures. Since GDI engines operate with lean mixture at part loads and stoichiometric mixture at full load, CO is not a problem for these engines. In GDI engine, due to the wetting of the piston and the cylinder walls with liquid fuel, HC emission can increase. Hydrocarbon (HC) emissions are a function of engine temperature and, therefore it can rise during cold start. The cold starts characteristics vary depending on the fuel distribution characteristics, the in-cylinder air motion, fuel vaporization, and fuel-air mixing (Gandhi et al., 2006).

During cold-start of a GDI engine, homogeneous operation can be employed due to a higher exhaust gas temperature resulting in a shorter time for catalyst light-off, and lower engine out HC emissions (Gandhi et al., 2006). Gasoline engines do not emit soot emission normally. Soot emission can occur at very rich mixtures. However, the GDI engines emit soot at stratified-charge operation, as in–cylinder can be areas with very rich mixtures. In addition, in GDI engine, if mixture formation do not realize at full loads due to rich mixture, the soot emission can increase. NOx emission is maximum at high cylinder temperatures and at λ =1.1. As torque output rises, temperatures rise and, in turn, the engine-out NOx emissions display an increase. NOx emissions increase especially at full load.

2.3 The Emission Control in GDI Engine

Environmental legislation determines the limits for exhaust emissions in the spark ignition engines. It is required the treatment of the exhaust gases to meet these limits. The three-way catalytic converter show high performance for converting the CO, HC and NOx in the engines with operation at $\lambda=1.0$. But, NOx cannot be completely converted harmless gases at lean mixture operation. Therefore, engines with lean mixture also require a NOx storage type catalytic converter to convert the NOx.

The two catalytic converters are successively used in GDI engine exhaust system. The one is Pre-catalytic converter (Three Way Converter -TWC). This converter has little volume and is connected close to the engine. The other is main catalytic converter which combines a NOx catalyst and a TWC. This converter has higher volume than the pre-catalytic converter and is connected not close to the engine. The Pre-catalytic converter convert the CO, HC and NOx to harmless gases (CO_2, H_2O and N_2) at $\lambda=1.0$. However, when engine operates at stratified mode with lean mixture, NOx cannot be converted to nitrogen. In such cases, NOx is sent to main catalytic converter (Anon, 2002).

In the NOx storage type catalytic converter, the components such as Ba and Ca are used for NOx conversion at lean mixtures. These components provide NOx to storage. At $\lambda=1.0$, the operation of the NOx converter resembles three way converter. At lean mixtures, NOx conversion is realized in three stages: NOx accumulation, NOx release and conversion. Nitrogen oxides reacts chemically with barium oxide (BaO) and thus barium nitrate ($Ba(NO_3)_2$ forms. (NOx storage stage). Then, to convert, engine is operated momentarily in the rich homogeneous mode. Thanks to rich mixture, there is CO in exhaust system. The barium nitrate reacts chemically with CO and, as a result of this CO_2, BaO and NO arise (NOx release stage). And then, NO reacts chemically with CO and, N_2 and CO_2 form (conversion stage). NOx storage converter can storage the NOx at temperatures of 250-500°C (Anon, 2002; Bauer, 2004). An exhaust gas recirculation system is necessary, as the NOx aftertreatment systems do not reach the conversion rates of $\lambda = 1$ concepts. With the exception at the highest loads, exhaust gas recirculation (EGR) is used extensively to control NOx emissions (Alkidas, 2007).

To meet the valid emission limits and diagnose the pre and main catalyst faults, and provide optimum engine operation 4 sensors (3 lambda sensor and 1 exhaust gas temperature sensor) are used in the exhaust system. The wide band lambda sensor upstream of pre-catalyst determines residual oxygen value in exhaust gas. The required λ for homogeneous lean operation can be controlled by this sensor. For each catalytic converter two lambda sensors (upstream and downstream sensor) are used. The faults of the pre and main converters can be diagnosed by signal of dual sensors. The temperature sensor is used to determine the temperature of the NOx catalyst (Küsell et al., 1999).

3. The Mixture Formation and Operation Modes in The GDI Engine

3.1 The Mixture Formation

The air-fuel mixture in the gasoline engines is prepared in-cylinder and out-cylinder. While the mixture in the engine with carburetor and port fuel injection is prepared out-cylinder, mixture in the gasoline direct injection engines is prepared in-cylinder, Figure 2.

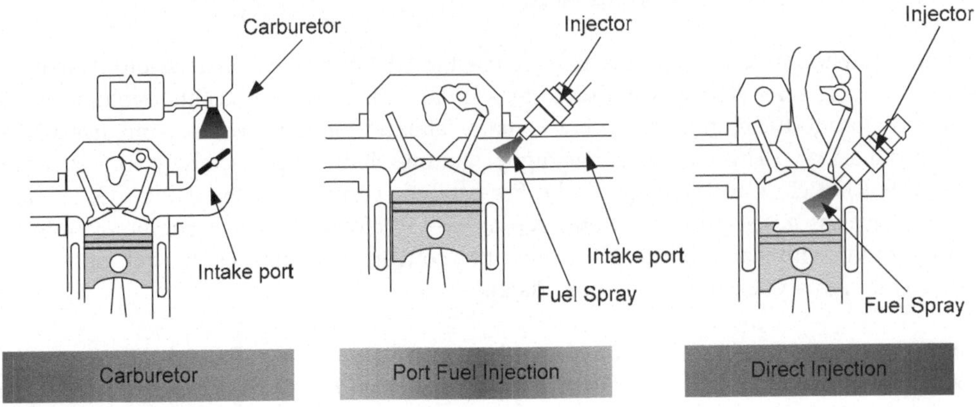

Fig. 2. The mixture formation systems in the gasoline engines.

In place of PFI engines where the fuel is injected through the port, in GDI engines, the fuel is injected directly into cylinders at a high pressure. During the induction stroke, only the air flows from the open intake valve and it enters into the cylinder. This ensures better control of the injection process and particularly provides the injection of fuel late during the compression stroke, when the intake valves are closed (Sercey et al., 2005). The acting of the intake system as a pre-vaporizing chamber is an advantage in the PFI engines (Rotondi, 2006). As the lack of time to fuel vaporize in GDI engines, the fuel is injected into the cylinder at a very high pressure to help the atomization and vaporization process. The duration for injection timing is little; advanced injection timing causes piston wetting and retarded injection timing decrease sufficient time for fuel-air mixing (Gandhi et al., 2006). In the PFI engine, a liquid film is formed in the intake valve area of the port, which causes delayed fuel vaporization. Especially during cold start, it is necessary to increase fuel amount for the ideal stoichiometric mixture. This "overfueling" leads to increasing HC emissions during cold start. Alternatively, injecting the fuel directly into the combustion chamber avoids the problems such as increasing HC and giving the excess fuel to engine (Hentschel, 2000).

To the GDI engines, it is implemented the two basic charge modes, stratified and homogeneous charge. At the partial load conditions, stratified charge (late injection) is used, that is, fuel is injected during the compression stroke to supply the stratified charge. The engine can be operated at an air-fuel ratio exceeding 100 and fully unthrottled operation is possible, but the engine is throttled slightly in this zone and the air-fuel ratio is controlled to range from 30 to 40 in order to introduce a large quantity of Exhaust Gas Recirculation (EGR) and to supply the vacuum for the brake system. A homogeneous charge (early injection) is preferred for the higher load conditions, that is, fuel is injected during the intake stroke so as to provide a homogeneous mixture. In most of this mode, the engine is operated under stoichiometric or a slightly rich condition at full load. In the lowest load conditions in this mode, the engine is operated at homogeneous lean conditions with a air-fuel ratio of from 20 to 25 for further improvement of fuel economy (Kume, 1996). During operation with homogeneous charge the adjustment of engine load is done by throttling while during operation with stratified charge the engine runs with unthrottled conditions and engine load is adjusted by fuel/air-equivalence ratio (Spicher et al., 2000). Fig.3 shows the homogeneous (early injection) and stratified-charge modes (late injection).

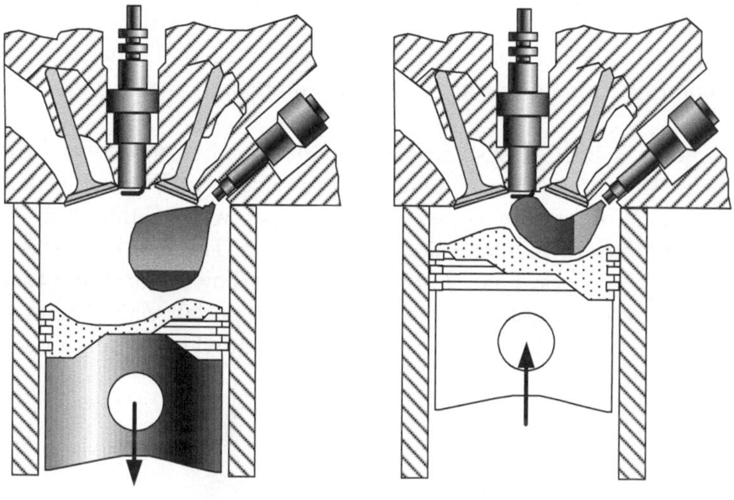

Fig. 3. Homogeneous and stratified-charge mode.

In the stratified operation, three combustion systems are used to form an ignitable mixture near spark plug at the instant ignition. These are the wall-guided, air-guided and spray-guided combustion systems, Fig. 4. The distinction between the different concepts is the used method with which the fuel spray is transported near the spark plug (Ortmann et al., 2001).

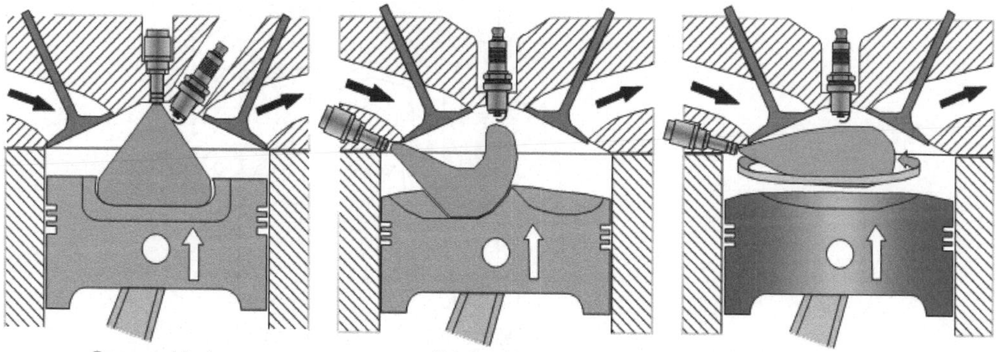

Fig. 4. The wall-guided, air-guided and spray-guided combustion systems at stratified charge (Stefan, 2004).

Wall-Guided combustion system: The fuel is transported to the spark plug by using a specially shaped piston surface. As the fuel is injected on the piston surface, it cannot completely evaporate and, in turn, HC and CO emissions, and fuel consumption increase. To use this system alone is not efficient.

Air-Guided combustion system: The fuel is injected into air flow, which moves the fuel spray near the spark plug. The air flow is obtained by inlet ports with special shape and air speed is controlled with air baffles in the manifold. In this technique, fuel does not wet the piston and cylinder. Most of stratified-charge GDI engines use a large-scale air motion (swirl or tumble) as well as specially shaped piston a surface in order to keep the fuel spray compact and to move it to the spark plug (Baumgarten, 2006). In the air-guided and wall-guided combustion systems the injector is placed remote to the spark plug.

VW direct injection combustion system is a combination of two systems– wall guided and air guided –by tumble flow. This system is less sensitive against the cyclic variations of airflow. This combustion system shows advantages as well in the stratified and in the homogenous mode. Injector is intake-side placed, Fig. 5. The fuel is injected to the piston under given angle. The piston has two bowls. The fuel bowl is on the intake-side; the air bowl is on exhaust-side. Tumble flow is obtained by special shaped intake port (Stefan, 2004). The fuel is guided simultaneously via air and fuel bowl to the spark plug.

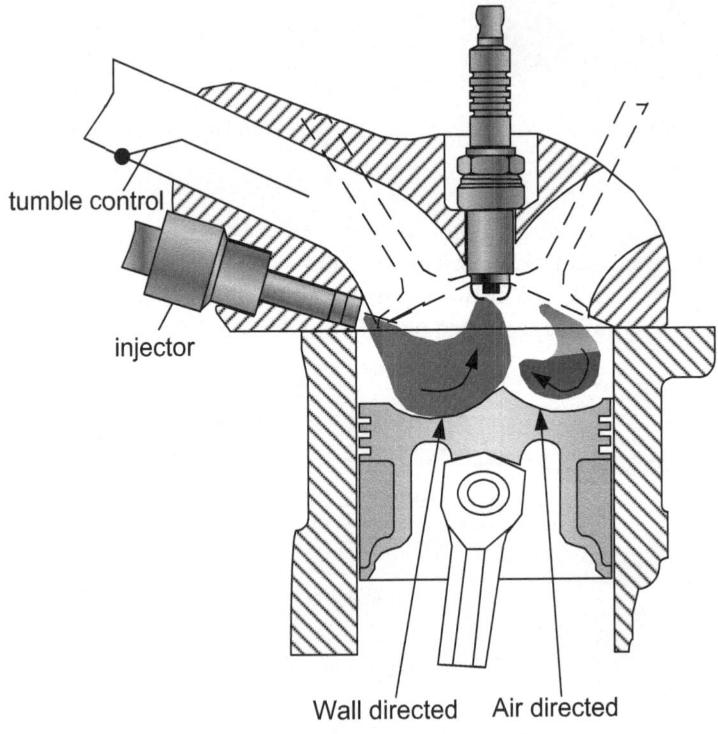

Fig. 5. Volkswagen FSI engine air-wall guided combustion system (Anon, 2002).

Spray-Guided combustion system: In the spray-guided technique fuel is injected near spark plug where it also evaporates. The spray-guided technique theoretically has the highest efficiency. The spray guided combustion process requires advanced injector systems such as piezo injection. This technique has some advantages: reduced wall wetting, increased stratified operation region, less sensitive to in-cylinder air flow, less sensitive to cylinder to cylinder variation and reduced raw HC emissions. Reported disadvantages are spark plug reliability (fouling) and poor robustness (high sensitivity to variation in ignition &injection timing) (Cathcart & Railton, 2001). Mercedes-Benz developed a new spray-guided combustion system. This system has the Stratified-Charged Gasoline Injection (CGI) engine with Piezo injection technology. The spray-guided injection achieves better fuel efficiency than conventional wall-guided direct injection systems. The main advantage of the CGI engine is obtained at the stratified operating mode. During this mode the engine is run with high excess air and thus excellent fuel efficiency is provided. Multiple injections extend this lean-burn operating mode to higher rpm and load ranges, too. During each compression stroke, a series of injections is made spaced just fractions of a second apart. This allows the better mixture formation and combustion, and lower fuel consumption (Website 1, 2010).

3.2 The Operating Modes

GDI engine operates at different operating modes depending on load and engine speed for a stable and efficient engine operation. These engines have three basic operating modes, stratified with an overall lean mixture, homogeneous with lean mixtures and homogeneous with stoichiometric mixtures. The engine is operated with the stratified, homogeneous lean and homogeneous stoichiometric modes; at low load and speed, at medium load and speed and at high load and speed, respectively. Fig. 6 shows an example of the GDI operating modes depending on engine load and speed.

The engine control unit continually chooses the one among the operating modes. Each mode is determined by the air-fuel ratio. The stoichiometric air-fuel ratio for petrol (gasoline) is 14.7:1 by weight, but ultra lean mode (stratified-charge) can involve ratios as high as 65:1. These mixtures are much leaner than conventional mixtures and reduce fuel consumption considerably. **Stratified-charge mode** is used for light-load running conditions, at constant or low speeds, where no acceleration is required. The fuel has to be injected shortly before the ignition, so that the small amount of air-fuel mixture is optimally placed near the spark plug. This technique enables the usage of ultra lean mixtures with very high air-fuel ratio, impossible with traditional carburetors or even port fuel injection (Website 2, 2010). The lean burn increases the NOx emissions. In this mode, EGR is actuated in order to decrease NOx. The area of stratified operation is limited by load and speed. At high load, the mixture in the stratified mode can be too rich, and thus soot can form. At high speed, it is impossible to provide sufficient stratification due to high turbulence in the cylinder. Therefore, at the higher load and speed range, the engine is operated in homogeneous mode to obtain low emissions and high torque (Küsell et al., 1999).

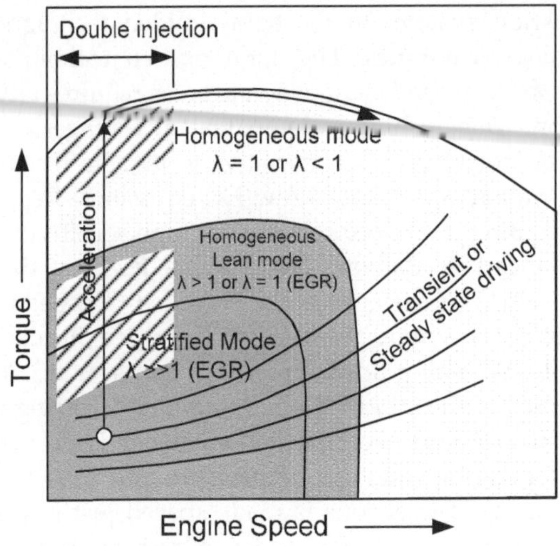

Fig. 6. GDI engine operating modes depending on load and speed (Küsell et al., 1999).

Homogeneous mode is used for acceleration, full load and high engine speeds. The air-fuel mixture is homogenous and the ratio is stoichiometric or slightly richer than stoichiometric. As the fuel is injected during the intake stroke, there is sufficient time for air-fuel mixture formation. In this mode, as engine operates with stoichiometric mixture, NOx emission decrease and therefore EGR is not activated.

In the transient areas the engine can be operated in homogeneous lean mode to optimize fuel consumption. **Homogeneous lean mode** is activated for moderate load and speed conditions. In this mode, fuel is injected during the intake stroke. The air-fuel mixture is homogeneous. The A/F ratio is lean or stoichiometric. As engine operates with lean mixture, NOx emission increase and therefore EGR is activated. The one another operating mode is homogeneous-stratified mode. This mode is used at acceleration conditions when passing from stratified to homogeneous mode. The two stage injection (double injection) is implemented. The primary injection is performed at intake stroke and majority of fuel is injected. The remaining fuel is injected at secondary injection and compression stroke. Double injection is made to reduce soot emissions and to decrease fuel consumption at low engine speeds in the transition area between stratified and homogeneous operation. The double injection can also be used to heat rapidly catalyst with a lean stratified operation mode. At low speed and high loads, combustion duration is long and temperature is high. Therefore, the engine tends to knock. In this homogeneous charge mode, by using dual injection at full load and by decreasing the ignition timing knock can be prevented.

4. The Fuel Supply and Engine Management System of the GDI Engine

4.1 The Fuel Supply System

The fuel systems for GDI engine require high fuel pressure levels. Fuel injection pressure is between 4 to 13 MPa (the actual trend is to increase the level of pressure). This pressure is higher than PFI engine pressure values ranging from 0.25 to 0.45 MPa. The higher pressures lead to a higher penetration and a better atomization. Although too high injection pressures increase atomization, but an over penetrating can cause the wall wetting problems (Rotondi, 2006).

In GDI engines, fuel supply system consists of the fuel tank, low-pressure pump, fuel filter, high-pressure pump, fuel rail, high-pressure sensor, injector and fuel pressure control valve (Figure 7). The fuel system is divided into: a low-pressure line and a high-pressure line. The pressure in low-pressure line is about 0-5 bar. While the pressure in high-pressure line is about 4-13 MPa (Anon, 2008).

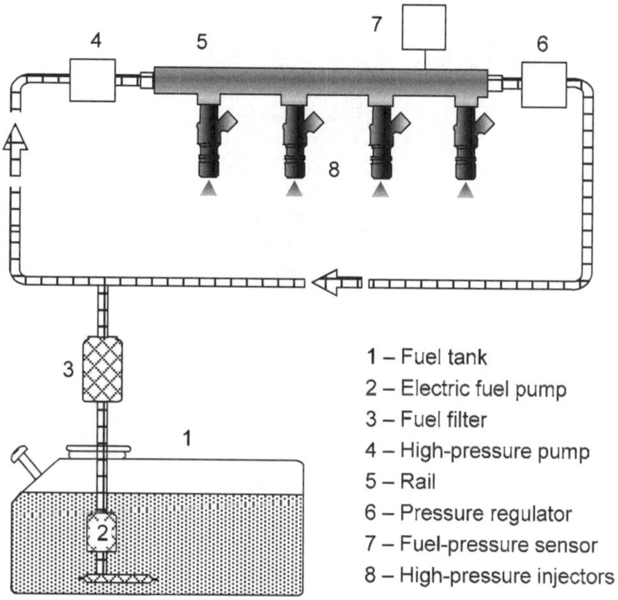

Fig. 7. The fuel system components for GDI engines.

The fuel tank is used to store the fuel. The fuel is delivered with the pressure of about 0,35 MPa from the tank to the high-pressure pump by means of an electric fuel pump (low-pressure pump). The electric pump is typically located in or near the fuel tank. Contaminants are filtered out by a high capacity fuel filter. The high-pressure pump driven by camshaft increases fuel pressure and sends the fuel to the rail. The high pressure pump increases the pressure up to 13 MPa. The fuel pressure can be set by application data depending on the operation point in the range from 4 MPa to 13 MPa. The pressure in the fuel rail is determined by the pressure sensor. To keep fuel pressure constant in the rail is very important in terms of the engine power, emissions, and noise. Fuel pressure is controlled in a special control loop. The deviations from adjusted value are compensated by an open-loop or closed loop pressure-control valve. In a closed loop control excessive fuel is returned by means of the pressure control valve. This valve allows just enough fuel to return to the tank. The fuel rail serves as fuel accumulator. The injectors, pressure control valve and high pressure sensor is mounted to the fuel rail. The injector is the central component of the injection system. Figure 8 illustrates a schematic view of the injector and its basic elements. The high pressure injector is located between the rail and combustion chamber. Injectors mounted on the rail are opened by Engine Control Unit (ECU) and, injectors inject the fuel into cylinder (Anon, 2006; Anon, 2008).

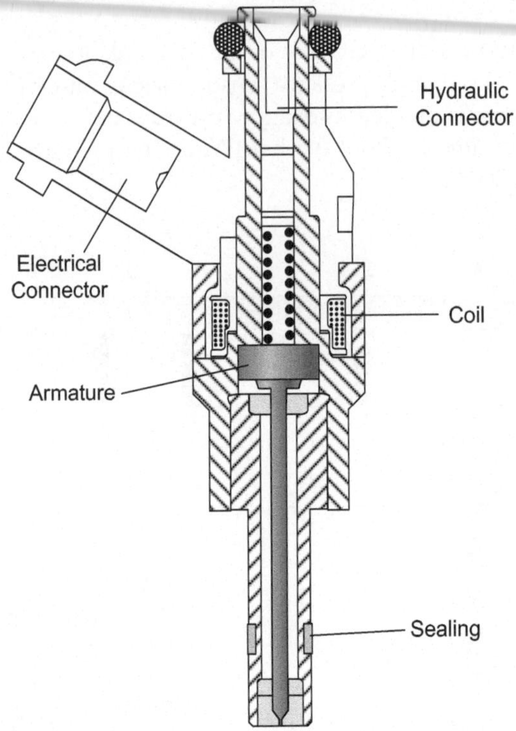

Fig. 8. The high pressure injector.

4.2 The Engine Management System

Engine management system consists of electronic control unit, sensors and actuators. The engine control unit continually chooses the one among operating modes depending on engine operating point and sensor's data. The ECU controls the actuators to input signals sent by sensors. All actuators of the engine is controlled by the ECU, which regulates fuel injection functions and ignition timing, idle operating, EGR system, fuel-vapor retention system, electric fuel pump and operating of the other systems. Adding this function to the ECU requires significant enrichment of its processing and memory as the engine management system must have very precise algorithms for good performance and drive ability.

Inputs (sensors): Mass air flow sensor, intake air temperature sensor, engine temperature sensor, intake manifold pressure sensor, engine speed sensor, camshaft position sensor, throttle position sensor, accelerator pedal position sensor, rail fuel pressure sensor, knock sensor, lambda sensor upstream of primary catalytic converter, lambda sensor downstream of primary catalytic converter, exhaust gas temperature sensor, lambda sensor downstream of main catalytic converter.

Outputs (actuators): Fuel injectors, ignition coils, throttle valve positioned, electric fuel pump, fuel pressure control valve, EGR valve, fuel-vapor retention system valve and fan control (Anon, 2002).

The engine load is mainly determined by a hot film air mass flow sensor as known from port injection systems. The determination of the EGR-rate and the diagnosis of the EGR-system are accomplished by the using of a manifold pressure sensor. The air/fuel ratio is controlled by means of a wide band lambda sensor upstream of primary catalytic converter. The catalyst system is diagnosed with a two point lambda sensor and an exhaust temperature sensor. An indispensable component is the electronic throttle device for the management of the different operation modes (Küsell et al., 1999). As an example of GDI engine management system, Bosch MED-Motronic system in Fig. 9 is given.

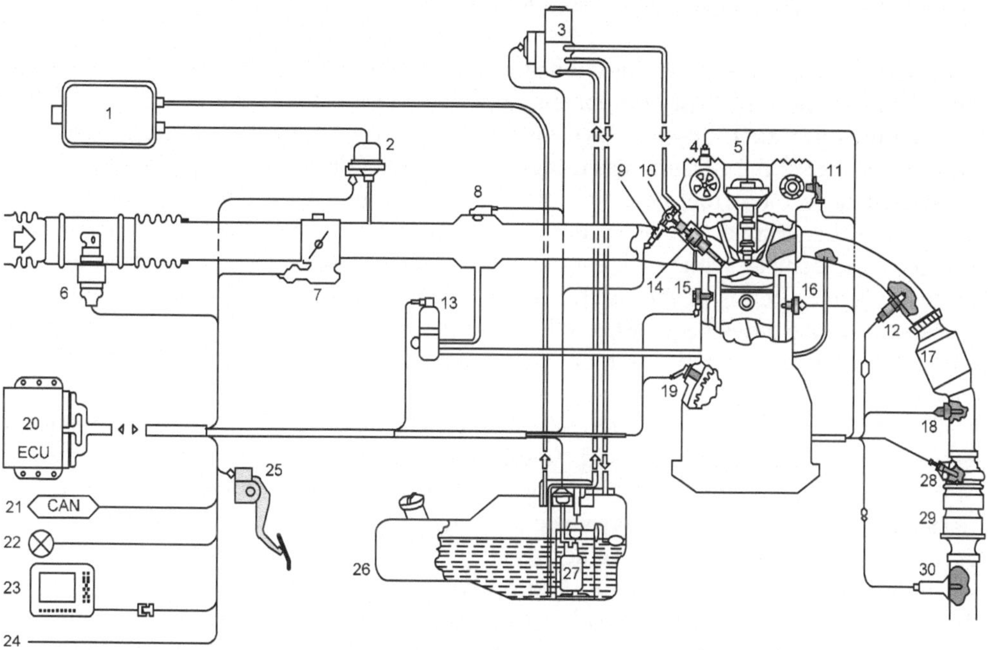

1 Activated charcoal canister
2 Regeneration valve
3 Type HDP2 high-pressure pump with integrated fuel-quantity control valve
4 Actuators and sensors for variable valve timing
5 Ignition coil and spark plug
6 Hot-firm air-mass meter with integrated temperature sensor
7 Throttle device (electronic throttle control EGAS with position sensor)
8 Intake-manifold pressure sensor
9 Fuel-pressure sensor
10 High-pressure fuel rail
11 Camshaft phase sensor
12 Lambda sensor upstream of primary catalytic converter
13 Exhaust-gas recirculation valve
14 High-pressure fuel injector
15 Knock sensor
16 Engine-temperature sensor
17 Primary catalytic converter (three-way catalytic converter)
18 Lambda sensor downstream of primary catalytic converter (optional)
19 Speed sensor
20 Engine ECU
21 CAN interface
22 Fault indicator lamp
23 Diagnostics interface
24 Interface with immobilizer ECU
25 Accelerator-pedal module with pedal-travel sensor
26 Fuel tank
27 In-tank unit comprising electric fuel pump, fuel filter and fuel-pressure regulator
28 Exhaust-gas temperature sensor
29 Main catalytic converter (Nox accumulator plus three-way catalytic converter)
30 Lambda sensor downstream of main catalytic converter

Fig. 9. Components used for electronic control in MED-Motronic system of the Bosch (with permission of Bosch) (Bauer, 2004).

5. Current trends and future challenges

At the present day, in the some gasoline engines are used port fuel injection system. This technique has achieved a high development point. As these engines operate with stoichiometric mixture, fuel economy and emissions of these engines can not be improved further. However, GDI engines have been popular since these engines have potential for reduction of toxic, CO_2 emissions and fuel consumption to comply with stringent Environmental Protection Agency (EPA) standards (Spegar et al., 2009). To attain this potential, it is required that use of the GDI engines with supercharging and/or turbo charging (Stan, 2009). The GDI engines with turbo charger enable the production of smaller displacement engines, higher fuel efficiency, lower emission and higher power (Bandel et al., 2006). The GDI engines also help eliminate the disadvantages conventional turbocharged engines (namely turbo lag, poorer fuel economy and narrowed emissions potential) to provide viable engine solutions (Spegar et al., 2009).

The primary drawback of direct injection engines is theirs cost. Direct injection systems are more expensive because their components must be well-made. In these engines, the high cost high-pressure fuel injection system and exhaust gas treatment components are required. The cost of the GDI engines is high at the present day, but GDI engines with turbo-charger that have more fuel economy are expected to be cheaper than diesel or hybrid engines in future. Thanks to mass production, if the prime cost of the GDI engines can be decreased, the vehicle with GDI engine that have turbo-charger can be leading on a worldwide level in terms of the market share. The firms such as Mitsubishi, Volkswagen, Porsche, BMW, Mercedes-Benz, Mazda, Ford, Audi, General Motors, Ferrari and Fiat prefer using GDI engine in their vehicles, today. Hyundai will start using the GDI engine in 2011.

Although different vehicles with alternative fuel have been come out, they are improbable to substitute conventional gasoline and diesel powered vehicles yet. Because the fuelling, maintenance infrastructure, cost, cruising distance and drive comfort of them are not satisfactory. Of the next-generation vehicles, only Hybrid Electric Vehicles (HEV) can be regarded as alternative energy vehicles. They have the potential to grade alongside conventional vehicles in terms of cost and convenience since their fuel costs are very low, although they cost more than conventional vehicles (Morita, 2003). It seems that large scale adoption of HEVs will not be realized unless their costs come down dramatically. GDI engine also doesn't force owner of motor vehicle to forgo luggage rack because of batteries, and doesn't make the car heavier. And it gives drivers lots of fun-to-drive torque very quickly.

The Spray-Guided Gasoline Direct Injection (SGDI) engine which has piezo injectors has showed a good potential in terms of the fuel economy and performance (Chang, 2007). Some GDI engines use piezoelectric fuel injectors today. The piezo-effect is used to provide opening and closing the injector in the direct injection systems. The piezo injectors are four-five times faster than conventional injectors. They can measure the fuel with greater precision. In addition, they can inject fuel between six and ten times during a combustion cycle. Precise piezo injection allows reducing the pollutants. GDI engines with piezo injectors can easily meet strictly emission limit changes ahead. Fuel consumption can be reduced by up to 15 percent and engine performance increased by about 5% (Website 3,

2010). Thanks to multiple injections, it is for the first time possible to extend lean-burn operating mode to higher rpm and load ranges, too. During each power stroke, a series of injections takes place. This improves mixture formation, combustion and fuel consumption. The injectors used in DI system have nozzles which open outwards to create an annular gap just a few microns wide. The peak fuel pressure in this system is up to 200 bar - around 50 times the fuel pressure in a conventional petrol injection system (Website 4, 2010). The firms such as Bosch, Delphi and Siemens have developed a piezo injection system for gasoline engines to automakers. The aim is to improve the performance of the direct injection systems. The Piezo injection with spray guided combustion system is used in the Mercedes-Benz CLS 350 CGI model vehicle (Website 5, 2010).

In GDI engine, as the spark plugs operate under high temperature, the fouling of them can cause the misfiring. To increase the life-time of the spark plug and engine efficiency, the system such as laser-induced ignition can be applied. Thus, engine efficiency can be more increased. The GDI engines are very suitable for the operating with alternative fuel. The studies on GDI engine with alternative fuel such as natural gas, ethanol, LPG have continually increasing at present day (Kalam, 2009; Teoh et al., 2008; Stein & House, 2009). If GDI engines with turbo charger use spray guided combustion process which has piezoelectric injector and high energy ignition system, the use of these engines are expected to increase more in short term.

6. References

Alger T., Hall M., and Matthews R. D., Effects of Swirl and Tumble on In-Cylinder Fuel Distribution in a Central Injected DISI Engine, SAE Paper 2000-01-0533.

Alkidas A. C., Combustion Advancements in Gasoline Engines, Energy Conversion and Management 48 (2007) 2751–2761.

Anon, Volkswagen AG, Bosch Motronic MED7 Gasoline Direct Injection, Volkswagen Self-Study Program 253, 2002, Wolfsburg.

Anon, Volkswagen AG, Twin Turbo Charger TSI Engine, Volkswagen Self-Study Program 359, 2006, Wolfsburg.

Anon, Volkswagen AG, TSI Turbocharged Engine, Volkswagen Self-Study Program 824803, 2008, U.S.A.

Anon, Volkswagen Passat TSI, Taşıt Tanıtım Kataloğu, 2009, Istanbul (in Turkish).

Bandel W., Fraidl G. K., Kapus P. E., Sikinger H. and Cowland C. N., The Turbocharged GDI Engine: Boosted Synergies for High Fuel Economy Plus Ultra-low Emission, SAE Paper 2006-01-1266.

Bauer H., Gasoline Engine Management-System and Components, Robert Bosch GmbH, Germany, 2004.

Baumgarten C., Mixture Formation in Internal Combustion Engines, Springer Verlag, Germany, 2006.

Cathcart G. and Railton D., Improving Robustness of Spray Guided DI Systems: The Air-assisted Approach, JSAE Annual Congress 2001, Vol. 40-01,p. 5-8.

Chang W. S., Kim Y. N. and Kong J. K., Design and Development of a Central Direct Injection Stratified Gasoline Engine, SAE Paper 2007-01-3531.

Çelik M. B., Buji İle Ateşlemeli Bir Motorun Sıkıştırma Oranının Değişken Hale Dönüştürülmesi ve Performansa Etkisinin Araştırılması, Doktora Tezi, Gazi Üniversitesi Fen Bilimleri Enstitüsü, 1999, Ankara.(in Turkish)

Çelik M. B., Performance Improvement and Emission Reduction in Small Engine with Low Efficiency, Journal of the Energy Institute, 80, 3, 2007.

Çınar C., Direkt Püskürtmeli Buji İle Ateşlemeli Motorlar, Selçuk-Teknik Online Dergisi, Cilt 2, No. 1-2001.(in Turkish)

Fan L., Li G., Han Z. and Reitz R. D., Modeling Fuel Preparation and Stratified Combustion in a Gasoline Direct Injection Engine, SAE Paper 1999-01-0175.

Ferguson C. R., Internal Combustion Engines, John Wiley&Sons, Inc., 1986, New York.

Gandhi A. H., Weaver C. E., Curtis E. W., Alger T. F., Anderson C. L., Abata D. L., Spray Characterization in a DISI Engine During Cold Start: (1) Imaging Investigation, SAE Paper 2006-01-1004.

Hentschel W., Optical Diagnostics for Combustion Process Development of Direct-Injection Gasoline Engines, Proceedings of the Combustion Institute, Volume 28, 2000/pp. 1119–1135.

Heywood J. B., Internal Combustion Engines Fundamentals, McGraw Hill Book, 2000, Singapore.

Kalam M. A., Experimental Test of a New Compressed Natural Gas Engine with Direct Injection, SAE Paper 2009-01-1967.

Karamangil M. İ., Direkt Püskürtmeli Benzin Motorları ve Mitsubishi Metodu, Uludağ Üniversitesi Mühendislik Mimarlık Fakültesi Dergisi, Cilt 9, Sayı 1, 2004.(in Turkish)

Kleeberg H., Dean T., Lang O. and Habermann K., Future Potential and Development Methods for High Output Turbocharged Direct Injected Gasoline Engines, SAE Paper 2006-01-0046.

Kume T., Lwamoto Y., Lida K., Murakami M., Akishino K. and Ando H., Combustion Control Technologies for Direct Injection SI Engine, SAE Paper 960600.

Küsell M., Moser W. and Philipp M., Motronic MED7 for Gasoline Direct Injection Engines: Engine Management System and Calibration Procedures, SAE Paper 1999-01-1284.

Lecointe B. and Monnier G., Downsizing a Gasoline Engine Using Turbocharging with Direct Injection, SAE Paper 2003-01-0542.

Morita K., Automotive Power Source in 21st Century, JSAE Review, 24 (2003) 3–7.

Muñoz R. H., Han Z., VanDerWege B. A. and Yi, J., Effect of Compression Ratio on Stratified-Charge Direct- Injection Gasoline Combustion, SAE Paper 2005-01-0100.

Ortmann R., Arndt S., Raimann J., Grzeszik R. and Würfel G., Methods and Analysis of Fuel Injection, Mixture Preparation and Charge Stratification in Different Direct Injected SI Engines, SAE Paper 2001-01-0970.

Rotondi R. and Bella G., Gasoline Direct Injection Spray Simulation, International Journal of Thermal Sciences, 45 (2006) 168–179.

Sercey G. D., Awcock G., Heikal M., Use of LIF Image Acquisition and Analysis in Developing a Calibrated Technique for in-Cylinder Investigation of the Spatial Distribution of Air-to-Fuel Mixing in Direct Injection Gasoline Engines, Computers in Industry 56 (2005) 1005–1015.

Smith J. D. and Sick V., A Multi-Variable High-Speed Imaging Study of Ignition Instabilities in a Spray-Guided Direct-Injected Spark-Ignition Engine, SAE Paper 2006-01-1264.

Spegar T. D., Chang S., Das S., Norkin E. and Lucas R., An Analytical and Experimental Study of a High Pressure Single Piston Pump for Gasoline Direct Injection (GDI) Engine Applications, SAE Paper 2009-01-1504.

Spicher U., Kölmel A., Kubach H. and Töpfer G., Combustion in Spark Ignition Engines with Direct Injection, SAE Paper 2000-01-0649.

Stan C. C., Analysis of Engine Performances Improvement by Down Sizing in Relationship with Super and Turbo Charging, Adapted Scavenging and Direct Injection, SAE Paper 2009-24-0075.

Stefan S., Optical Diagnostics on FSI Transparent Engine, FISITA World Automotive Congress, Barcelona 23-27 May, Barcelona Spain, 2004.

Stein R. and House C., Optimal Use of E85 in a Turbocharged Direct Injection Engine, SAE Paper 2009-01-1490.

Stoffels H., Combustion Noise Investigation on a Turbocharged Spray Guided Gasoline Direct Injection I4-Engine, SAE Paper 2005-01-2527.

Stone R., Introduction to Internal Combustion Engines, SAE, Inc., 1999, Warrandale.

Teoh Y. H., Gitano H. W. and Mustafa K. F., Performance Characterization of a Direct Injection LPG Fuelled Two-Stroke Motorcycle Engine, SAE Paper 2008-32-0045.

Website 1: http://www.greencarcongress.com/2006/02/mercedesbenz_pr.html, (17.04.2010).

Website 2: http://germancarwiki.com/doku.php/fsi, (17.04.2010).

Website 3: http://www.epcos.com/web/generator/Web/Sections/Components/Page, locale=en,r=263288,a=263380.html, (17.04.2010).

Website 4: http://www.schwab-kolb.com/daimler/en/dc000259.htm, (17.04.2010).

Website 5: http://www.mercedes-benz.com.tr/content/turkey/mpc/mpc_turkey_website/tr/home_mpc/passengercars/home/new_cars/models/cls-class/c219/overview/drivetrain_chassis.0002.html, (17.04.2010).

Zhao F., Lai M. C., Harrington D. L., Automotive Spark-Ignited Direct-Injection Gasoline Engines, Progress in Energy and Combustion Science, Volume 25, Issue 5, October 1999, Pages 437-562.

2

Liquid Sprays Characteristics in Diesel Engines

1. Introduction

For decades, the process of injecting an active fluid (diesel fuel) into the thermodynamic behaviour of a working fluid (air or gas) has been a priority in the research of the phenomena that occur in combustion systems. Due to technological improvements it's possible in present times to characterise the injection fuel process in such conditions that match those happening when the engine is running under standard conditions, hence the purpose of these studies, which focus in the achievement of a perfect mixture between the working and active fluids; as a result of this, a series of consequences are triggered that lead to an optimum combustion, and therefore in the improvement of the engines capabilities. In Diesel engines the combustion process basically depends on the fuel injected into the combustion chamber and its interaction with the air.

The injection process is analysed from this point of view, mainly using as basis the structure of the fuel spray in the combustion chamber, making this study of high importance for optimizing the injection process, and therefore reducing the pollutant emissions and improving the engines performance. Because of these, the importance to obtain the maximum control of the diesel spray structure using electronic control systems has become vital. To reduce pollutant emissions and achieving a high engine performance, it's necessary to know which parameters influence these ratings the most. It is consider being several meaningful factors that have an influence, but the most important one is the diesel spray, more specifically the penetration of the liquid length of the spray thru the combustion chamber or piston bowl. The analysis of the liquid length penetration is very useful to determine the geometric design of high speed Diesel engine combustion chambers with direct injection. For example, in a low speed regime and light load conditions, the unburned hydrocarbon emissions will be reduced greatly if contact between the spray of fuel (liquid length) and the combustion chamber wall is avoided. If now we consider a high speed regime and heavy load, the emission of fumes is reduced if there is contact between the spray of fuel and the combustion chamber wall, hence the importance of measuring the liquid phase penetration of the fuel in Diesel engines with direct injection, using sophisticated and complex measuring techniques.

2. Diesel spray characteristics

Depending on the mechanism to characterise, diesel spray can be analysed in a macroscopic or microscopic point of view. With the purpose of understanding in detail this process, the various physical parameters involved during the transition of a pulsed diesel spray will be expressed in this chapter, however it is essential to know the systems that make possible for an injection process to take place. These are the injection nozzle, active fluid to inject (liquid), and the working fluid on which the liquid is injected, as seen in figure 1.

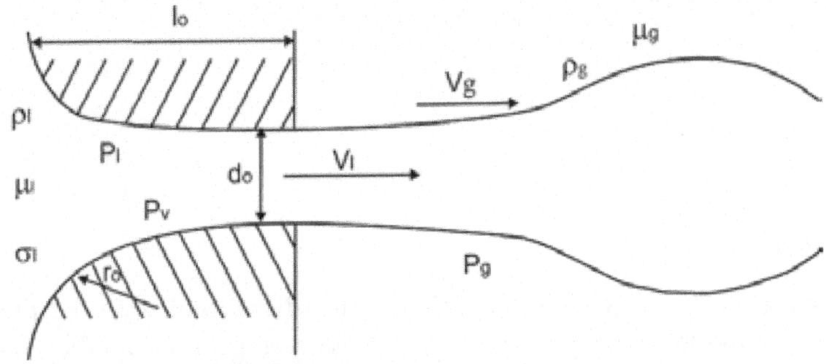

Fig. 1. Meaningful variables of the injection process.

For a Newtonian fluid with constant temperature distribution and an injection nozzle with a completely cylindrical orifice, the variables that influence the dispersion of the spray are:

Nozzle Geometry
- Orifice Diameter (do)
- Length (lo)
- Orifice entrance curvature radius (ro)
-Superficial Roughness (∈)
Injection Conditions
-Pressure of Liquid Injected Fluid (P_l)
-Pressure of Gas Working Fluid (P_g)
-Pressure increasing ($\Delta P = P_l - P_g$)
-Medium velocity of the injected Liquid fluid (V_l)
- Medium velocity of the working gas fluid (V_g)
-Duration of the injection (t_{inj})
Injected Fluid Properties (Liquid)
-Density (ρ_l)
-Kinematic Viscosity (μ_l)
-Vapour Pressure (P_v)
-Superficial Tension (σ)
Working Fluid Properties (Gas)
-Density (ρ_g)
- Kinematic Viscosity (μ_g)

All these variables can be, can be fitted into a dimensionless form that allows us to have much simpler relations and better defined. The dimensionless variables used in most cases are:

Relation of densities:

$$\rho^* = \frac{\rho_l}{\rho_g} \tag{1}$$

Relation of viscosities:

$$\mu^* = \frac{\mu_l}{\mu_g} \tag{2}$$

Reynolds Number, relation between inertial and viscous forces:

$$Re = \frac{\rho d \upsilon}{\mu} \tag{3}$$

Weber Number, relation between superficial tension force and inertial force:

$$We = \frac{\rho d \upsilon^2}{\sigma} \tag{4}$$

Taylor Viscosity Parameter:

$$Ta = \frac{Re}{We} = \frac{\sigma}{\mu \upsilon} \tag{5}$$

Ohnesorge Number:

$$Oh = \frac{\sqrt{We}}{Re} = \frac{\mu}{\sqrt{\rho \sigma d}} \tag{6}$$

Length/diameter relation of the Nozzle (l_o/d_o)

Nozzle radius entrance/diameter relation (r_o/d_o)

Discharge coefficient of the nozzle:

$$C_d = \frac{\upsilon l}{\sqrt{\frac{2\Delta P}{\rho_l}}} \tag{7}$$

Cavitation Parameter:

$$K = \frac{2(P_l - P_\upsilon)}{\rho_l \upsilon^2} \tag{8}$$

Reynolds Number: Density and kinematic viscosity must be particularised for liquid or gas, furthermore these properties can be evaluated for intermediate conditions between both fluid film conditions. These parameters can be divided into two groups:

1. External flow parameters (relation of densities, Weber number, Taylor parameter), these parameters control the interaction between the liquid spray and the surrounding atmosphere.
2. Internal flow parameters (Reynolds number, cavitation parameter, length/diameter relation, nozzle radius entrance/diameter relation, discharge coefficient): these parameters control the interaction between the liquid and the nozzle.

2.1. Macroscopic Characteristics

The macroscopic description of a diesel spray generally emphasise the interaction of the latter and the control volume where it is injected and mixed, and because of this the diesel spray can be defined with the following physical parameters (Figure 2.2):

1. Spray tip penetration
2. Spray angle
3. Breack up length

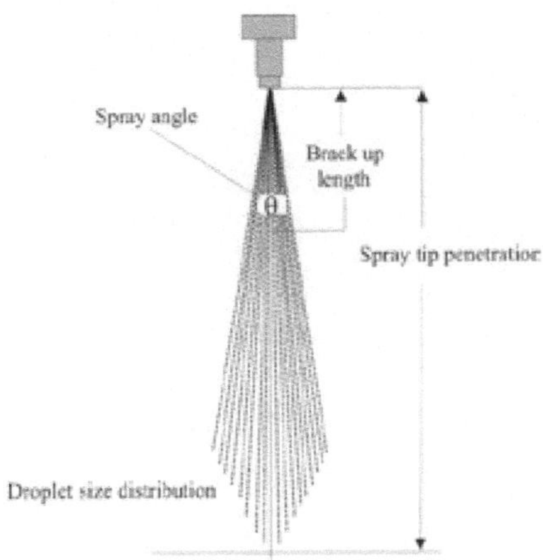

Fig. 2. Physical parameter of a diesel spray (Hiroyasu & Aray, 1990).

2.1.1. Front Penetration

The injection front penetration (S) is defined as the total distance covered by the spray in a control volume, and it's determined by the equilibrium of two factors, first the momentum quantity with which the fluid is injected and second, the resistance that the idle fluid presents in the control volume, normally a gas. Due to friction effects, the liquids kinetic energy is transferred progressively to the working fluid. This energy will decrease continuously until the movement of the droplets depends solely on the movement of the working fluid inside the control volume. Previous studies have shown that a spray penetration overcomes that of a single droplet, due to the momentum that the droplets located in the front of the spray experiment, accelerating the surrounding working fluid, causing the next droplets that make it to the front of the spray an instant of time later to have less aerodynamic resistance. We must emphasise that diesel fuel sprays tend to be of the compact type, which causes them to have large penetrations.

Several researchers have studied the front penetration and have found a series of correlations that allow us to establish the main variables that affect or favour the penetration of a pulsed diesel spray. The following are some of the most relevant:

From the theory of gaseous sprays, (Dent, 1971) was one of the pioneers in the study of spray phenomena. The author proposed an experimentally adjusted correlation which is applicable to pulsed diesel sprays; this correlation was the compared by (Hay & Jones, 1972) with other correlations, finding certain discrepancies between them. However, this correlation is considered to be applicable in a general form to diesel sprays:

$$S(t) = 3{,}07\left(\frac{\Delta P}{\rho_a}\right)^{\frac{1}{4}}\left(\frac{294}{T_a}\right)^{\frac{1}{4}}\sqrt{d_o t} \quad (9)$$

(Hiroyasu & Arai, 1990) proposed two expressions to determine the sprays penetration as a function of the time of fracture (t_{rot}), and so defining the fracture time can fluctuate between 0,3 y 1 ms depending on the injection conditions.

$$t_{rot} = 28{,}65\frac{\rho_l d}{\sqrt{\rho_g \Delta P}} \quad (10)$$

$$S = 0{,}39\sqrt{\frac{2\Delta P}{\rho_l}}\cdot t \quad (11)$$

$$t = t_{rot} \quad (12)$$

$$S = 2{,}39\left(\frac{\Delta P}{\rho_g}\right)^{0{,}25}\sqrt{d_o t} \quad (13)$$

$$t = t_{rot} \quad (14)$$

An empirical equation considering the dimensionless parameter $\rho^* = (\rho_a/\rho_l)$ was developed by (Jiménez et al., 2000) obtaining the following expression:

$$S(t) = 0{,}6^{-3}U_o(t)^{0{,}9}\left(\frac{\rho_a}{\rho_l}\right)^{-0{,}163} \quad (15)$$

Where Uo is the medium velocity at the beginning of the injection in [m/s] and t is injection time duration in [m/s]. In this equation the behaviour of the sprays penetration is considered for temperature variations in the working fluid between 293 K and 423 K. Although the equation considers the atmospheric pressure values of the working fluids (low density), it is also valid for high densities.

Penetration according to (Jaward et al., 1999):

$$S = C_1 (\Delta P)^{0,25} \sqrt{t\rho_l^{0,25} \rho_g^{-0,14}} \tag{16}$$

From the derivation of the expressions developed by (Dent, 1971) and (Arai et al., 1984), (Bae et al., 2000) proposes this expression for the penetration of the spray:

$$S = C\left(\frac{\Delta P}{\rho_g}\right)^{0,25} \sqrt{d_o t} \tag{17}$$

$$t = t_o = \sqrt{\left(\frac{\rho_l}{\rho_g}\right)\left(\frac{d_o}{V_{iny}}\right)} \tag{18}$$

Penetration according to (Correas, 1998):

$$S = C_2 U_o^{0,5} \sqrt{d_{eq} t} \tag{19}$$

$$d_{eq} = d_o \sqrt{\frac{\rho_l}{\rho_g}} \tag{20}$$

Considering C1 and C2 experimental constants, d_{eq} to be the equivalent diameter, and C another experimental constant as a function of the discharge coefficient, it can be said that the discharge coefficient and the constant C have a direct dependence on the injector type used and in less measure on the working conditions. Therefore and according to (Hiroyasu & Dent, 1990) proposal, the discharge coefficient (Cd) for a determined injector does not modify the constant C value. Other works of great importance concerning the penetration of spays in VCO nozzles were presented by (Bae & Kang, 2000), in which he classifies different types of sprays for different densities of the working fluid.

As a summary it can be said that the penetration of the spray basically depends on the following parameters:

-Injection pressure increasing ΔP: Increasing the injection pressure in relation to the control volume where the fuel is injected (ΔP), increases the velocity of the penetration of the spray and hence the development of the latter will be easier at the beginning, (Hiroyasu et al., 1980) and (Arai et al., 1984).

According to (Ahmadi et al., 1991), because a part of the liquid advances rapidly through the internal spray area where the aerodynamic interaction is poor, the injection pressure fluctuations are not related to the injections velocity. On the other hand, at the tip of the spray the high aerodynamic interaction causes the latter to lose velocity, making the recently injected liquid to reach and pass this slower moving tip, taking its place as the new spray tip and afterwards being slowed down as well by the control volumes surroundings. As well, (Nishida et al., 1992) and (Tinaut et al., 1993) suggest that the velocity of the droplets at the tip is usually slower than in other regions of the spray, so the simple fact that the velocity of the droplets is slower than the velocity of penetration demands a constant droplet renewal in the tip of the spray.

-Density ratio (ρ^*): this dimensionless parameter ρ^* or relation of densities, according to (Hiroyasu et al., 1980), (Arai et al., 1984) and (Payri et al., 1996), considerably affects the penetration of the spray, due to the fact that increasing the relation of densities causes the penetration to reduce considerably, this is because of the increase or reduction of the aerodynamic interaction, according to the respective parameter scale.

-Working fluid temperature (Tg): density reduction can be caused by the increase of the working fluids temperature, hence, the decrease of spray penetration. However, previous studies show that the spray's temperature doesn't produce significant effects in the penetration in relation to other parameters, (Hiroyasu et al., 1980) and (Arai et al., 1984).

2.1.2. Cone angle

The cone angle is defined as the angle formed by two straight lines that stat from the exit orifice of the nozzle and tangent to the spray outline (sprays morphology) in a determined distance. The angle in a diesel spray is formed by two straight lines that are in contact with the spray's outline and at a distance equivalent to 60 times de exit diameter of the nozzles orifice. This angle usually is between 5 and 30 degrees. This determines greatly the fuels macroscopic distribution in the combustion chamber. In one hand, the increase in angle decreases the penetration and can cause interference between sprays (when sprays are injected using multi-orifice nozzles) in the same chamber favouring the merging of droplets. On the other hand, an excessive penetration is favoured when the angle decreases lower than certain values, causing the spray to collide with the piston bowl or the combustion chamber.

In previous studies there have been a series of proposals to determine the cone angle, some of the most important are as follows:

$$\tan\frac{\theta}{2} = 0{,}13\left(1+\frac{\rho_a}{\rho_l}\right) \tag{21}$$

This expression is considered for densities of the working fluid lower than (ρ_g) 15 kg/m^3, but the dimensionless injector relation is not considered(l_o/d_o). However, (Reitz & Braco, 1979) and (Arai et al., 1984) do consider this dimensionless parameter in their investigations to determine the maximum aperture of the cone angle, proving that it indeed has great influence on the opening of the cone angle.

Cone angle according to (Hiroyasu et al., 1980):

$$\theta = 0,05\left(\frac{d^2\rho_a\Delta\rho}{\mu_a^2}\right)^{0,25} \tag{22}$$

The droplets size related to the wavelengths of the most unstable waves was established by (Ranz & Marshall, 1958) and therefore, the cone angle is defined by the combination of the injection velocity and the radial velocity of the waves of greater growth in their superficial unstableness, defining the cone angle with the following expression:

$$\tan\frac{\theta}{2} = \frac{1}{A}4\pi\left(\frac{\rho_g}{\rho_l}\right)^{\frac{1}{2}} f(\Gamma) \tag{23}$$

$$\Gamma = \frac{\rho_l}{\rho_g}\left(\frac{Re_l}{We_l}\right)^2 \tag{24}$$

$$A = 3,0 + 0,277\left(\frac{l_o}{d_o}\right) \tag{25}$$

Where: A is a constant determined experimentally in function of the relation length/diameter of the nozzle (l_o/d_o), which is represented by the equation (24) according to (Reitz & Braco, 1979). Figure 3 shows the dependence of the cone angle in function of aerodynamic forces, (Ranz & Marshall, 1958) cited by (Heywood, 1988) y (Ramos, 1989), and for concepts on droplet evaporation, (Ranz & Marshall, 1952).

Cone angle proposed by (Hiroyasu & Arai, 1990):

$$\theta = 83,5\left(\frac{l}{d}\right)^{-0,22}\left(\frac{d}{D_o}\right)^{0,15}\left(\frac{\rho_g}{\rho_l}\right)^{0,26} \tag{26}$$

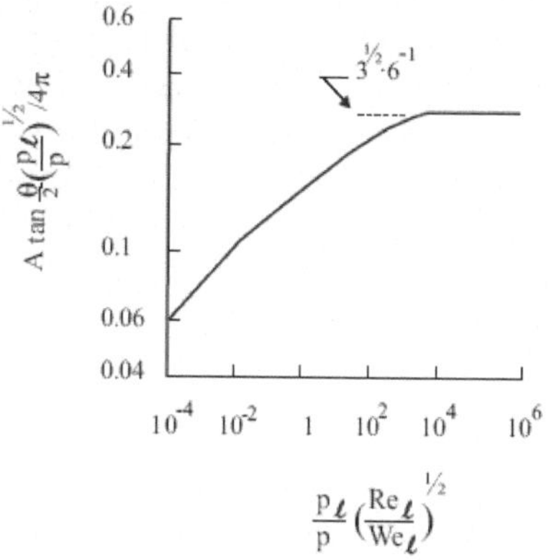

Fig. 3. Cone angle dependence in function of aerodynamic forces (Ramos, 1989).

Where: Do represent the diameter of the nozzles jacket. With this expression it's possible to determine the angle of opening of the fully developed spray, where the angle is practically a function of the nozzles orifice geometry and the dimensionless term of the relation of densities (ρ^*). Others parameters such as cinematic viscosity can in some way modify the limits of the developed spray, but not the angle of the cone.

The cone angle is mainly affected by the geometric characteristics of the nozzle, the density ratio (ρ^*), and the Reynolds number of the liquid, (Reitz & Bracco, 1979, 1982), apart from depending on other variable such as those described as follows:

-Increasing pressure (ΔP): An increase in the injection pressure causes an increase in the cone angle up to a maximum value, above decrease gradually.

-Density ratio (ρ^*): An increase in the relation of densities is a factor that causes an increase in the cone angle due to an increase in the aerodynamic interaction, according to (Arrègle, 1998) and (Naber & Siebers, 1996), for values greater than ($\rho^* > 0.04$) the cone angle tends to be independent of this parameter.

-Working fluid temperature (Tg): Increasing working fluid temperature, increases the evaporation process in the sprays exterior zone, consequently a decrease in the angle of the cone, (Hiroyasu et al., 1980).

2.1.3. Liquid Length

The liquid length of the spray is a very important characteristic to define the behaviour of the spray in the combustion chamber. This zone of the spray is also called continuous or stationary and it is understood as being from the nozzle exit to the point were the separation of the first droplets occur. To define this zone the use of diverse measurements methods and techniques is of vital importance. In the literature we find some of the most useful measurement methods and techniques in the analysis of the liquid length, (Hiroyasu & Arai, 1990), (Chehroudi et al., 1985), (Arai et al., 1984), (Nishida et al., 1992), (Gülder et al., 1992), (Christoph & Dec, 1995), (Zhang et al., 1997) and (Bermúdez et al., 2002, 2003).

To analyze the internal structure of the spray, (Hiroyasu & Aray, 1990) identified two zones inside the atomizing regime, the zone of the incomplete spray and the zone of the complete spray. Figure 4 shows structure in a general way. The difference between them is due to the fact that with the incomplete sprays the disintegration of the surface of the spray begins at a certain distance from the point of the nozzle of the injector, indicating a distance Lc, while in the case of the incomplete sprays distance Lc is nearly cero and Lb is maintained virtually constant on increasing speed. Furthermore (Hiroyasu & Aray, 1990) show that cavitation greatly favours the atomization process in the complete spray regime.

To define liquid length a series of expressions have been proposed which have been suggested in specific conditions according to each case and among the most relevant the following can be cited:

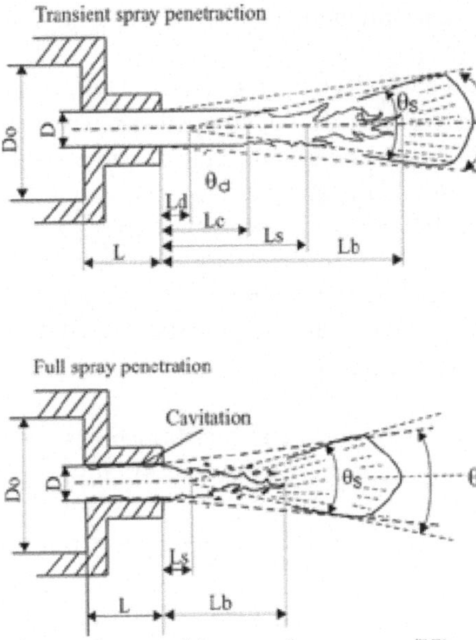

Fig. 4. Internal structure of complete and incomplete spray (Hiroyasu & Aray, 1990).

Based on experimental results of the measurement of the liquid length in complete sprays (Hiroyasu & Aray, 1990) proposed the following equation:

$$L_b = 7d\left(1+0,4\frac{R}{D}\right)\left(\frac{\rho_g}{\rho_l U_o^2}\right)^{0,05}\left(\frac{L}{D}\right)^{0,13}\left(\frac{\rho_l}{\rho_g}\right)^{0,5} \tag{27}$$

Liquid length according (Bracco, 1983):

$$L_b = 7,15\left(\frac{\rho_l}{\rho_g}\right)^{0,5} \tag{28}$$

Liquid length according (Yule & Salters, 1995):

$$L_b = 2{,}65^{-3}(d)We_l^{-0,1}Re_l^{-0,3}\left(\frac{\rho_l}{\rho_g}\right)^{-0,08} \tag{29}$$

The most important parameters on liquid length penetration are the following:

1. The ratio of work fluid densities/liquid (ρ^*): an increase on the ratio of densities produces a decrease in liquid length due to an increase in the aerodynamic interaction between the spray and the environment in which this is developed as shown by (Arai et al., 1984), (Chehroudi et al., 1985), (Hiroyasu & Arai, 1990), (Christoph & Dec, 1995), (Cannan et al., 1998), (Naber & Siebers, 1996) and (Siebers, 1998).
2. The relationship between length/nozzle diameter (l_o/d_o): this relationship influences the liquid length penetration when the volume of control where the combustible is injected at atmospheric conditions. However, when the control volume pressure is high, the influence of this parameter in liquid length penetration decreases, according to investigations made by (Ha et al., 1983) and (Xu & Hiroyasu, 1990).
3. Nozzle orifice diameter (d_o): the liquid length has a linear behaviour with the nozzle diameter. Liquid length penetration decreases to minimum values when the nozzle diameter is reduced to minimum values, in other words, a change in the diameter of the nozzle orifice results in a directly proportional change in the penetration of liquid length as recent research shows, (Siebers, 1998), (Verhoeven et al., 1998) and (Schmalzing et al., 1999).
4. Working fluid temperature (Tg): working fluid temperature is one of the thermodynamic properties that strongly affect liquid length penetration, since the rate of combustible vaporization is directly related to the energy content of the working fluid in the inside of the cylinder (e.g., high temperatures) and in the degree of the mixture of both fluids (injected fuel-gas or air) (Christoph & Dec, 1995). However, working fluid temperature has no relevant effect at high pressure injection because both, an increase in the speed of injection and the amount of fuel injected, ease the effect with respect of low pressures, (Zhang et al., 1997). An increase in working fluid temperature at constant density causes and increase in the specific energy of the latter and therefore a decrease in liquid length during spray penetration is a consequence of high drag of vaporization energy towards the fuel, (Siebers, 1999).
5. Fuel temperature (T_f): fuel temperature is a variable that greatly affects liquid length penetration in such a way that on increasing the temperature of the latter liquid length tends to decrease lineally. It has been proven that at under conditions of low temperature and working fuel density there are more significant effects that under high conditions of temperature and density, because in the latter case the effect witch respect an absolute scale is insignificant, (Siebers, 1998).
6. Physical-Chemical properties of the fuel: these properties of the fuel (i.e., density, viscosity and volatility) have a considerable impact on liquid length penetration

with volatility being the most influential property on penetration. (Siebers, 1998, 1999) observes that a low volatility fuel requires more energy to be heated and then evaporate than a high volatile fuel. Therefore, for a low volatile fuel, liquid length penetrates much more than a more volatile fuel because the amount of energy dragged towards the fuel depend basically on the process of evaporation.

Liquid length of a diesel spray is a parameter of much interest in the study of the injection-combustion process. In later topics in this same chapter we will discuss this parameter where a complete experimental analysis of the characterization of the liquid length of a diesel spray is approached.

3. Microscopic Characteristics

The macroscopic description is characterized by the content of droplets of diverse sizes and the changes on the changes in their special kinetics. For example, the atomization mechanism is responsible for distributing the droplets in the injection process and to a great extent the good distribution of the droplets in relation to their size depend on it. Generally the quality of the atomization of a liquid spray can be estimated on the medium diameter of the droplets. A determined medium diameter represents the equivalent diameter that characterizes the entire group of the droplets of the spray. Equation (30) establishes the general form based on which all the correlations that determine Sauters medium diameter have been defined.

$$D_{mn} = \sqrt[m-n]{\frac{\sum_{i=1}^{k} D_i^m N_i}{\sum_{i=1}^{k} D_i^n N_i}} \tag{30}$$

Where Ni is the number of droplets of the group with diameter Di. Generally speaking, medium diameters are used to simplify calculation and analysis of data. Medium diameter is that which defines the characteristics of a population of drops present in a sample. In some processes Sauters medium diameter is used, which represents the diameter of droplets which have the same volume/surface relation in the totality of the spray, as well as the arithmetic average diameter (D10) which are represented by the following respective equation:

$$SMD = \frac{\sum_{i=1}^{k} D_i^3}{\sum_{i=1}^{n} D_i^2} \tag{31}$$

$$D_{10} = \frac{\sum_{i=1}^{k} D_i}{\sum_{i=1}^{k} N_i} \tag{32}$$

It must take into account that using medium diameters is very useful to simplify droplet populations existing in an atomizing process. For this reason it is essential to use the distribution of droplet size.

3.1. Droplet size distribution

The diameter of the droplets obtained as a result of atomization is based on a series of parameters as follows:

1. Rate of injection: the diameter of the droplet increases with the rate of injection as an increase in the volume of the injected liquid produces a greater drag of the working fluid, the aerodynamic interaction grows and the critical size of the droplets increases. Apart from this, increasing the numeric population of droplets intensifies de coalescence, resulting in a growth in the geometry of the droplets.

2. Density ratio (ρ^*): the relation of densities has two opposing effects on the size of the droplets, intensification of atomization and the possibility that there will be coalescence. On increasing the relationship of densities a greater aerodynamic interaction exists, which causes the droplets to slow down and an increase in the numerical population in their field.

3. Working fluid temperature (Tg): on increasing working fuel temperature their is an increase on the rate of evaporation, due to which at the beginning of this the droplets with small diameters tend to evaporate completely while those droplets with greater diameters maintain a stable geometry until they evaporate completely.

4. Spatial evolution of the size of the drops: the average size of the droplets tends to grow in relation to the increase of the distance between the drops and the injector point. In some studies it has been suggested that the average diameter of the drops is greater in the direction of the radius of the spray while other suggest the opposite, that is the medium diameter is reduce in relation to the distance from it.

5. Evolution of the diameter of droplets during time: It's generally considered that the medium diameter of the droplets decreases at the point of the spray and increases at the tail, while in areas distant from the injector they maintain a rate of constant values. Generally speaking, the sizes of the droplets tend to diminish at the beginning of the injection and grow at the end.

The most common formulas to determine Sauters medium diameter are:

Sauters medium diameter according to (Hiroyasu & Kadota, 1974):

$$\mathrm{SMD} = 4{,}12d\left(\mathrm{Re}\right)^{0,12}\mathrm{We}^{-0,54}\left(\frac{\mu_l}{\mu_g}\right)^{0,54}\left(\frac{\rho_l}{\rho_g}\right)^{0,18} \tag{33}$$

Where A being an experimental constant (A = 2330) and Q the injected volume [m^3]

Sauters medium diameter according to (Hiroyasu & Arai, 1990) and (Hiroyasu et al., 1989)

1. For incomplete spray

$$SMD = 0,38d\left(Re\right)^{0,25} We^{0,32}\left(\frac{\mu_l}{\mu_g}\right)^{0,37}\left(\frac{\rho_l}{\rho_g}\right)^{-0,47} \tag{34}$$

2. For complete spray

$$SMD = 8,7\left(Re_l We_l\right)^{-0,28} d_o \tag{35}$$

These formulae have been the most used to determine Sauters medium diameter, even though these correlations experimentally obtained have been modified over the years, they maintain a very important basis in which to determine Sauters medium diameter. Each of these formulae may experience further modifications and better approximations according to the quality of the specific model or experiment.

4. Measurement techniques

Some problems of fluid mechanics are complex where multiphase systems are concern and when combustion phenomena are produced. In many cases current knowledge is still incomplete due to the complexity of the physical-chemical processes: (non-stationary processes, irreversible processes and out-of-balance chemical reactions) that occur at the limits of different scientific disciplines such as fluid mechanics, thermodynamics and chemistry. In order to progress in its study we need available experimental data that provide information of the different processes and degrees of interest for the study, such as for example, mass and energy transport, movement and the size of particles, concentration of the different species, thermodynamic properties, and chemical composition among others.

The physical phenomena of interaction matter-radiation (absorption, dispersion, interference, diffraction, among others) are very sensitive to small variations in the localize physical parameters of the fluid, and furthermore they do not interact with the physical processes in the environment of fluid mechanics, and so are useful in the analysis of these problems. Technological advance in diverse fields basically optics, electronics and information technology have allowed for this development of equipment able to measure some localized physical parameters of fluids in a very precise way, and are the basis for the development of optical techniques of measurement and visualization used in studies of fluid mechanics.

4.1. Classical visualization techniques

The classical visualization methods are based on the variations of the refraction rate that are produced in the fluids heart due to the changes in its physical properties. When an beam of light propagates through a fluid, the variations of the refraction rate causes variations in both the intensity and in wave phase, therefore the emerging light contains information of the fluid properties in the light beam trajectory propagation. Basically these optical techniques can be divided in 3 types: Shadowgraphy, Schlieren and Interferometry, which have been used since the 1860's, (Foucault, 1859) in France and (Toepler, 1864) in Germany gave the first insights of the Schlieren technique. Toepler was the first to develop this technique for the study of liquids and gas flow, and later on used by (Hayashi et al., 1984) and (Konig & Sheppard, 1990), among others.

-Shadowgraphy: the environment is illuminated with a straightening of a light beam and the image is taken after the emerging light propagates freely through the space. The visualization technique with diffused rear illumination is a similar technique but the environment is lit up with a diffuse beam light. The difference between these techniques consists on placing a diffuser between the beam and the environment to illuminate. These techniques allow visualizing the liquid phase of the fuel spray and are greatly used in the study of the injection process of combustion internal engines. The visualization with rear diffused illumination technique allows the estimation of the different macroscopic parameters in an injection process. (Zaho & Ladommatos, 2001) have studied the spray penetration and consider this technique to be reliable and easy to use for this type of analysis.

-Schlieren photography: this technique is similar to that of the shadowgraphy, the difference is that the image is taken after a spatial filtering in the image plane of the light source. Adjusting adequately the spatial filtering dimensions it is possible to visualize both the liquid and vapour phase of the fuel spray, but not to quantify them. These techniques have been used in the injection and combustion processes of the internal combustion engine (Preussner et al., 1998), (Spicher & Kollmeire, 1986) and (Spicher et al., 1991), as well as in the analysis of propulsion systems (Murakamis & Papamoschou, 2001) and (Papampschou, 2000).

4.2. Scattering techniques

The classical visualization techniques incorporate the information throughout the beams propagation trajectory, by which the information about the existing three-dimensional structures in the vessel of the fluid is lost. This information can be obtained illuminating the fluid with planes of light and taking pictures of the dispersed light by the environment, normally in the perpendicular direction of the plane. This kind of visualization techniques can be included in a much general group which is the scattering technique. The light scattering phenomena can be of two types, elastic or inelastic, depending on if the process produces or not the radiation frequency.

4.2.1. Elastic scattering techniques

The elastic dispersion phenomena of light are studied within the theory of Lorenz-Mie. There are basically two approximations depending on the size of the particles: Mie scattering and Rayleigh scattering.

-The Mie scattering is an interaction of the elastic type of light with particles of much greater size than that of its wave length (droplets, ligaments, among others). The characteristics of the scattered light are related to the form, size, refraction rate and number of scattering particles. These properties are the basis of the different optical techniques of measurement described as follows:

1. Visualization with a laser sheet the fluid is illuminated with a laser sheet beam obtaining images of the scattered light (Mie regime), normally on the perpendicular direction of the sheet. This technique allows estimating the macroscopic characteristics of fuel sprays and analysing the existence of internal structures, ligaments, among others. This technique is one of the most used in the study of the injection process in an internal combustion engine (Dec, 1992) and (Preussner et al., 1998).
2. Technique of laser anemometry: it is based on the interaction of coherent light with the existing particles in movement inside the heart of the fluid in such a way that the sizes of these particles allow them to be treated in Mie scattered imaging. These interactions produce a change in the frequency of radiation (Doppler Effect) that can be related to both the speed and size of the particles. In the so called Laser Doppler Anemometry (LDA), two coherent light beams interact in one region (control volume) with the existing moving particles in the fluid and the fluctuation of the disseminated light intensity allows the estimation of the particles speed. (The obtained light intensity is basically intensity with a background modulated by a cosine function, whose temporal variation depends solely on the frequencies of the dispersed beams. The frequency of modulation for this signal can be related to the velocity of the particles). The Phase Doppler Anemometry (PDA) is based on the same principle but it uses several photo sensors placed in different spatial positions. With which it's possible to estimate the diameter of the diffusive particles considering them spherical by the temporal phase lag between signals received by each photo detector. This technique requires a series of optical accessories that difficult its use in measurement of a real thermo engine. Although some investigators (Auriemma et al., 2001), (Corcione et al., 1998), (Cossali et al., 1996), (Georjon et al., 1997) and (Guerrassi & Champoussin, 1996) have used the phase Doppler anemometry to develop very specific analysis, the mayor usage is still the characterization of the distribution of diameters and velocities of fuel droplets in accessible optical models that simulate similar conditions of those found in real thermal engines (Arrègle, 1998) and (Jiménez et al., 2000).
3. The velocimetry imaging techniques allow velocity field measuring in a fluids plane that is illuminated with a screen of light. There are several ways to use these techniques, depending on the method selected to register and to process information, however all of them are very important: in Particle Image Velocimetry (PIV) the fluid is illuminated with several light pulses and the instant images are registered using multiple exposure techniques. The instant velocities are obtained dividing the particles displacement in each time consecutive image by two pulses. In Particle Shadow Velocimetry the fluid is illuminated in a long period of time in which the displacement of the particles are registered as lines on the image and the velocities are calculated dividing the line length by time interval. In Particle Tracking Velocimetry a series of consecutive exposures take place (several light pulses) in one image and the velocity is estimated by tracking the particles. The velocimetry techniques are used mainly to analyse flow of gases en the thermal engine. Some of the most recent applications for this technique can be found in the literature (Choi & Guezennec, 1999), (Kakuhou et al., 1999), (Nauwerck et al., 2000), (Neussert et al., 1995), and (Trigui et al., 1994), where the main application is focussed to the study of mixture formation inside the combustion chamber of a thermal engine, furthermore it considered to be one of the best techniques for this kind of analysis.

4. Rayleigh scattering is of the elastic kind, where the size of particles is much smaller of that of the lights wavelength, for example the gas molecules. The intensity of the scattered light is proportional to the total density of all kinds of existing particles inside the illuminated zone and provides images of global concentration of all the species, although it doesn't allow discrimination between them. Furthermore for example, the Rayleigh signal for a particle approximate 1 μm is close to twenty orders of magnitude lower than the Mie signal, for which the signal is highly affected by both the presence of large particles and by the background light. The two most commonly used procedures to reduce the interference of particles are shown by (Zhao et al., 1993). The main researchers using the Rayleigh technique (Espey & Dec, 1994), (Lee & Foster, 1995) and (Zhao et al., 1991) have been basically to determine concentrations of vapour and liquid phases and mainly in zones with high flame presence. As well as in the temperature measure and species concentration for the combustion diagnostic.

4.2.2. Inelastic scattering techniques

On the other hand, the inelastic scattering of light is studied in the quantum mechanics field, specifically in the study of matter-radiation interaction phenomena. These phenomena are extremely sensitive to the frequency of radiation and the species chemical composition because they depend on electronic transitions between molecular energy levels caused by the absorption of photons of defined frequency that stimulate the molecules to higher energetic conditions. After which the molecules come to stable conditions releasing radiant energy where its spectral characteristics are also very well defined. Different optical techniques of measure are bases on these phenomena, detailed as follows:

-Laser Induced Incandescence is a technique based on the thermal emission that is produced when the carbon particles are stimulated with a very intense electromagnetic radiation. The obtained signal is proportional to the volume fraction of the carbon particles concentrated in the measured zone. Because of this, the technique is very useful for the study of combustion processes (Dec, 1992), (Dec et al., 1991), (Dec & Espey, 1992), (Winklhoefer et al., 1993) and (Zhao & Ladommatos, 1998), mainly to determine the qualitative distribution of soot in the high radiation zone during a injection-combustion process.

-Laser Induced Fluorescence (LIF) is a technique based on the fluorescent properties that some molecules present. When these molecules absorb electromagnetic energy of a determine frequency they acquire a higher energetic condition (stimulation) and afterwards they return to their original energetic state releasing this energy (fluorescence). The spectral characteristics of this radiation are determined by the molecules characteristics. If the fluid doesn't have fluorescent molecules, molecular tracers that present fluorescence can be added. For example: NO, NO_2, acetone, biacetyl, rodamina, or different colorants. The fluorescent signal is proportional to the density of the tracers inside the illuminated zone. In many cases the environment is illuminated using laser beam sheet and the technique is then known as planar induced laser fluorescence (PLIF). In the planar laser induced exciplex fluorescence (PLIEF) tracers called exciplex (complex excitation), like for example: naphthalene mixtures and TMPD (tetramethyl-1,4-phenylenediamine) that allow to separate spectrally the corresponding liquid and vapour phase fluorescence of a biphasic system, and therefore measure simultaneously each ones concentration (Juliá, 2003). Although this

technique has much application in injection-combustion processes (Felton et al., 1995), (Fujimoto et al., 1997), (Hiroshi et al., (1997), (Kido et al, 1993) and (Kim & Ghandhi, 2001), it is not considered to be the most appropriate to detect species when compared to other, like for example: Mie-Scaterring. This is due to the incoherencies presented when detecting species in these types of processes (Preussner et al., 1998) and (Takagi et al., 1998)
Phosphorescent particle tracking (PPT) is a similar technique to that of particle tracking velocimetry (PTV). The phosphorescence is an inelastic diffusion of light characterized by it long temporal duration, much higher than that of fluorescence, which makes it ideal to track the movement of particles in the fluid.

5. Experimental characterization of the liquid length penetration

5.1. Introduction

The main objective of this section is to carry out the characterization of the liquid length penetration of a diesel spray. To achieve this it has been necessary to consider a group of experiments which allow the determination of the influence that the injection parameters and the thermodynamic variables have upon the penetration of a diesel spray in evaporative conditions. The first developed study is based on the analysis of the penetration of the spray in its liquid phase, where it is expected to define the degree of influence that the following have over this phenomena: thermodynamic variables (pressure, temperature and density) present in the combustion chamber at the moment when the fuel is injected, the injection pressure and the geometry of the nozzle. To make this study it's necessary to use the ombroscopy technique for the taking of digital images, as well as an acquisition system to process data. It is to point out that the ombroscopy has been the most used technique in the macroscopic characterization of diesel sprays, specifically in the study of the liquid phase penetration. As mentioned in section 4, the techniques of measure to carry out studies of the liquid phase of diesel sprays are very diverse. The most used until know are expressed in this chapters literature. (Cambell et al., 1995), (Canaan et al., 1998), (Christoph & Dec, 1995), (Felton et al., 1995), (Hiroyasu & Miao, 2002) and (Knapp et al., 1999).

5.2. Experimental work approach

A working plan that groups the different experiments to carry out has been structured in such a way to analyse qualitatively the injection process. To achieve this, the experimental work has been planned as follows:

The use of the experimental in system with the inert atmosphere method and through the ombroscopy technique analyse the penetration of the liquid phase of the diesel spray.
- Parametric analysis to consider:
1. Influence of the injection process on the liquid length penetration.
2. Influence of the diameter of the nozzle on the liquid length penetration.

The analysis of the liquid length penetration is useful to determine the geometric design of combustion chambers for high speed regime diesel engines with direct injection. For example, in low speed regime and light load the hydrocarbon emissions will be reduced if the contact of the spray (liquid length) with the combustion chambers wall is avoided. For high speed regimes and heavy loads, the reduction of fumes can be achieved by contact between the spray and the chamber wall. Because of these, the necessity to measure the liquid penetration in diesel engines of direct injection emerges, motivating the use of measure techniques even more complex and sophisticated.

In previous studies (Christoph & Dec, 1995) investigated the effects that temperature and the fluids density have on the liquid phase penetration. In this study they used a Diesel engine witch optical access views, and through the elastic-scatter technique they obtained images of the spray. (Zhang et al., 1997) analyzed the effects that the injection pressures, diameter of the nozzle and admission air temperature have on liquid length penetration. For this they used a compression machine which had an equivalent compression ratio to that found in a Diesel engine. In this analysis an argon laser beam was used as the light source and an E-10 camera was also used to capture the images. (Siebers, 1998) investigated the maximum axial penetration of the liquid phase of an evaporated diesel spray in a chamber of constant volume, using the Mie-scattered technique for image capturing. The main altered parameters where the injection pressure, orifice diameter of the nozzle, temperature and density of the working fluid in the inside of the chamber.

The investigation of the sprays liquid phase for a common rail system at high temperatures was made by (Bruneaux & Lemenand, 2002). The variation in parameters in this investigation where: the injection pressure, the temperature of the working fluid and the diameter of the nozzle. This study was made in a chamber similar to the one used by (Verhoeven et al., 1998), in which it was possible to maintain high pressures and temperatures inside the chamber and so simulating similar conditions found in a real Diesel engine. The technique of measure used was based on a light source supported by a planar laser induced exciplex fluorescence system and a charged-coupled device (CCD) camera to capture images. It's evident that each investigator uses in his experiments defined and heterogeneous techniques of measure. However occasionally and in some complexity degree the final results tend to be very similar independently of the used, reason why the motivation to develop the basis for the experiments presented in this chapter arose with one of the most flexible visualization techniques, the ombroscopy.

The characterization of the liquid length penetration of an evaporated diesel spray was done under the following methodology:

1. Experimental system configuration: to undertake the experiments that lead to obtain information about the liquid length penetration of the spray without flame, it has been necessary to form the experimental system in an inert atmosphere. Furthermore to conceive as a first phase the use of ombroscopy technique to obtain images of the liquid phase of the spray (Figure 5 shows the schematics diagram of the global experimental setup configuration).

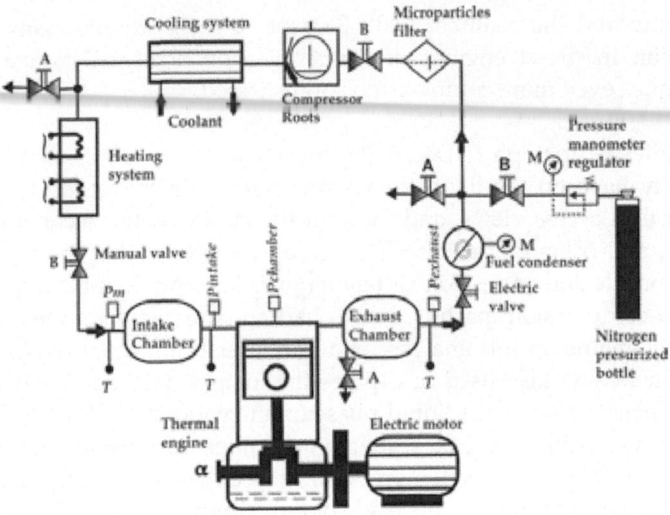

Fig. 5. Schematic diagram of the experimental setup.

2. Configuration of the group of experiments: The considered group of experiments defines the variables to be analysed, as well to determine their influence on the liquid length penetration of the spray. The main variables for study are:
   -Injection pressure.
   -Orifice diameter of the nozzle.
   -Working fluid density constant.

Figure 6 shows the schematics of the nozzle that has been used in the experiments. It has been experimented with five nozzles of similar geometry with single axisymetric orifice and same kind of jacket.

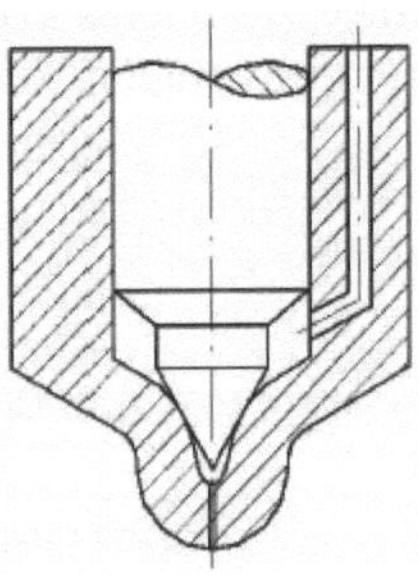

Fig. 6. Scheme of the nozzle used in the experiments.

Four nozzles were tested at four different injection pressures, while the intake temperature and pressure were kept constant at 70 °C and 1.3 bar, respectively. The four nozzles have single axisymmetric holes with 115, 130, 170 and 200 μm in diameter, and the injection pressure was 300, 700, 1100 and 1300 bar. Table 1 shows the estimated mass flow rates and discharge coefficients for each nozzle and injection pressure. A diagnostic thermodynamic model developed by (Martínez et al., 2007) was employed to calculate the working fluid properties (temperature and density) in the cylinder. Cylinder pressure was measured with a transducer installed on a lateral wall. The pressure at bottom dead center was measured with a resistive transducer located between the prechamber intake and the chamber itself. A temperature sensor was also installed in the prechamber intake to measure the working fluid temperature at bottom dead centre. Since pressure and temperature data were available, thermodynamic conditions were characterized at top dead center ± 3 crank angle degrees, which is considered the most stable region during the fuel injection process (Martínez et al., 2007).

Injection pressure (bar)	Nozzle diameter (μm)	Measured mass flow rate (g/s)	Theoretical mass flow rate (g/s)	C_d
300	115	1.53	2.04	0.746
700	115	2.52	3.38	0.745
1100	115	3.13	4.32	0.725
1300	115	3.34	4.72	0.708
300	130	2.27	2.61	0.870
700	130	3.50	4.32	0.810
1100	130	4.05	5.52	0.734
1300	130	4.42	6.03	0.733
300	170	3.36	4.46	0.753
700	170	5.32	7.38	0.721
1100	170	6.47	9.43	0.686
1300	170	6.87	10.30	0.666
300	200	3.63	6.18	0.587
700	200	6.74	10.20	0.660
1100	200	8.53	13.10	0.653
1300	200	9.29	14.30	0.651

Table 1. Injection parameters and their corresponding mass flow rates and discharge coefficients.

5.3. Mathematical correlation

Liquid phase penetration of a jet injected into an inert environment has well defined stages. The first stage begins with the injection and ends when the jet breaks up. This is the intact length stage or the first break-up regime, (Hiroyasu & Aray, 1990) suggested the following correlation to estimate the time for the first break-up regime to occur:

$$t_b = \frac{15.8\rho_f d_n}{C_d \sqrt{2\rho_a \Delta P}} \quad (36)$$

where Cd is the discharge coefficient, dn (μm) is the nozzle diameter, ΔP (Pa) is the pressure drop through the nozzle, and ρ_f and ρ_a (kg/m^3) are the fuel and working fluid densities, respectively. For the particular conditions studied here, Equation (36) predicts times for the first break-up regime between 25 and 30 μs, and our experimental measurements indicate an average time of 50 μs. Experimental evidence (Ahmadi et al., 1991), (Auriemma et al., 2001), (Christoph & Dec, 1995) and (Martínez et al., 2007) indicates that the liquid penetration length, LL, increases proportionally to the square root of time from the injection onset until the second break-up regime is reached at time tr. Thereafter the liquid penetration length varies little and hence it is considered constant from a macroscopic point of view. Therefore, a mathematical correlation suitable to model the liquid penetration length is:

$$0 < t < t_r : \qquad LL(t) = \alpha\sqrt{t} \tag{37}$$

$$t > t_r : \qquad LL(t) = Cte = LL_{max} \tag{38}$$

which is illustrated in Figure 7. Coefficients α and LL depend on numerous parameters, such as the fluid thermodynamic conditions and geometrical parameters of the injection system. A satisfactory mathematical correlation must take into account the effect of the nozzle diameter, the discharge coefficient, the injection pressure, and the working fluid density. These parameters have been previously found to be enough to characterize the liquid penetration length (Bae & Kang, 2000), (Bae et al., 2000), (Bermúdez et al., 2002, 2003), (Bracco, 1983), (Canaan et al., 1998) and (Chehroudi et al., 1985). It is therefore expected that a detailed analysis of these parameters can yield an accurate correlation that can be of assistance in the successful designing of combustion chambers required by modern heavy duty diesel engines. In this paper we attempt power law correlations for α and LLmax (Equations 39 and 40).

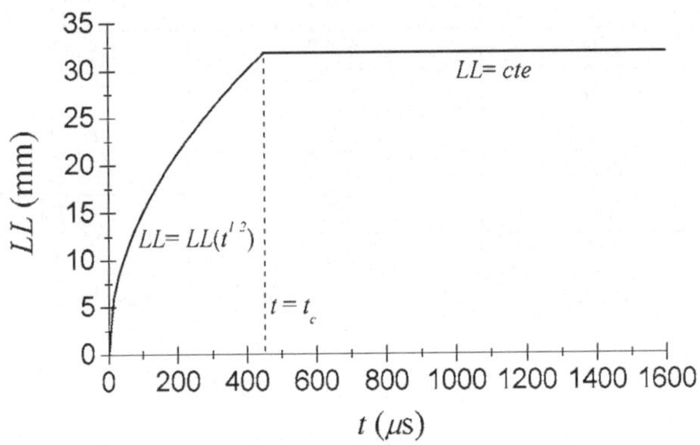

Fig. 7. Plot showing different stages of the considered model.

$$\alpha \ \mu \ d_n^A \ \rho_a^B \ P_{iny}^C \ C_d^D \tag{39}$$

$$LL_{max} \ \mu \ d_n^E \ \rho_a^F \ P_{iny}^G \ C_d^H \tag{40}$$

5.4 Determination of the fuel injection onset

The fuel injection onset can be determined assuming that LL increases proportionally to the square root of time until the second break-up regime is reached at tr, i.e. LL = α $t^{1/2}$ for 0 < t < tr. Time tr is defined as the time when the ratio between LL to $t^{1/2}$ with a correlation coefficient R^2 = 99 %. Coefficient α is estimated by fitting experimental data measured before the second break-up regime is reached, as shown in Figure 8, where the experimental data can be approximated by LL = 1.07 $t^{0.497}$ with a correlation coefficient R^2 = 99.8 %.

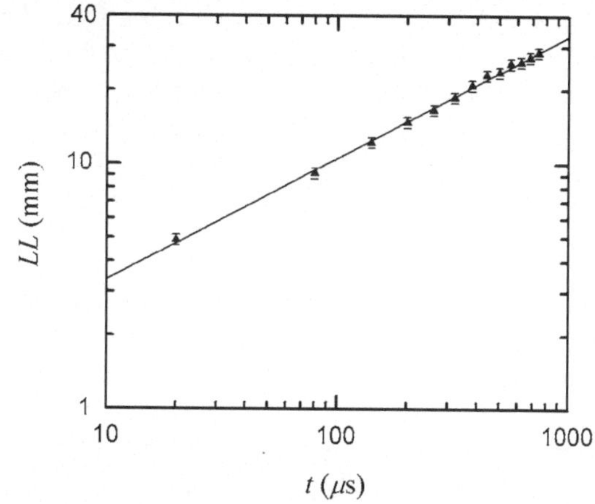

Fig. 8. Estimation of α and the fuel injection onset

5.5. Determination of the discharge coefficient

The discharge coefficients of each nozzle hole at the injection pressures studied here were estimated using the following correlation:

$$C_d = \frac{m_f}{An\sqrt{2\Delta P \rho_f}} \tag{41}$$

where the discharge coefficient C_d is defined as the ratio of the mass flow rate injected in the cylinder and the theoretical mass flow rate computed from the Bernoulli equation. The mass flow rate of fuel injection was measured by a fuel rate indicator (EVI-IAV). Experimental measurements provided enough data to estimate the discharge coefficient for each nozzle and injected condition, which are shown in Table 1.

6. Results and discussion

Equation 42 is the best fit for predicting penetration length in the fuel injection process before the second break-up regime, when : $0 < t < t_r$

$$LL(t) = 6.47 d_n^{0.56} \rho_a^{-.027} P_{iny}^{0.23} C_d^{0.08} t^{1/2} \tag{42}$$

Figures 9 (a, b) and 10 (a, b) show a comparison between calculated (Equation 42) and experimental liquid penetration lengths. In all cases curves and experimental data are in good agreement and the correlation coefficient is R^2 = 93.3 %, which means only 6.7 % of all data are not accounted by the proposed correlation. Analyzing Equation 42 we find that the liquid length penetration is strongly affected by the nozzle diameter whose exponent in Equation 42 is greatest. The density of the working fluid and the injection pressure have comparable and inverted effects on the liquid penetration length, $\partial LL/\partial \rho a \approx -(P_{inj}/\rho_a)(\partial LL/\partial P_{inj})$ or $\partial \rho a/\partial P_{inj} \approx -(\rho_a/P_{inj})$. Additionally we notice from Equation 42 that the liquid velocity penetration, $\partial LL/\partial t$, is proportional to $P_{inj}^{0.23}$, which is the same proportionality as for LL itself. On the other hand, an increase in the working fluid density causes the liquid penetration resistance to rise, which yields a shortening in the liquid penetration length. It is worth mentioning that the effect of ρa on LL reported here is in good agreement with experimental data presented by (Dent, 1971), who suggested the following correlation:

$$LL(t)\,\mu\,\rho_a^{-0.25}\,t^{\frac{1}{2}} \tag{43}$$

Equation 42 reveals that under the experimental conditions studied here, 0.58 < Cd < 0.87, the liquid penetration length is very insensitive to the value of the discharge coefficient, which causes a maximum variation of the liquid penetration length of only about 3 %.

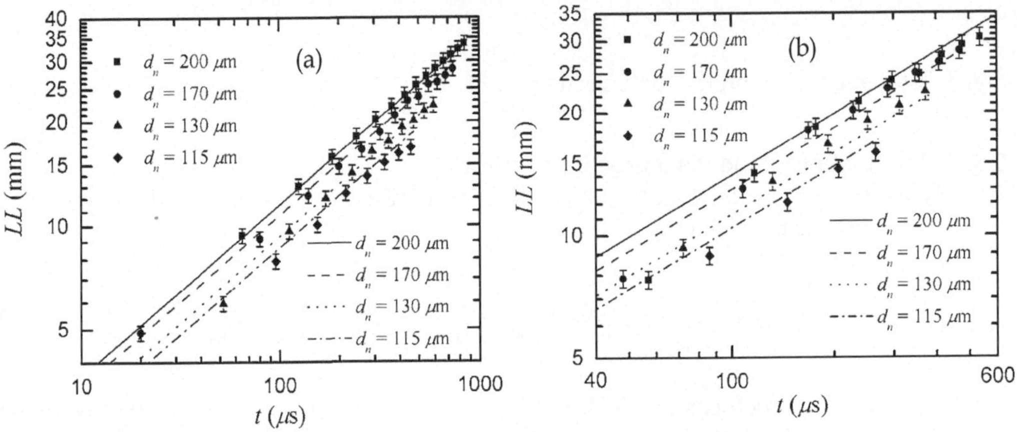

Fig. 9. Comparison between experimental data and the proposed correlation, equation 42. (a): P_{inj} = 300 bar and (b): P_{inj} = 700 bar, ρ_a = 26 kg/m³ and Tg = 906 K.

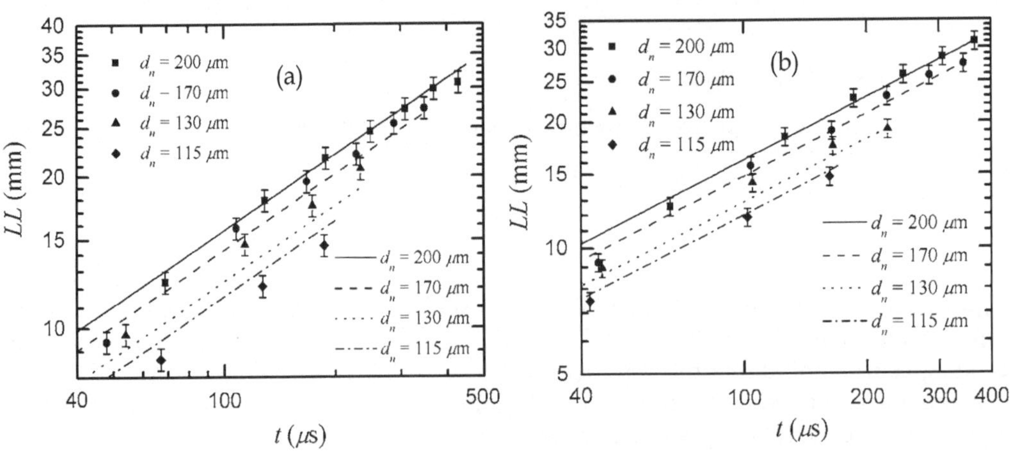

Fig. 10. Comparison between experimental data and the proposed correlation, equation 42. (a): P_{inj} = 1100 bar and (b): P_{inj} = 1300 bar, ρ_a = 26 kg/m³ and Tg = 906 K.

7. Conclusions and remarks

Experimental measurements were carried out to estimate the liquid penetration length of a diesel fuel jet injected in an inert environment. The effects of the characteristic parameters, i.e. the nozzle diameter, discharge coefficient, injection pressure, and working fluid density were analyzed. The transient fuel injection process was recorded using optical access, and the liquid penetration length before the second break-up regime was measured using the ombroscopy technique. The aim of the present research is to generate a correlation that accurately predicts liquid penetration length at conditions typical of modern Heavy Duty common rail diesel engines operating with direct fuel injection. A statistical analysis of our experimental measurements suggests a power function correlation to model the liquid penetration length. The proposed model is in good agreement with experimental data and yields a correlation coefficient $R^2 = 93.3$ %. Furthermore, the suggested correlation illustrates important details about how the main parameters affect the fuel injection process. The nozzle diameter has the greatest effect on liquid penetration length. A reduction in nozzle diameter yields a shorter penetration length because it causes an earlier start of the second break-up regime. Increasing the injection pressure provokes premature droplet break-up within the jet, which results mainly due to cavitation at the nozzle exit. If the working fluid density in the combustion chamber increases the penetration length is shorter and the second break-up regime is delayed due to the free-share flow between the working fluid and the fuel jet, which produces higher evaporation rates of droplets from the diesel jet. Finally, under the experimental conditions studied here, the discharge coefficient has a negligible effect on the liquid penetration length. However, the discharge coefficient influences the cavitation phenomenon at the nozzle exit and modifies the droplet velocity within the jet.

8. References

Ahmadi Befrui, Wieseler B. y Winklhofer E. (1991) "The propagation of Fuel Spray in a Research Diesel Engine A Joint Numerical and Experimental Analysis". SAE Technical Paper 910181.

Arai M., Tabata M., Shimizu M. y Hiroyasu H. (1984) "Disintegrating Process and Spray Characterization of Fuel Jet Injected by a Diesel Nozzle". SAE Technical Paper 840275.

Arrègle J. (1998) Análisis de la Estructura y Dinámica Interna de Chorros Diesel. Tesis Doctoral, E.T.S. Ingenieros Industriales. Universidad Politécnica de Valencia, Spain.

Auriemma M., Corcione F. E., DIMartino U. y Valentino G. (2001) "Analysis of the Intake Flow in a Diesel Engine Head Using Dynamic Steady Flow Conditions". SAE Technical Paper 2001-01-1307.

Bae Ch. y Kang J. (2000) "Diesel Spray Characteristics of Common-Rail VCO Nozzle Injector". Congreso THIESEL-2000, Valencia, Spain.

Bae Ch. y Kang J. (2000) "Diesel Spray Development of VCO Nozzles for High Pressure Direct-Injection". SAE Technical Paper 2000-01-1254.

Bae Ch., Yu J., Kang J., Cuenca R. y Lee O. (2000) "The Influence of Injector Parameters on Diesel Spray". Congreso THIESEL-2002, Valencia, Spain.

Bermúdez V., García J. M., Juliá E. y Martínez S. (2002) "Instalación Experimental para el Estudio del Proceso de Inyección-Combustión en Motor Diesel de Inyección Directa". XV Congreso Nacional de Ingeniería Mecánica, Cádiz, Spain.

Bermúdez V., García J. M., Juliá E. y Martínez S. (2003) "Engine with Optically Accessible Cylinder Head: a Research Tool for Injection and Combustion Processes". SAE Technical Paper 2003-01-1110.

Bracco F. V. (1983) Structure of High Speed Full Cone Sprays. Recent Advances in Gas Dynamics, Plenum Publishing Corporation, N.Y.

Bruneaux G. y Lemenand C. (2002) "A Study of Liquid Phase Structure in Very High Pressure Common Rail Diesel Injection Using Optical Diagnostics". Congreso THIESEL-2002, Valencia, Spain.

Cambell P., Sinko K. y Chehroudi B. (1995) "Liquid and Vapour Phase Distributions in a Piloted Diesel Fuel Spray". SAE Technical Paper 950445.

Canaan R. E., Dec J. E. y Green R. M. (1998) "The Influence of Fuel Volatility on the liquid-Phase Fuel Penetration in a Heavy-Duty D.I. Diesel Engine". SAE Technical Paper 980510.

Chehroudi B., Chen S. H., Bracco F. V. y Onuma Y. (1985) "On the Intact Core of Full-Cone Sprays". SAE Technical Paper 850216.

Choi W. Ch. y Guezennec Y. G. (1999) "Study of the Flow Field Development During the Intake Stroke in an I.C. Engine Using 2-D PIV and 3-D PTV". SAE Technical Paper 1999-01-0957.

Christoph E. y Dec J. E. (1995) "The Effect of TDC Temperature and Density on the Liquid-Phase Fuel Penetration in a D.I. Diesel Engine". SAE Technical Paper 952456.

Corcione F. E., Vaglieco B. M. y Valentino G. (1998) "A Study of Physical and Chemical Delay in a High Swirl Diesel System Via Multiwavelength". SAE Technical Paper 980502.

Correas D. (1998) Estudio Teórico Experimental del Chorro Libre Diesel Isotermo. Tesis Doctoral, E.T.S. Ingenieros Industriales. Universidad Politécnica de Valencia, Spain.

Cossali G. E., Gerla A., Coghe A. y Brunello G. (1996) "Effect of Gas Density and Temperature on Air Entrainment in a Transient Diesel Spray". SAE Technical Paper 960862.

Dec J. E. (1992) "Soot Distribution in a D.I. Diesel Engine Using 2-D Imaging of Laser-Induced Incandescence, Elastic Scattering, and Flame Luminosity". SAE Technical Paper 920115.

Dec J. E., Axel O., Loye Z. y Siebers D. L. (1991) "Soot Distribution in a D.I. Diesel Engine Using 2-D Laser-Induced Incandescence Imaging". SAE Technical Paper 910224.

Dec J. E. y Espey Ch. (1992) "Soot and Fuel Distribution in a D.I. Diesel Engine via 2-D Imaging". SAE Technical Paper 922307.

Dent J. C. (1971) "A Basis for the Comparison of Various Experimental Methods for Studying Spray Penetration". SAE Technical Paper 710571.

Espey C. y Dec J. (1994) "Quantitative 2-D Fuel Vapor Concentration Imaging in a Firing D.I. Diesel Engine Using Planar Laser Induced Rayleigh Scattering". SAE Technical Paper 940682.

Felton P. G., Kyristsis D. C. y Fulcher S. K. (1995) "LIF Visualization of Liquid Fuel in the Intake Manifold during Cold Start". SAE Technical Paper 952464.

Foucault L. (1859) Memoiré sur la Construction des Télescopes en Verre Argenté, Vol. 5, pp. 197-237. Ann. Observ. Imp. París.

Fujimoto M., Tabata M. y Tanaka T. (1997) "Planar Measurements of NO in an S.I. Engine Based on Laser Induced Fluorescence". SAE Technical Paper 970877.

Georjon T., Chalé H. G., Champoussin J. C., Marié J. L. y Lance M. (1997) "A Droplet Tagging Method to Investigate Diesel Spray". SAE Technical Paper 970351.

Gülder Ö. L., Snelling D. R. y Smallwood G. J. (1992) "Diesel Spray Structure Investigation by Laser Diffraction and Sheet Illumination". SAE Technical Paper 920577.

Guerrassi N. y Champoussin J. C. (1996) "Experimental Study and Modelling of Diesel Spray/Wall Impingement". SAE Technical Paper 960864.

Ha J., Sato G. T., Tanabe H., Fujimoto H. y Kuniyoshi H. (1983) "Investigation on the Initial Part and the Spray Formation Delay of Diesel Spray". SAE Technical Paper 830451.

Hay N. y Jones P. L. (1972) "Comparison of the Various Correlations for Spray Penetration". SAE Technical Paper 720776.

Hayasi T., Taki M., Kojima S. y Kondo T. (1984) "Photographic Observation of Knock With a Rapid Compression and Expansion Machine". SAE Technical Paper 841336.

Heywood J. B. (1988) Internal Combustion Engine Fundamentals, pp. 522-536. McGraw-Hill International Editions.

Hiroshi N., Hiroyuki E., Yoshihiro D., Matsuhei N., Hiroshi O. y Taizo S. (1997) "NO Measurement in an Diesel Spray Flame Using Laser Induced Fluorescence". SAE Technical Paper 970874.

Hiroyasu H. y Arai M. (1990) "Structures of Fuel Sprays in Diesel Engines". SAE Technical Paper 900475.

Hiroyasu H., Arai M. y Tabata M. (1989) "Empirical Equations for the Sauter Mean Diameter of a Diesel Spray". SAE Technical Paper 890464.

Hiroyasu H. y Kadota T. (1974) "Fuel Droplet Size Distribution in Diesel Combustion Chamber". SAE Technical Paper 740715.

Hiroyasu H., Kodata T. y Arai M. (1980) Fuel Spray Characterization in Diesel Engines. Combustion Modelling in Reciprocant Engines, Mattavi and Amann, Plenum Press.

Hiroyasu H. y Miao H. (2002) "Optical Techniques for Diesel Spray and Combustion". Congreso THIESEL-2002, Valencia, Spain.

Jaward B., Gulari E. y Heinen N. A. (1999) Characteristics of Intermittent Fuel Spray. 1999.

Jiménez J., Castro F. y Giménez B. (2000) "The Tip Evolution of an Evaporative Intermittent Fuel Spray". Congreso THIESEL-2000, Valencia, Spain.

Juliá E. (2003) Medida de Concentraciones de Combustible en Chorros Diesel Mediante Técnicas de Fluorescencia Inducida por Láser. Tesis Doctoral, E.T.S. Ingenieros Industriales. Universidad Politécnica de Valencia, Spain.

Kakuhou A., Urushihara T., Itoh T. y Takagi Y. (1999) "Characteristics of Mixture Formation in a Direct-Injection S.I. Engine With Optimized In-Cylinder Swirl Air Motion". SAE Technical Paper 1999-01-0505.

Kido A., Ogawa H. y Miyamoto N. (1993) "Quantitative Measurements and Analysis of Ambient Gas Entrainment into Intermittent Gas Jets By Laser-Induced Fluorescence of Gas (LIFA)". SAE Technical Paper 930970.

Kim T. y Ghandhi J. B. (2001) "Quantitative 2-D Fuel Vapor Consentration Measurements in an Evaporating Diesel Spray Using the Exciplex Fluorescence Method". SAE Technical Paper 2001-01-3495.

Knapp M., Luczak A., Beushausen V., Hentschel W. y Andresen P. (1999) "Vapour/Liquid Visualization with Laser Induced Exciplex Fluorescence in an SI Engine for Different Injection Timings". SAE Technical Paper 961122.

Konig G. y Sheppard C. G. W. (1990) "End Gas Autoignition and Knock in a Spark Ignition Engine". SAE Technical Paper 902135.

Lee K. y Foster D. (1995) "Cycle-by-Cycle Variation in Combustion and Mixture Concentration in the Vicinity of Spark Plug Gap". SAE Technical Paper 950814.

Martínez S., Sánchez F., Rodríguez G., Riesco J y Gallegos A. (2007) "Simultaneous Measurement of Evaporating Fuel Spray Using Laser Induced Exciplex Flourescence" International Journal of Kones.

Martínez S., Sánchez F., Riesco J., Gallegos A y Aceves S. (2007) "Liquid penetration length in direct diesel fuel injection" Applied Thermal Engineering.

Murakamis E. y Papamoschou D. (2001) Experiments on Mixing Enhancement in Dual-Steam Jets. Department of Mechanical and Aeorospace Engineering, University of California at Irvine, Irvine, CA.

Naber J. D. y Siebers D. L. (1996) "Effects of Gas Density and Vaporization on Penetration and Dispersion of Diesel Sprays". SAE Technical Paper 960034.

Nauwerck A., Gindele J., Spicher U., Rosskamp H. y Landwehr G. (2000) "Investigation of the Transient In-Cylinder Flow Inside a Two-Stroke Engine With Particle Image Velocimetry". SAE Technical Paper 2000-01-0902.

Neussert H. J., Spiegel L. y Ganser J. (1995) "Particle Tracking Velocimetry A Powerful Tool to Shape the In-Cylinder Flow of Modern Multi-Valve Engine Concepts". SAE Technical Paper 950102.

Nishida M., Nakahira T., Komori M., Tsujimura K. y Yamaguchi I. (1992) "Observation of High Pressure Fuel Spray With Laser Light Sheet Method". SAE Technical Paper 920459.

Papamoschou D. (2000) Mixing Enhancement Using Axial Flow. Department of Mechanical and Aeorospace Engineering, University of California, Irvine, CA.

Payri F., Desantes J. M. y Arrμegle J. (1996) "Characterization of D.I. Diesel Sprays in High Density Conditions". SAE Technical Paper 960774.

Preussner C., Döring C., Fehler S. y Kampmann S. (1998) "GDI Interaction Between Mixture Preparation, Combustion System and Injector Performance". SAE Technical Paper 980498.

Ramos J. I. (1989) Internal Combustion Engine Modeling, pp. 150-158. Hemisphere Publishing Corporation.

Ranz W. E. y Marshall W. R. (1952) Evaporation from Drops, Vol. 48 parte I, pp. 141-146. Canad. J. Chemical Engineering Progress.

Ranz W. E. y Marshall W. R. (1952) Evaporation from Drops, Vol. 48 parte II, pp. 173-179. Canad. J. Chemical Engineering Progress.

Ranz W. E. y Marshall W. R. (1958) Some Experiments on Orifice Sprays, Vol. 36, pág. 175. Canad. J. Chemical Engineering Progress.

Reitz R. D. y Bracco F. B. (1979) "On the Dependence of Spray Angle and Other Spray Parameters on Nozzle Design and Operating Conditions". SAE Technical Paper 790494.

Reitz R. D. y Bracco F. V. (1979) Ultra-High-Speed Filming of Atomizing Jets, Physics of fluids, Vol. 22, pp. 1054-1064. Physics of Fluids.

Reitz R. D. y Bracco F. V. (1982) Mechanism of Atomization of a Liquid Jet, Vol. 25. Physics of Fluids.

Schmalzing C. O., Stapf P., Maly R. R., Renner G., Stetter H. y Dwyer H. A. (1999) "A Holistic Hydraulic and Spray Model - Liquid and Vapor Phase Penetration of Fuel Sprays in DI Diesel Engines". SAE Technical Paper 1999-01-3549.

Siebers D. L. (1998) "Liquid-Phase Fuel Penetration in Diesel Sprays". SAE Technical Paper 980809.

Siebers D. L. (1999) "Scaling Liquid-Phase Fuel Penetration in Diesel Spray Based on Mixing-Limited Vaporization". SAE Technical Paper 1999-01- 0528.

Spicher U. y Kollmeire H. (1986) "Detection of Flame Propagation During Knocking Using Simultaneously High Speed Schlieren Cinematography and Multi-Optical Fibre Techique". SAE Technical Paper 861523.

Spicher U., Kroger H. y Ganser J. (1991) "Detection of Knocking of Combustion Using Simultaneously High Speed Schlieren Cinematography and Multi- Optical Fibre Techique". SAE Technical Paper 912312.

Takagi Y., Itoh T., Muranaka S., Liyama A., Iwakiri Y., Urushihara T. y Naitoh K. (1998) "Simultaneous Attainment of low Fuel Consumption, High Output Power and Low Exhaust Emissions in Direct Injection SI Engines". SAE Technical Paper 980149.

Tinaut F. V., Castro F., Melgar A., Sanchez M. L. y Gimenez B. (1993) Visualization and Measurement of Speed and Size in Diesel Sprays, pp. 78-83. Societée de Ingenieurs de l´Automovile.

Toepler A. (1864) Memoiré sur la Construction des Télescopes en Verre Argenté. Cohen M. and Sohn, Bonn.

Trigui N., Kent J. C., Guezennec Y. G. y Choi W. C. (1994) "Characterization of Intake-Generated Flow Fields in I.C. Engine Using 3-D Particle Tracking Velocimetry (3-D PTV)". SAE Technical Paper 940279.

Verhoeven D., Vanhemelryck J. L. y Baritaud T. (1998) "Macroscopic and Ingnition Characteristics of High-Pressure Sparys of Single-Component Fuels". SAE Technical Paper 981069.

Winklhofer E., Fuchs H. y Philipp H. (1993) "Diesel Spray Combustion an Optical Imaging Analysis". SAE Technical Paper 930862.

Xu M. y Hiroyasu H. (1990) "Development of a New Optical Technique for Measuring Diesel Spray Penetration". SAE Technical Paper 902077.

Yule A. J. y Salters D. G. (1995) The Break-up Zone of a Diesel Sprays. Part 1: Length of Zone and Volume of Unatomized Liquid, Vol. 5, pp. 157-174. Atomization and Sprays.

Zhang L., Tsurushima T., Ueda T., Ishii Y., Itou T., Minamia T. y Yokota K. (1997) "Measurement of Liquid Phase Penetration of Evaporating Spray in a DI Diesel Engine". SAE Technical Paper 971645.

Zhao F. Q., Kadota T. y Takemoto T. (1991) "Temporal and Cyclic Fluctuation of Fuel Vapor Concentration in a Spark Ignition Engine". SAE Technical Paper 912346.

Zhao F. Q., Taketomi M., Nishida K. y Hiroyasu. (1993) "Quantitative Imaging of the Fuel Concentration in a S.I. Engine with Laser Rayleigh Scattering". SAE Technical Paper 932641.

Zhao H. y Ladommatos N. (1998) Optical Diagnostics for Soot and Temperature Measurement in Diesel Engines, Vol. 1, pp. 244-254. Department of Mechanical Engineering, Brunel University, U.K.

Zhao H. y Ladommatos N. (2001) Engine Combustion Instrumentation and Diagnostics, pp. 457-470. Society of Automotive Engineers, Inc.Warrendale, Pa.

3

Experimental Cells for Diesel Spray Research

1. Introduction

The study of the combustion process in Diesel engines has been going even deeper into depth with the application of new techniques of measure and more rigorous methodologies. This has taken into new expectations in the development of parametric studies and in the construction of tools (physical models or experimental setup) that allow the reproduction of similar thermodynamic conditions to the ones present in the inside of a cylinder in a real thermal engine, making it possible to obtain greater approximations between the theoretical relation and the experimental one.

2. Experimental setup classification

The physical models or experimental setup used to study the injection-combustion process usually are of very specific characteristics, depending on the phenomena to analyse. These models can be classified in the three following groups:

2.1. According to the type of the working fluid

-Models with cold working fluid are used to study sprays in non-evaporating conditions, but many times in conditions similar to the existing ones in a real thermal engine.

- Models with hot working fluid, this kind of model is the best way to study sprays in evaporating conditions and in the majority of times it's possible to simulate a real thermal engine in its temperature and density conditions.

2.2. According to the atmosphere type

-Models with inert atmosphere are used to study diesel spray without the presence of flame. Its greater application is basically the biphasic study of the spray (liquid and vapour).

-Models with reactive atmosphere, this kind of model is generally applicable to combustion studies and normally atmospheric air is used to achieve the reaction.

2.3. According to the geometry of the chamber

Models of constant volume vessel are used to study the diesel sprays in non reactive atmosphere and reactive atmosphere. This kind of model is not of common use, most of all on the development of studies of the sprays by means of reactive atmospheres since they require previous ignitions of mixtures of combustible gases to achieve high pressures and temperatures in the inside of the injection chamber.

Historically the first developed studies about combustion process in a thermal engine were by means of direct recording of the luminosity of the flame, and afterwards by the Schieren technique for studies of auto-ignition and the knocking combustion. The visualization involves the formation of visible images directly or indirectly by the action of light over sensible materials to the latter. Therefore the light and the recording of the formed images are essential to any form of visualization. The light can be generated by the object itself if it's luminous (auto-ignition or combustion). If it's not, the light source has to be deployed to illuminate the object to reflect (free injection). Traditionally the photographic films were the only way to register images for the chemical processes induced by the action of light, however in some applications; the photoelectric devices are now substituting the photographic film. Due to the nature of the engines combustions, each image must be taken in a very short amount of time to freeze any fluid movement or of the flame during the exposure. This can be obtained by intermittent recording with high speeds of the shutter to register an image from a cycle, and a sequence of images in function of the turning angle of the crankshaft along many engine cycles or by means of the recording of hundreds of images per second to obtain a image sequence within the same engine cycle (relating to the first recording case as a projection of a high speed monostable image, and the last one as high speed continuous, correspondingly). These recording methods and other techniques of measure have been evolving gradually and increasing their application to the different kinds of existing models, such as it will be mentioned in the subsequent sections.

The analysis of diesel sprays has had very important advancements beginning with the physical model implementation because during time, very specific studies have been developed. In the last decade of the 20th century, (Hiroyasu et al., 1989) was one of the pioneers in the use of injection models and in the implementation of visualization methods (Fraunhofer diffraction). Hiroyasu, in his investigations to study Sauter's medium diameter (SMD), used an injection model with the only possibility of doing experiments under low atmospheric conditions (pressures of the order of 3 bar and temperature of 285 K), which allowed relatively good microscopic studies to be made but with limitations to study the diesel spray under the thermodynamic variables that can be achieves in a real thermal engine.

To develop his experiment of SMD, (Minami et al., 1990) used a similar model to that one employed by (Hiroyasu et al., 1989), where the objective was to obtain the density conditions similar to the existing ones inside the combustion chamber of a real thermal engine. This author to develop his experiments used as the working fluid pressurized nitrogen at 20 bar and the injection pressures analysed were of the order of 2000 bar. These pressures could not be analysed using the model of Hiroyasu due to the low densities that were achieved in his model, so that it resulted irrelevant to do certain kinds of studies, mostly in high injection pressures. Minami employed the Fraunhofer technique to visualize the diameter of the droplets, using a lighting source based in a ruby laser beam pulsed at 30 ns and with a wavelength of 694 nm. Unlike Hiroyasu, Minami expanded the beam making it parallel to the spray through lenses, which afterwards is attenuated by the own fuel spray, where the transmission of light is focused by a parabolic mirror with which an image of the filmed spray on the 35 mm film is obtained. It was possible in this model to analyse the penetration of the spray by installing a high speed photographic camera to take the images and replacing the beam of light by halogen lamp as the light source.

There where basically two problems to clear up the interference of stripes in the photographic plane when using this technique:

- The first issue was due to the interferences caused by the thickness of the visualization window which was approximately 50 mm. This problem was solved by changing the trajectory of the beam between the reference and the object, using a intensity rate of 9:1 and fixing the point that focuses the lens of the relay as close as possible to the photographic plate.

- The second problem was generated by the collision of waves, which were produced in the interior of the chamber in a short amount of time during the period of injection. This was a consequence of the fuel injection velocity that surpassed the sound velocity. The collision of waves inside the combustion chamber made it difficult to take pictures because of the reflection inside it. The problem was solved by changing the time of synchronization of the injection time in reference to the picture taking, changing to times of 0,2 to 0,4 ms after start of injection.

The evaporated diesel spray studied by (Tabata et al., 1991) was made using a model similar to that presented by others scientist (Hiroyasu et al., 1989) and (Minami et al., 1990), with a small difference in its configuration. A heat system was installed in the interior of the chamber which had the objective of heating up the working fluid (nitrogen) to conditions that were optimal to achieve fuel evaporation. The main drawback studying evaporated diesel spray was the low range of available densities for high permissible temperatures. Due to this inconvenience it was impossible to reproduce similar thermodynamic conditions to those found in a thermal engine. It was limited to predict behaviours at low pressures (pressures inside the injection chamber in the order of 20 bar). (Higgins et al., 2000) studied the ignition of the spray and the behaviour of the pre-mixture of burn in a model of constant volume, where it was possible to achieve pressures inside the chamber of 350 bar and a temperature range of 800 K to 1100 K for densities between 7.27 kg/m3 y 45 kg/m3. This model, unlike those used by other authors (Hiroyasu et al., 1989), (Minami et al., 1990) and (Tabata et al., 1991) has two spark plugs and a ventilator to equalize the atmospheric conditions in the inside of the chamber, besides of having the walls electrically conditioned to simulate the conditions of the wall temperature found in a thermal engine. Another objective for this conditioner was to avoid the vapours of the fuel to condensate on the visualization windows. This model didn't require any kind of modifications to develop burning or free injection studies, unlike the one used by (Minami et al., 1990), which did need very sophisticated adaptations to be able to make more complex studies, as the burning study.

The effects on the diesel spray caused by the geometry of the chamber were studied by (Montajir et al., 2000). To develop the investigations they based themselves on an experimental installation which consisted in a small thermal engine of slow speed regime and with a combustion head of rectangular geometry adapted with a window for visualization. In this model as in the ones described previously, nitrogen was used as the working fluid (injected in the cylinder a temperature of 293 K and with a pressure inside the chamber of 45 bar) to visualize the diesel spray without combustion. To achieve the development of studies in different thermodynamic scopes with this model, the main option was to change the compression ratio. This was achieved by placing spacer rings between the cylinder and the combustion head. This modification used to be also very complex and expensive. Furthermore it didn't allow continuity in the experimental session because of the constant stops required to carry out mechanical changes in the installation.

The stratification effects of the combustion were studied by (Plackmann et al., 1998) using a model of constant volume, also called combustion constant volume vessel. Where the visualization took place through three quartz windows. One of the windows was located in the back side and the other two in the middle section of the pump. This model was pressurized with an air-propane mixture with pressures of 40 bar; these fluids were used as oxidant and test combustible respectively. The ignition took place by the means of two diametrically opposite electrodes which function as a spark plug to provoke the spark. One of the electrodes was connected to a high tension source that came from a discharge capacitive ignition system and the other one to mass, being this last system to be the one that caused the difference in electric potential to achieve the ignition of the mixture.

In their studies about the behaviour of gasoline direct injection (GDI), (Shelby et al., 1998) used an engine with three optical accesses made out of quartz. To carry out the visualization of burning the planar laser induced fluorescence (PLIF) as the technique of measure and a high speed camera to film the combustion process were used. The camera was placed perpendicularly to the beam, having with this the advantage of taking images directly. This thermal engine unlike the one used by (Montjair et al., 2000) had completely transparent walls, which allowed a complete capture of the phenomena (injection-combustion). Furthermore, it required complex routines at the moment of doing any changes in the installations configuration.

With the purpose of visualizing the atomization process in a gasoline direct injection system, (Preussner et al., 1998) used a model pressurized with oxygen (pressure and temperature of 20 bar and 673 K correspondingly). Where the combustion products of each cycle were constantly evacuated by the use of nitrogen and the visualization of the atomization process was done through windows located on the far ends of the injection chamber. Mie-Scattering was the technique of measure used during the experiments. In addition to this technique, the scientists developed tests with the laser induced fluorescence technique, which proved not to be the most appropriate for some measures in the gasoline direct injection systems, because it is an optical method of measure somewhat incoherent to detect concentration of species compared to Mie-Scattering. However, it was possible to use other techniques of measure. One disadvantage presented by this model was the way of evacuating the burn gases, because it was required to introduce nitrogen to the injection chamber when the cycle ended to expel these gases and was not recoverable, which made expensive the experimentation, besides of needing very long test routines because of discontinuity in the tests due to the type of cleaning used.

In his experiments to study the effects of vaporization, (Takagi et al., 1998) used an experimental installation based on a thermal engine of 2.0 litres provided with sapphire walls and only one optic access for image capturing. The optic access was placed in one section of the piston head, being the most appropriate collation place due to the geometric characteristics of the model. In the same way as Preussner, laser induced fluorescence was used as the technique of measure, obtaining extremely satisfying results in addition to confirm (Preussner et al., 1998) theories, who postulated that the laser induced fluorescence technique was no the best option to study the combustion effects due to the problems generated to detect species. Something that does not exist in the case of vapour concentration studies, as long as the fuel is mixed with tracers to make it behave as a fluorescent source (the combustible by its own nature tends to illuminate when its subject to high pressure and temperature, which is why tracers are required to carry out concentration studies, because the presence of tracers allows to define the different wavelengths for each element that constitutes the mixture). This model wasn't sufficiently flexible enough for the use of other techniques of measure due to the fact that the sampling process of images was of the intrusive kind, therefore making it an inappropriate method for the study of internal processes in an internal combustion engine.

To study the effects of the injection pressure and the diameter of the nozzle on the fumes emissions, (Siebers & Pickett, 2002) used a model consisting in a chamber of constant volume where it is possible to simulate similar thermodynamic conditions to the one in a Diesel engine (Figure 1). Other applications for this model are described in (Naber & Siebers, 1996), (Siebers, 1998), (Siebers & Higgins, 2001) and (Siebers et al., 2002). The chamber has four optic accesses placed orthogonally and equidistant, having the advantage of using different optical techniques simultaneously making more than one analysis in each experiment. But it also has disadvantages, mainly the way of obtaining the appropriate thermodynamic conditions in the chamber's interior. These conditions such as the temperature and density are achieved by causing the ignition (by means of a spark plug) of the combustible gases in the inside of the chamber, the method is unusual and in many cases even dangerous. The range of operation for the temperature as for the density is in the order of (600 a 1400) K and (3,6 a 60) kg/m^3 respectively. Furthermore, it's possible to regulate the concentration of oxygen in the interior of the chamber from cero (e.g., for inert conditions) to values greater than 21% in volume, depending on the type of study aiming for.

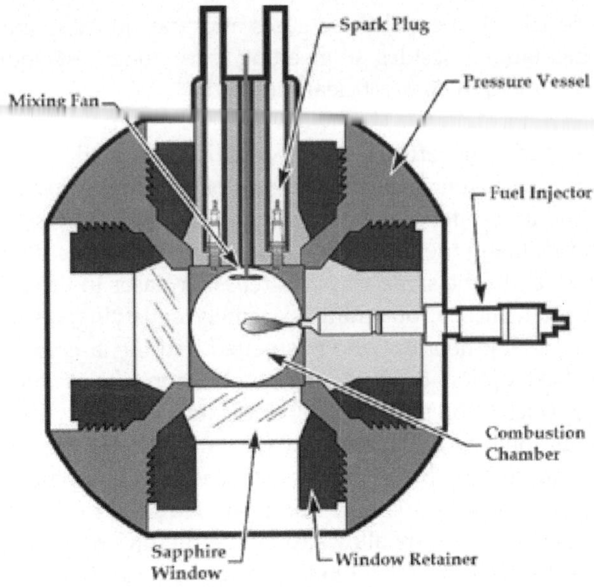

Fig. 1. Experimental setup: constant volume vessel (Siebers, 1998).

3. Proposal of the physical model

To achieve the optimal feasibility the physical model must always obey the following equivalence requirements:

-Reproduce inside the injection chamber, the thermodynamic properties that the working fluid has in a real thermal engine in normal operation.

-Easiness to sweep the atomized fuel in the injection chamber and maintain the visualization accesses clean.

-To make possible the studying of different phenomena it's indispensable to have versatility in the cases of the density magnitudes and in the working fluids thermodynamic properties, besides having flexibility in the geometry of the injection chamber (the geometry of the injection chamber can vary with small arrangements or by making modification of the supplementary type in the physical model).

- Allow repetitiveness in the tests when constant thermodynamic are maintained.

3.1. Operational Conditions

The operational conditions depend fundamentally on the characteristics of the working fluid in the injection chamber at the moment when the injection of fuel is initiated. These conditions commonly vary from one engine to another. When a physical model is designed, where the thermodynamic properties of the working fluid are not similar to those of atmospheric air, (generally heavy molecular gases are used) the model must be reproduced based on analogies, an equivalent atmosphere to that in a real thermal engine in low pressure and temperature conditions and these must be easily achieved, so that the development of the experimental study is allowed (Figure 2). It is to be noted that model of hot fluid that will be presented in the next sections doesn't need analogical processes because this physical model is in fact a real thermal engine. However, the thermodynamic variables that are desired to combine for each particular study must be correctly defined. On the one hand, it's a safety routine and on the other hand, it gives knowledge of the high pressure range that can be achieved in the combustion chamber. This last argument is fundamental for the defining of the thermodynamic conditions inside the cylinder (Arrègle, 1998).

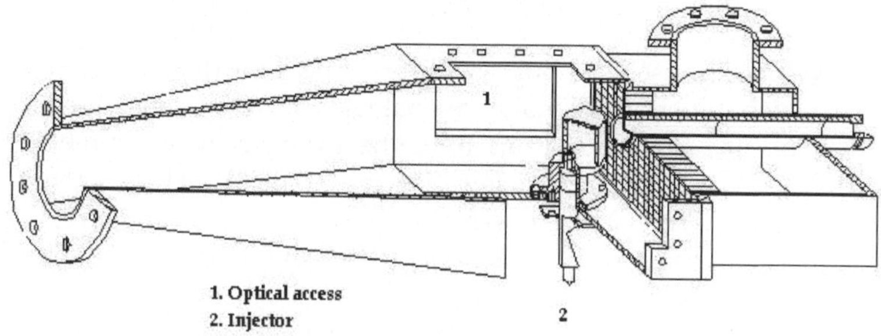

Fig. 2. Experimental setup: the work fluid is SF6 (test with low pressure and temperature)

1. Reynolds analogy: The viscous forces have a mayor role in the penetration of the spray in a controlled atmosphere, in this case a gaseous atmosphere and a diesel fuel spray. The Reynolds number and the relationship between the inertial and viscous forces must be conserved always so the penetration of the spray is not modified.

$$\mathrm{Re} = \frac{\rho_g \phi_{32} \upsilon}{\mu_g} \tag{1}$$

$$\mathrm{Re}^* = \frac{\rho_l^* \phi_{32}^* \upsilon^*}{\mu_l^*} \tag{2}$$

$$\mathrm{Re} = \mathrm{Re}^* \tag{3}$$

Maintaining constant spray characteristics: Sauter's medium diameter, the penetration velocity and the Reynolds number can be simplified as follows:

$$\frac{\rho_g}{\mu_g}[\text{engine}] = \frac{\rho_g^*}{\rho_g^*}[\text{experimental setup}] \tag{4}$$

2. Weber's analogy: in the atomization process of a liquid vane, the aerodynamic forces and the superficial tension are directly involved in the breaking transition of the droplets, which is why the conservation of Weber's number that associates these forces, is a main factor to consider in the injection process. Weber's analogy can be expressed as follows:

$$We = \frac{\rho_g \phi_{32} \upsilon^2}{\sigma_l} \tag{5}$$

$$We^* = \frac{\rho_g^* \phi_{32}^* \upsilon^{2*}}{\sigma_l^*} \tag{6}$$

$$We = We^* \tag{7}$$

The results can be directly used as long as Sauter's medium diameter and the injection velocity are kept, which implies the equality of analogies. It can be considered that during a very short amount of time of the atomization, the heat transfer between the working fluid and the fuel in the combustion chamber is low. The fuel starts to heat up when it is found as a droplet. Therefore, if a same fuel is used the superficial tension variation that occurs between the real thermal engine and the physical model will be very small. This hypothesis is confirmed by the results obtained by (Hiroyasu & Arai, 1990), in which differences during the atomization process at temperatures of the working fluid between 293 K and 573 K aren't seen. This is why Weber's analogy is simplified to a density conservation of the working fuel between the real thermal engine and the physical model.

3. Analogy equality: the equality of analogies between Weber and Reynolds is simplified as shown. This analogy between the real thermal engine and the physical model is in accordance to the theoretical and experimental results on atomization and spray penetration. These results are presented by the authors previously cited.

-Fuel Characteristics

$$\rho_l = \rho_l^* \tag{8}$$

$$\sigma_l = \sigma_l^* \tag{9}$$

$$\mu_l = \mu_l^* \tag{10}$$

-Injection pressure

$$\Delta p = \Delta p^* = P_{iny} - P_g \tag{11}$$

-Working fluid characteristics

$$Re: \left[\frac{\rho_g}{\mu_g} = \frac{\rho_g^*}{\mu_g^*} \right] \tag{12}$$

$$We : \left[\rho_g = \rho_g^* \right] \quad (13)$$

The most critical parameters that are usually present at the moment of insuring the analogy between the physical model and the real thermal engine are: the density and viscosity of the working fluid. Furthermore, it's necessary to calibrate the injector to achieve a pressure increasing and therefore, an exit velocity of the injected fuel similar to that happening in a real thermal engine in normal operation.

3.2. Geometric characteristics of the physical model

The geometric characteristics of the physical model must obey certain requirements. The most important one is in the injection fuel chamber, which must have optical accessibility to make proper measurements and observations. As it has been previously explained, to carry out the majority of experiments more than one technique of measure is required. This is why it is convenient to have several optic accesses (these access are generally windows, mainly out of quartz). The geometry of these accesses is in function of the test aimed to do, being in the majority of cases the front longitude of the spray the basis of study. In a real thermal engine the diesel spray length is conditioned by the diameter of the combustion pre-chamber (in a thermal engine of indirect injection) or by the piston bowl (in the case of thermal engines of direct injection). This length in a diesel spray of an internal combustion engine of self-propulsion can measure between 18 to 50 mm before the collision with the wall. However, in the majority of cases the sprays manages to collide with the wall of the bowl before the atomization process has ended. It's because of this reason the importance of studying the spray up to lengths in the order of 100 mm, sufficiently long for the spray to achieve complete atomization, having special attention to the injection chamber. A second requirement that the physical model's geometry must fulfil is the easiness of evacuation of the injected fuel inside the injection chamber to reduce the dirtiness of the optic accesses. Many scientists as seen in the previous section (Hiroyasu & Arai, 1990), (Hiroyasu et al., 1989) and (Minami et al., 1990) have used chambers of the volumetric type were the atmosphere is renewed after every injection. In using this methodology of measure, an injection system is required that allows having certain flexibility to inject discontinuously and making possible the admission of clean fluid before each injection cycle. This system usually doesn't allow the use of many optical techniques because in very particular cases, it's necessary the use of tenths of injection cycles to do a complete spray study, having very long, arduous and costly test sequences.

In this section the development, designed and tune-up of an experimental installation is presented. The main objective is to carry out studies of the evaporated diesel spray in inert and reactive conditions. To achieve this purpose a model has been brought up based on a two stroke thermal engine and a group of auxiliary equipment that allow the adjustment of operational parameters of the installations and of control for the various independent variables that interact in the injection-combustion process.

3.2.1. Objectives of the experimental installation

-Simulate the process of direct injection of diesel fuel in high pressure, density and temperature conditions.

-Study the microscopic and macroscopic characteristics of the diesel spray in evaporative conditions.

-Carry out experiments of the diesel spray in an inert and reactive atmosphere.

3.2.2 Design requirements and determining factors

The experimental installation systems must allow the following functions and capacities:

1. Control and regulate the working fluid's temperature and pressure.
2. Optic accesses to visualize the diesel spray by means of visualization techniques.
3. Modification of characteristic parameters of the injection.
4. The injection chamber must have a minimum number of movements in relation to the visualization equipment.

3.2.3. Observations relative to system design

The most relevant aspects that must be considered in the design of the main systems that constitute the experimental installation based on a thermal engine are described next:

1. To be able to have high pressures and temperatures of the working fluid in the admission system.
2. Having a good regulation of the exhaust backpressure.
3. The thermal engine must be thermally fitted by an external system that allows it to have optimum operational conditions.
4. There must be an injection system for high pressure independent of the thermal engine.
5. The vibration control to reduce vibrations transmitted by the thermal engine to its surroundings must be controlled by a limitation vibration system.

The experimental installation allows switching between two different configurations to study the diesel spray in evaporative conditions. These configurations are detailed as follows:

1. Open loop configuration, reactive atmosphere: this configuration allows studying the diesel spray injection in an oxidant atmosphere. The working fluid used to fill the control volume where the fuel is injected is atmospheric air. When the combustion is executed, the gases and the product species of the latter are then evacuated to the atmosphere after each operational cycle. Afterwards, fresh air is introduced in the cylinder to carry out a new injection.

2. Closed loop configuration, inert atmosphere: This configuration allows studying the diesel spray in an inert atmosphere in which there's no oxidation of the injected fuel. To achieve this, a nitrogen supply system has been incorporated to the installation with which the injection chamber is filled.

Experimental installation characteristics

-Repetitiveness of the various tests so an excellent stability in the different operation points is achieved.

-The precision in the measures demands quality of the measuring equipment and an accurate calibration of them to avoid systematic errors during the tests.

-Having reliability and safety assumes a rigorous fitting of the room where the experimental installation is housed.

-Good accessibility and compatibility to facilitate the control, verification and proceeding tasks on all the thermal engine systems.

3.2.4. Reactive atmosphere configuration

The reactive atmosphere configuration or open loop is useful for carrying out studies of the injection, self-ignition and combustion processes, in the same way to carry out analysis in the combustion process and measures of contaminant emissions. Figure 3 shows the schematic diagram of the experimental installation that is used in the study of diesel sprays in a reactive atmosphere (Martínez et al., 2007) and (Bermúdez et al., 2000, 2003).

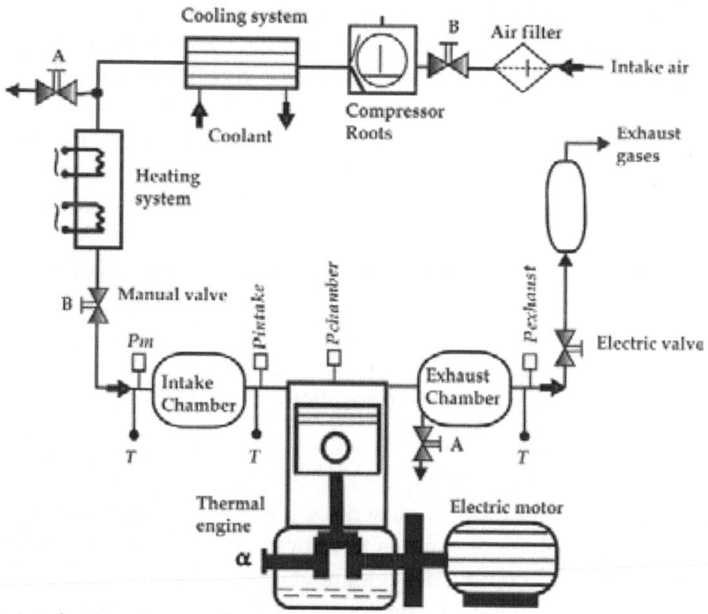

Fig. 3. Experimental setup in open loop configuration.

Description of the elements that constitute the experimental installation, impurity filter: A commercial filter is used to purify air from the impurities of the environment:

-Compressor: A Roots compressor is used to boost air from the environment towards the engine's admission, as well to achieve the sweep of gases produced by the combustion.

-Heat exchanger: it is of the armour and pipe type (water-air). Its function is to regulate the exit air temperature.

-Heater: prepares the air of admission through a resistance heater which is installed in series with the water-air exchanger.

-Pm: the medium pressure sensor absolute scale is in charge of sending a voltage signal to the PID controller so it can regulate the opening of the electro valve. The aperture of the electro valve is in function of the target pressure established in the admission.

-Pintake: is the instantaneous pressure sensor in absolute scale that measures the pressure in the admission chamber, which is used as a reference for the medium pressure in the cylinder.
-Pexhaust: is the instantaneous pressure sensor in absolute scale that measures the pressure in the exhaust chamber. This pressure measurement is used as an operating reference for the exhaust backpressure electro valve.

-T: are the thermocouples type k: the thermocouple is located downstream of the admission chamber and is responsible of measuring the air's temperature just at the entrance of the admission chamber. This signal is sent to a PID that rules over the heater based on the established target temperature. The thermocouple is located upstream from the admission chamber and measures the temperature just before it enters the admission.

-Electro valve: to increment the pressure of the working fluid in the injection chamber it's necessary to establish a backpressure in the exhaust to avoid a short cut of the working fluid from admission to exhaust. This is achieved by closing the exhaust systems electro valve.

-A: represents the valves used to drain the vapour from the circuit, as of residual fuel as well of the oil that accumulates on the ducts.

-B: represents the sphere valves used to free the circuit elements. These are mainly used when the installation operates with inert atmosphere.

3.2.5. Inert atmosphere configuration

The inert atmosphere configuration or closed loop is used for carrying out studies of the injection process in liquid phase as well as in vapour phase. This configuration allows the spray to evolve through the combustion chamber presenting the two phases (liquid and vapour) without permitting flame during the process. This is achieved by a very simple way, the spray is injected in a control volume (combustion chamber) were the working fluid is a inert gas (e.g. nitrogen), therefore a volume free of oxygen molecules but with similar properties to those of air.

Figure 4 shows the schematic diagram of the experimental installation used to study diesel sprays in an inert atmosphere.

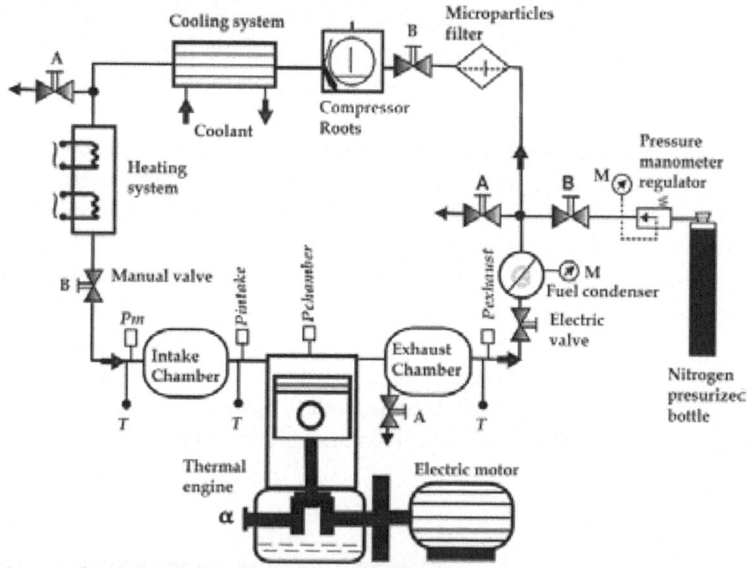

Fig. 4. Experimental setup in close loop configuration.

Description if the elements that constitute the experimental installation. All of the elements that constitute the adapted experimental installation to study diesel sprays in a reactive atmosphere are used in inert atmosphere, with the exception of the expansion chamber and the filter of particles. Additionally to these, two elements which are indispensable for this configuration are included. These elements are described next:

-Condensate separator of the collision type: the configuration in a closed loop must have nitrogen recirculation without the presence of fuel, therefore the injected fuel in the combustion chamber and the gas in which the latter has been injected to, must be separated once they have been expelled from the cylinder. The condensate separator is in charge of the first phase of separation. Upon circulating the mixture (fuel-gas) through the discharge system and arriving at the separator, where the fuel is then condensed into small drops that go to the bottom of the vessel and the gas and other particles that are still mixed are separated by a filter f micro particles (see next section). The liquid fuel concentrated at the bottom of the vessel is purged by a system of volumetric valves and disposed thereafter.

-Filter of micro particles: the second phase of gas purification is carried out through a filtering of high purity that allows the gas to be separated from possible fuel residues that haven't condensate in the previous phase, or from possible solid particles dragged from the circulation system or other particles loosed by the engine. The filter of micro particles allows the separation of particles up to 0.001 μm.

3.2.6. Operation mode

The complexity of operation and precision of the measures is of the same magnitude as on the reactive atmosphere configuration and also the mechanism for the circulation of the working fluid. When the installation is operating in an inert atmosphere, the nitrogen is boosted by the Roots compressor through the conditioner system and afterwards introduced into the cylinder. After the fuel is injected and mixed with gas, this mixture (fuel-nitrogen) is expelled towards the condensate separator system and to the high efficient filter, where the larger drops of fuel are first condensed and then the solid particles dragged by the gas through the circuit elements or the engine are eliminated. Once the nitrogen has been purified, it is driven up to the Roots compressor admission where it is boosted again towards the motor to carry out a new injection cycle. A pressurised nitrogen deposit is used to fill up the close loop circuit before initiating operations of the installation. Once the installation is operating, the deposit with help of an electro valve and a PID regulator are responsible of maintaining a constant volume of nitrogen circulating through the system since some small gas leaks towards the engines crankcase through the looseness between the piston and the cylinder may exist. The Figure 5 (a) shows a schematic diagram of the combustion chamber through which is possible to visualize the injection-combustion process. Figure 5 (b) shows an example of an image sequence of a multiple spray. Figure 6 (a) and (b) show a sequence of the development of an axi-symetric spray in an inert and reactive atmosphere respectively.

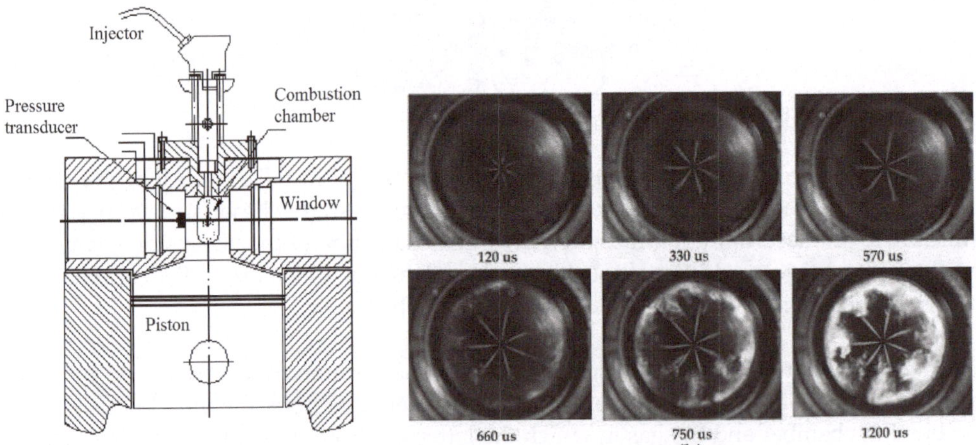

(a) (b)

Fig. 5. (a) Schematic diagram of combustion chamber and (b) Example of multi diesel spray.

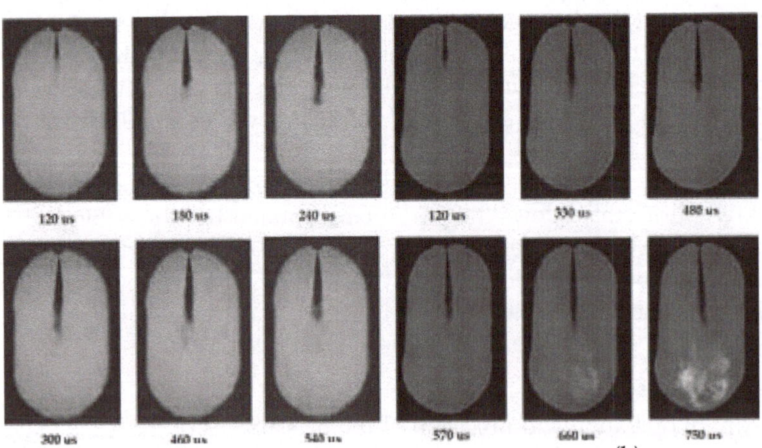

(a) (b)

Fig. 6. (a) Example of diesel spray in inert atmosphere and (b) example of diesel spray in reactive atmosphere.

4. Conclusions and remarks

In this chapter, several experimental setup based on different models have been presented. This experimental setup allows a basic study of the injection and combustion processes. The models have diverse applications and their use depends on the kind of study that is intend to carry out, these models are limited in some aspects, chiefly in the work fluid renewal, likewise is used to being complicated to reproduce the thermodynamic conditions that are presented in the combustion chamber of a real thermal engine. The experimental setup that is presented and based on a thermal engine is a excellent tool that permits to reproduce thermodynamic conditions and of operation as the existing in a engine in standard conditions. The cylinder head that has arranged the engine has three optical accesses which is allow visualization and the use of optical techniques. Diverse techniques have been utilized in this experimental setup; like LDA, PDA, PLIEF, PIV, Spectroscopy, Raman and Rayleigh Scattering, Two colors, and Shadowgraph, principally. These studies have permit to characterize the diesel spray in evaporating conditions to obtain relevant information that permit to improve the combustion processes in modern diesel engines.

5. References

Arrègle J. (1998) Analysis of the Structure and Internal Dynamic of the Diesel Spray. PhD Dissertation, Universidad Politécnica de Valencia, Spain,.

Bermúdez V., García J. M., Juliá E. and Martínez S. (2002) "Instalación Experimental para el Estudio del Proceso de Inyección-Combustión en Motor Diesel de Inyección Directa". XV Congreso Nacional de Ingeniería Mecánica, Cádiz, Spain.

Bermúdez V., García J. M., Juliá E. and Martínez S. (2003) "Engine with Optically Accessible Cylinder Head: a Research Tool for Injection and Combustion Processes". SAE Technical Paper 2003-01-1110.

Hinggins B., Siebers D. L. and Aradi A. (2000) "Diesel-Spray Ingnition and Premixed-Burn Behavior". SAE Technical Paper 2000-01-0940.

Hiroyasu H. y Arai M. (1990) "Structures of Fuel Sprays in Diesel Engines". SAE Technical Paper 900475.

Hiroyasu H., Arai M. and Tabata M. (1989) "Empirical Equations for the Sauter Mean Diameter of a Diesel Spray". SAE Technical Paper 890464.

Minami T., Yamaguchi I., Shintani M., Tsujimura K. y Suzuki T. (1990) "Analysis of Fuel Spray Characteristics and Combustion Phenomena under High Pressure Fuel Injection". SAE Technical Paper 900438.

Martínez S., Sánchez F., Rodríguez G., Riesco J and Gallegos A. (2007) "Simultaneous Measurement of Evaporating Fuel Spray Using Laser Induced Exciplex Flourescence" International Journal of KONES.

Martínez S., Sánchez F., Riesco J., Gallegos A and Aceves S. (2007) "Liquid penetration length in direct diesel fuel injection" Applied Thermal Engineering.

Montjair R. M., Tsunemoto H. and Minami T. (2000) "Fuel Spray Behavior in a Small DI Diesel Engine Effect of Combustion Chamber Geometry". SAE Technical Paper 2000-01-0946.

Naber J. D. y Siebers D. L. (1996) "Effects of Gas Density and Vaporization on Penetration and Dispersion of Diesel Sprays". SAE Technical Paper 960034.

Plackmann J. D., Kim T. and Ghandhi J. B. (1998) "The Effects of Mixture Stratification on Combustion in a Constant Volume Combustion Vessel". SAE Technical Paper 980159.

Preussner C., Döring C., Fehler S. and Kampmann S. (1998) "GDI Interaction Between Mixture Preparation, Combustion System and Injector Performance". SAE Technical Paper 980498.

Shelby M. H., VanDerWege B. A. and Hochgreb S. (1998) "Early Spray Development in Gasoline Direct Injected Spark Ignition Engines". SAE Technical Paper 980160.

Siebers D. L (1998) "Liquid-Phase Fuel Penetration in Diesel Sprays". SAE Technical Paper 980809.

Siebers D. L. and Higgins B. (2001) "Flame Lift-Off on Direct-Injection Diesel Sprays Under Quiescent Conditions". SAE Technical Paper 2001-01-0530.

Siebers D. L., Hinggins B. S. y Pickett L. M. (2002) "Flame Lift-Off Length on Direct-Injection Diesel Fuel Jets: Oxygen Concentration Effects". SAE Technical Paper 2002-01-0890.

Siebers D. L. and Pickett L. M. (2002) "Injection Pressure and Orifice Diameter Effects on Soot in DI Diesel". Congreso THIESEL-2002, Valencia, Spain.

Tabata M., Fujii H., Arai M. and Hiroyasu H. (1991) *Mean Drop Diameter of a Diesel Spray in Vaporizing Process*. 1991.

Takagi Y., Itoh T., Muranaka S., Liyama A., Iwakiri Y., Urushihara T. and Naitoh K. (1998) "Simultaneous Attainment of low Fuel Consumption, High Output Power and Low Exhaust Emissions in Direct Injection SI Engines". SAE Technical Paper 980149.

4

Experimental study of spray generated by a new type of injector with rotary swinging needle

1. Introduction

Numerous papers have indicated that the present standard systems of injecting fuel directly into the combustion chamber, in the most economical, direct-injection diesel engines, have reached the limits of development. In order to maintain the emission of toxic components of exhaust gases within the ranges defined by both EURO V and projected EURO VI standards, various modifications to the combustion system will be necessary. It is known that injection through a conventional multi-hole nozzle, in combination with induced swirl in the air in the chamber, does not ensure that particulates and oxides of nitrogen are not formed. These toxic components are among the most difficult to subsequently remove from the exhaust gases (Hiroyasu & Arai, 1990; Peake, 1997; Kuszewski & Lejda, 2009). Effective limitation of NO_X and PM emission can be achieved by providing proper macro structure parameters of the spray. First of all, it is very important to distribute fuel effectively in the combustion chamber, considering the rotary motion of the air and the spray shape (Beck et al., 1988; Dürnholz & Krüger, 1997; Kollmann & Bargende, 1997). Next, a larger range of the spray front and apex angle of spray is needed (Kollmann & Bargende, 1997; Kuszewski & Lejda, 2009). Moreover, it is important to know the distribution of fuel in the spray, because then it is possible to select the correct rotary motion of the charge in the cylinder, which will guarantee as effective a mix of the air and the fuel as is possible. A fuel spray with a different macro structure might have an important effect on these processes; however it is necessary to use a different type of injector, a new type of construction and modus operandum (Varde and Popa, 1983; Kuszewski, 2002). The design for such a device forms the subject of this paper (Metz & Seika, 1998; Szlachta & Kuszewski, 2002; Szymański & Zabłocki, 1992). The special feature of the spray nozzle of this injector is the variability of the fuel-spraying holes during injection. The variability of the cross-sections of these holes is achieved by the rotary swinging movement of the needle (RSN injector). The results of investigations described below show that a spray generated by this injector design has macro-structural parameters that differ from those of a standard traditional injector.

2. The construction and principle of operation of an injector with rotary swinging needle movement

The construction of an injector with a rotary swinging needle movement is shown in Fig. 1. Fuel is supplied through a high-pressure tube to the upper part of the injector body. The fuel pressure acts on the working area 'b' of a piston, 7, joined by a pin, 13, to an injector needle driver, 14. This causes the driver to rotate simultaneously with the rotation of the needle, 17, which is joined to the driver.

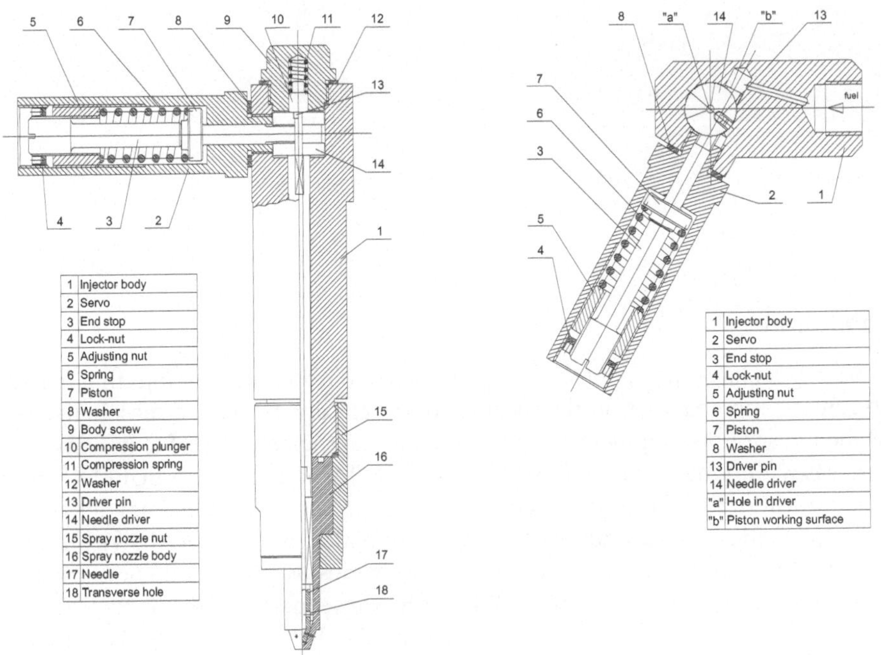

Fig. 1. The construction of an RSN injector

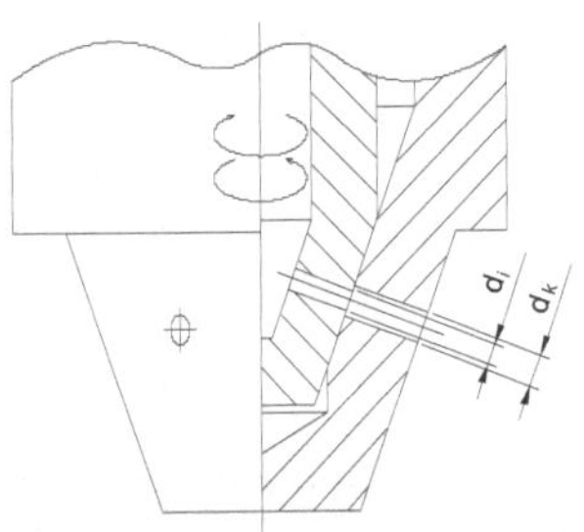

Fig. 2. Magnified part of RSN spray nozzle, d_k: outlet hole diameter in the spray nozzle body; d_i: outlet hole diameter in the needle

Fuel flows to the needle through a hole 'a', in the needle driver and through a fissure between the driver and the injector body. From there the fuel flows through an axially-symmetrical hole in the needle, both to the holes in the needle (spraying holes) and in the injector body (outlet holes), on the cone side surface which seals the needle (Fig. 2). The release of fuel from the sprayer starts when the needle-performing a rotary swinging movement-causes a misalignment of the spray and the outlet holes, so that they are momentarily occluded, resulting in a non-circular area of flow (Fig. 3). This area increases in proportion to the increasing angle of needle rotation. It should be mentioned that during the initial phase (low rotary angle), the needle has a 'dead' movement, which prevents increase of the flow area. The rotation of the needle in one direction continues so long as the axes of the spraying and outlet holes overlap. In this position, the piston of the driver rests on the end stop, 3, which acts as a regulatory element, limiting needle movement. The release of fuel causes a decrease in fuel pressure, which causes the spring, 6, to move the piston, 7 (moving in the servo, 2), screwed into the injector body, until it reaches a position corresponding to the closure of the injector. During this movement the flow area from the injector decreases to zero. The unique feature which distinguishes the new construction from the standard construction is that fuel flows out through a non-circular area, quasi-lenticular in shape, during the beginning and end phases of injection. During injection, fuel is throttled only in this non-circular area.

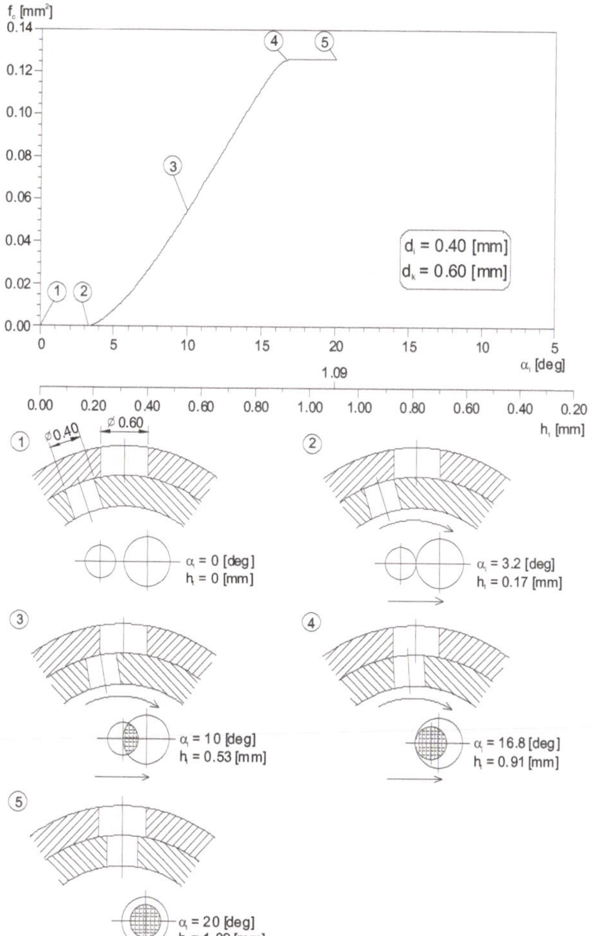

Fig. 3. The diagram of outlet hole in the spray nozzle body and outlet hole in the needle position and growing of geometrical flow area, f_c: geometrical flow area; α_i: angle of needle rotation (part 17 on Figure 1); h_t: piston stroke (part 7 on Figure 1).

3. Method of investigations

The parameters of the macrostructure of the stream of sprayed fuel were determined on the basis of measurements carried out using specially constructed equipment, which enabled both the direct observation of the development of the spray during the fuel injection to a chamber of fixed volume (Metz & Seika, 1998; Szlachta & Kuszewski, 2002; Szymański & Zabłocki, 1992) and the measurement of the fuel distribution within the spray of droplets. The scheme of test stand for visual studies is shown in Fig. 4.

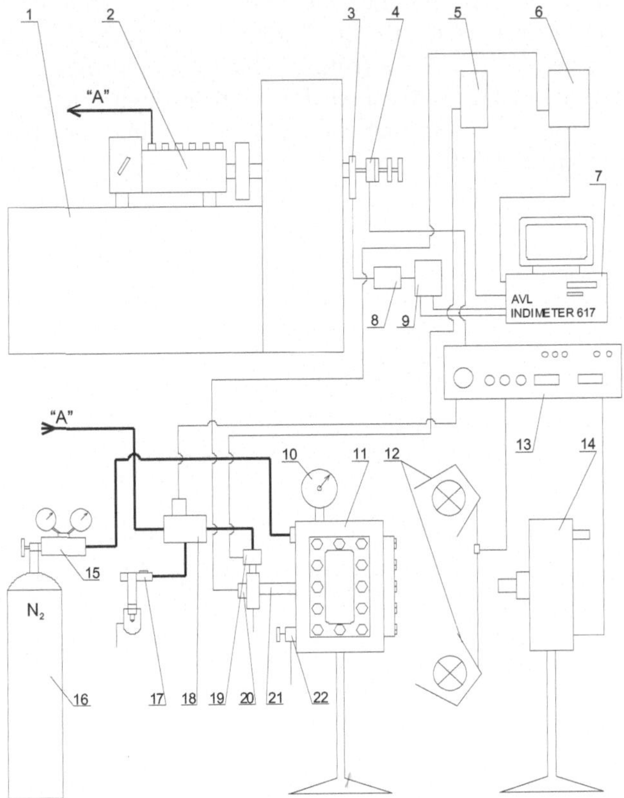

Fig. 4. Scheme of test stand for visual studies, 1: test bench Hansmann EFH 5008; 2: injection pump P56-01; 3: position and rotational speed sensor of pump shaft AVL; 4: synchronizing plates; 5: charge amplifiers AVL 3056; 6: tensometric bridge; 7: computer with data acquisition device for high-speed courses recording AVL Indimeter 617; 8: signals decoder; 9: feeder; 10: manometer; 11: visualization chamber of constant volume; 12: lights; 13: driver of high-speed camera; 14: high-speed camera; 15: pressure regulator with manometers; 16: nitrogen cylinder; 17: auxiliary injector; 18: switching valve; 19: piezoquartz pressure sensor AVL 5QP6002; 20: inductive position sensor of injector piston (or needle); 21: tested injector; 22: release valve

The essential elements of test stand were: visualization chamber of constant volume, where a fuel was injected by using tested injector, 21, test bench ,1, injection pump, 2, for pumping the fuel to injector, the high-speed camera, 14, with driver, 13, and the computer with data acquisition devices for high-speed courses recording 7. Inside the chamber to which fuel was injected, there was fixed backpressure verified with manometer, 10. For safety reasons, gas filling the chamber was nitrogen, passed from the cylinder 16. The fuel for tested injector, 21, was supplied from the injection pump, 2, driven by an electric motor of test bench, 1. Switching valve, 18, referred fuel from injection pump to the tested injector, 21, (at the time of visualization test of stream development) or to the auxiliary injector, 17, (in the intervals between fundamental tests). To record of fuel stream images, it was necessary to synchronize the work of high-speed camera, 14, the test bench, 1, the lights, 12, and switching valve, 18. It was obtained using a special camera driver, 13, co-operating with synchronizing plates, 4, fixed to the driving shaft of the test bench engine. Additionally, during work of the high-speed camera, using piezoquartz pressure sensor, 19, the values of pressure before the injector were recorded. Furthermore, using inductive position sensor, 20, it was recorded displacement of injector needle (or a piston in the injector of new type) and using an optical rotation speed sensor, 3, – it was recorded rotational speed of injection pump camshaft. The record of these parameters was possible thanks to the computer, 7, with a special data acquisition device. The stream images were recorded at a speed of 5000 fps. The successive images of a developing fuel stream were recorded every 0.0002 s (0.2 ms). Next, the films of width 16 mm were scanning and digital images were analyzed using a computer.

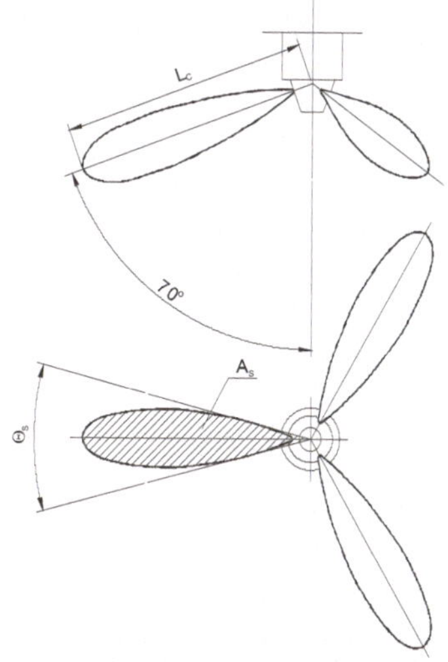

Fig. 5. The method of determining the values of the parameters of the fuel spray, L_C: range of the spray front; Θ_S: apex angle of the spray; A_S: area of the spray projection in a plane perpendicular to the injector axis

The visual studies enabled the following to be analysed: the range of the spray front – L_C, the apex angle of the spray – Θ_S, and the area of the spray projection in a plane perpendicular to the injector axis – A_S. The last criterion of the spray macrostructure estimation was introduced because of the irregular shape of the spray generated by the RSN injector. The method of determination of the values of the analysed parameters of the spray of injected fuel is depicted in Fig. 5. A standard injector with a D1LMK 140/M2 pattern sprayer, and the new RSN injector were compared. Both sprayers had three outlet holes, the diameter of the holes in the standard injector body and the RSN type being equal.

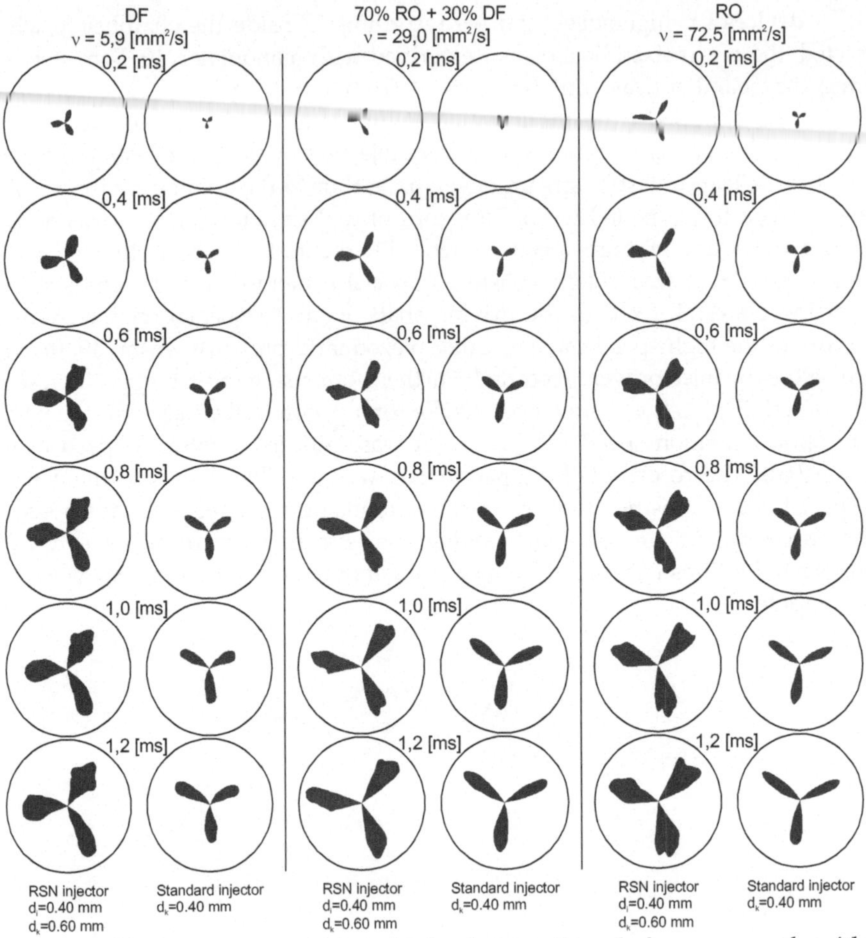

Fig. 6. Pictures of fuel spray propagation obtained at use three-hole spray nozzle with rotary swinging needle movement and standard (p_o = 170 bar, p_b = 20 bar, q = 130 mm^3/injection, n_p = 600 rpm, fuel: DF, 70% RO + 30% DF, RO), d_k: outlet hole diameter in the spray nozzle body; d_i: outlet hole diameter in the needle; v: kinematic viscosity; DF: diesel fuel; RO: rape oil

In Fig. 6 the pictures of fuel spray propagation obtained at use three-hole spray nozzle with rotary swinging needle movement and standard are shown. From the figure 6 it is clear that the spray generated by injector with rotary swinging movement of the needle is developing in a different way than in the standard nozzle. It causes differences in values of used parameters to assess the macrostructure of fuel spray. In particular it is clear, that the spray forming by injector with a rotary swinging needle movement is irregular in shape, and its area (area of the spray projection in a plane perpendicular to the injector axis), the apex angle of the spray and the front range are usually significantly higher compared to the classical sprayer. Particularly noteworthy are the results presented in Figure 6 concerning rape oil, which is characterized by a high value of kinematic viscosity (v = 72.5 mm^2/s). As we can see, the area occupied by rape oil spray from the RSN needle movement is much larger than achieved for classical sprayer.

4. The range of the spray front

In both injectors the following values were set, being the same for each type: line pressure at the opening of the sprayer p_o = 170 bar, fuel dose q = 130 mm^3/injection and rotary velocity of the camshaft of the injection pump n_p = 600 rpm. Fuels of different viscosity (Diesel Fuel (DF), Rape Oil (RO), and 70/30 RO/DF mixture), were injected into the observation chamber, which was filled with nitrogen at pressures of 15, 20, and 25 bar. During all investigations of the range of the spray front, the apex angle and surface area of the spray, a time scale of 0–1.4 ms was chosen. Beyond this range, for some injection parameters, the spray front reached the walls of the observation chamber. The range of the spray front for DF, formed by the RSN sprayer under various values of the background pressure in the observation chamber, is presented in Fig. 7. For RSN injector, it can be seen that an increase of nitrogen pressure in the observation chamber caused – as was expected – a reduction in the range of the spray front. This phenomenon is characteristic of standard sprayers, and may be ascribed to the effect of the aerodynamic resistance on droplets of variable size. An increase in the background pressure (gas density), causes an increase in aerodynamic resistance, and reduced dynamic pressure of the gas into which the injection is made, creating adverse conditions for the disintegration of secondary droplets. Therefore, larger droplets with greater penetrative capability are formed (obviously a larger droplet has greater kinetic energy and will therefore travel further).

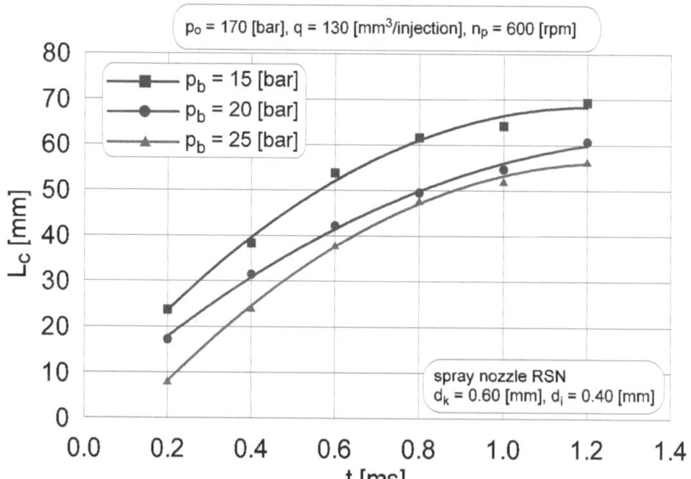

Fig. 7. The range of the Diesel Fuel spray front formed by the RSN type, at various background pressures in the observation chamber

The greatest range of the front of the DF spray formed by both the standard injector (Fig. 8) and the RSN injector, was observed at p_b = 15 bar. However, at p_b = 20 bar, the range of the front was less than at p_b = 25 bar. Most probably this was because during the analysed (single) injection, at p_b = 25 bar, the initial velocity of fuel at the sprayer outlet was higher than that at p_b = 20 bar. This was caused by the greater difference between the line pressure and the pressure in the observation chamber. Therefore, the greater kinetic energy of the spray at p_b = 25 bar, had a stronger influence on the movement of the front of the spray than the enhanced aerodynamic resistance of the environment. A comparison of Fig. 7 and 8 shows that, generally, the range of the front of the spray generated by the RSN sprayer is greater than that of the standard injector.

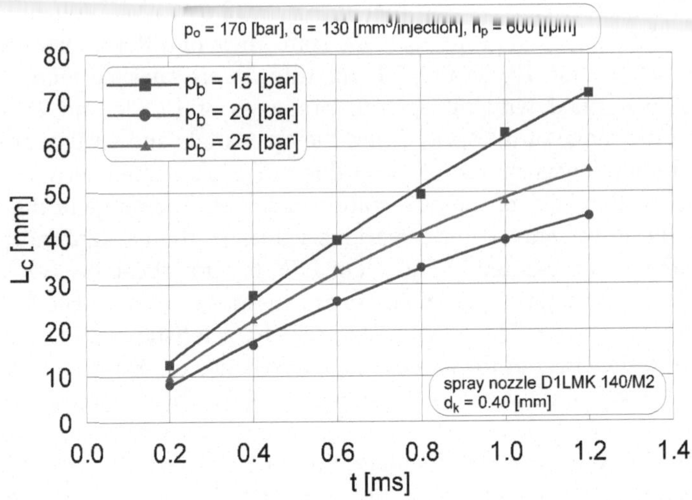

Fig. 8. The range of the Diesel Fuel spray front formed by the standard injector, at various background pressures in the observation chamber

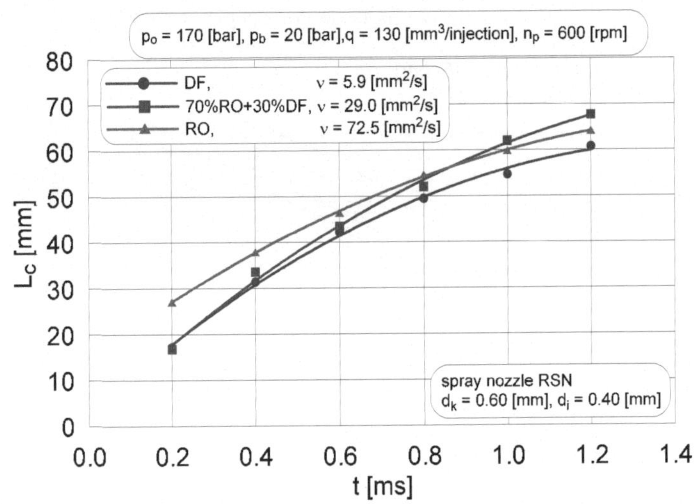

Fig. 9. The range of the front of the spray, formed by the RSN injector for fuels differing in physical properties

As could be expected, the use of fuels of considerably greater viscosity affected both types of injectors by considerably increasing the injection pressures. This was caused by a reduction in the value of the index of fuel outflow from the sprayer holes. These changes were the main contributors to the increased spray front range for fuels of increased viscosity (RO – $\nu = 72.5\ mm^2/s$; 70/30 RO/DF – $\nu = 29.0\ mm^2/s$), in relation to (DF – $\nu = 5.9\ mm^2/s$) – see Figures 9 and 10. An additional reason for the increased range of the spray front when using higher viscosity fuels (observed for both types of injectors), was probably due to the increase in droplet size, when conditions conducive to their disintegration became worse.

From a comparison of Fig. 9 and 10, it may be seen that – as in the case of DF – the spray range of other fuels was greater for the RSN injector over the entire time of spray development.

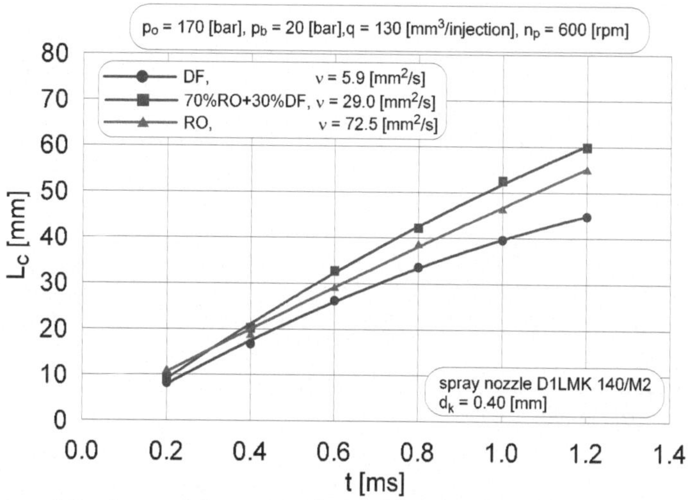

Fig. 10. The range of the front of the spray, formed by the classical injector for fuels differing in physical properties

5. The apex angle and surface area of the spray

In Fig. 11 it may be seen that, in the case of the RSN sprayer, a change in background pressure did not significantly affect the values of the apex angles of the spray over the whole period of its development. However, the spray surface area varied, the greatest area being observed for p_b = 15 bar, i.e., at the background pressure at which the range of the spray was greatest.

Conversely, in the case of the standard injector, the effect of p_b on the apex angle Θ_S was more visible – cp. Fig. 12. As could be expected, the largest apex angles occurred at maximum background pressure. The values of the apex angles of the spray diminished during its development, i.e., the penetration of the spray in a direction perpendicular to its axis was reduced; this has a negative effect on mixing. It may be only partly compensated by the fact that the spray surface area increases with its development. The smallest surface area of the spray was recorded during the intermediate background pressure, p_b = 20 bar, i.e., for a value corresponding to the shortest range of the spray front.

From a comparison of Fig. 11 and 12 it will be seen that the values A_S, achieved by the RSN injector, were greater than for the standard injector. It may also indicate the superior properties of the spray from the RSN injector, due to improved air/fuel mixing processes. The larger area of the spray allows distribution of the fuel around the combustion chamber of DI engine much effectively. In this case it is possible to reduce a rotary motion of the charge. Too strong rotary motion of the charge can lead to sprays overlapping and can cause the coalescence of fuel drops. It is unfavourable on account of PM formation.

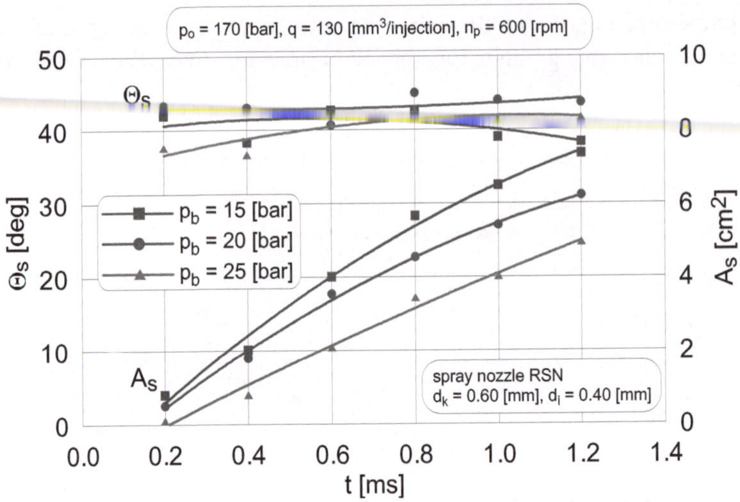

Fig. 11. The apex angle and surface area of the spray formed by the RSN type at various background pressures levels

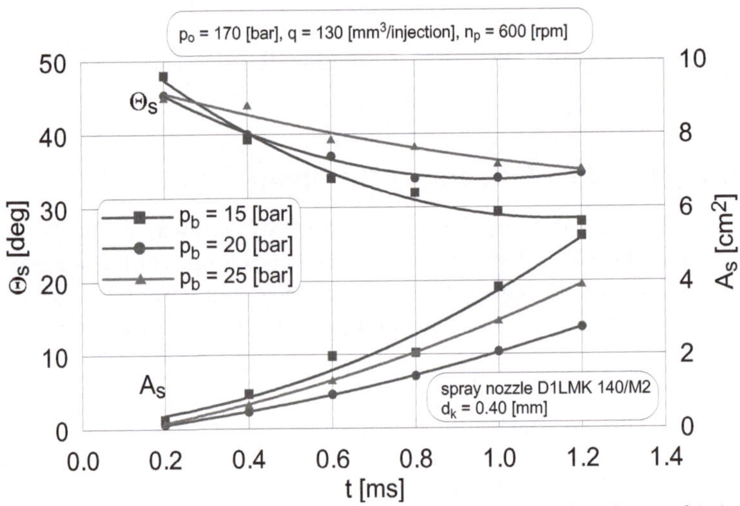

Fig. 12. The apex angle and surface area of the spray formed by the classical injector at various background pressures levels

The application of fuels with increased kinematic viscosity had little effect on the surface area of the spray, A_S (Fig. 13 and 14). At the same time, it may be noted that the dimensions of this area are much greater for the RSN-type than for the standard injector.
The value of the spray angles generated by the standard injector decreased inversely as the sprays developed. The value of the angle was virtually independent of the type of fuel used. On the other hand, in the case of the RSN sprayer, the apex angle of the spray was dependent not only on the time of the spray development, but also on the type of fuel. It is significant that the largest values of these angles were found in fuels with the lowest viscosities and surface tension (DF). They did not change during the spray development period. It is very likely that the smaller drops deviated more acutely towards outside the

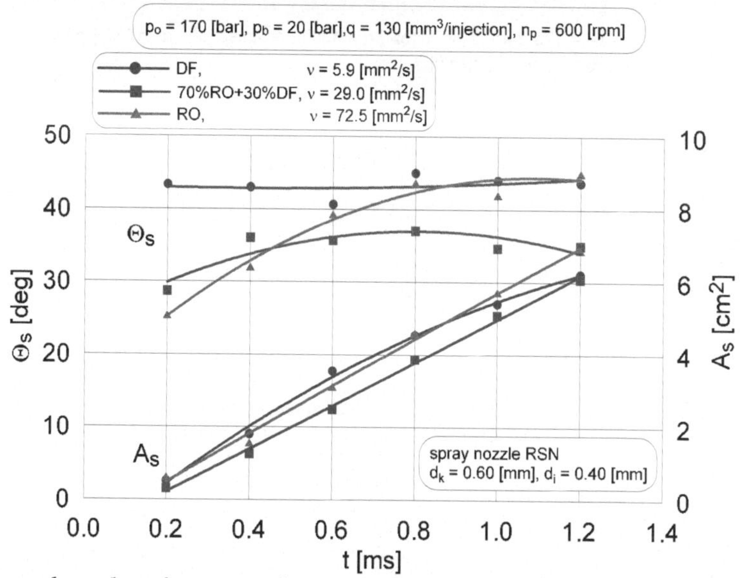

Fig. 13. Apex angle and surface area of spray formed by the RSN model when spraying fuels differing in physical properties

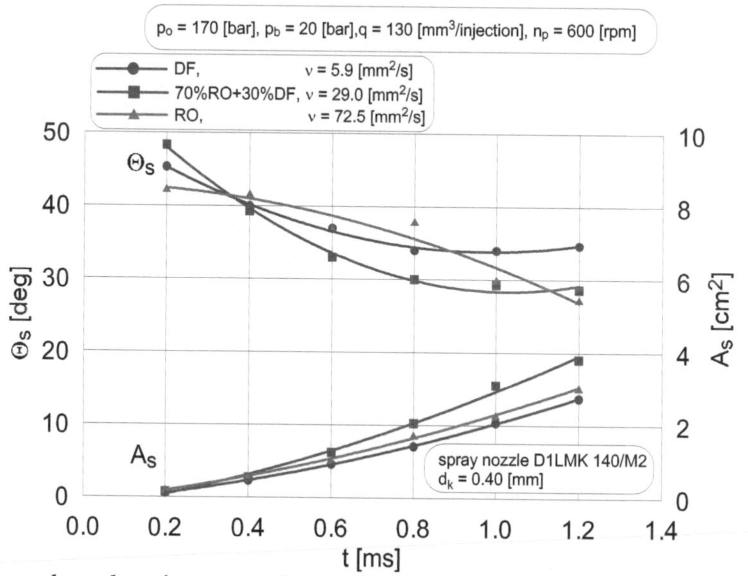

Fig. 14. Apex angle and surface area of spray formed by the standard injector, spraying fuels with different physical properties

spray. RO, with the highest viscosity, behaved differently. The apex angle of the spray increased steadily, and for time t = 1.2 ms (the end of the analysed fuel injection), it was greater than for DF. Presumably, in this case the apex angle of the spray resulted from the additional factor which increased the turbulence of outflow from the sprayer, caused by the variability of cross-sections of the spraying holes, and the resulting permanent change in the ratio of the length of the outlet hole to its sectional area.

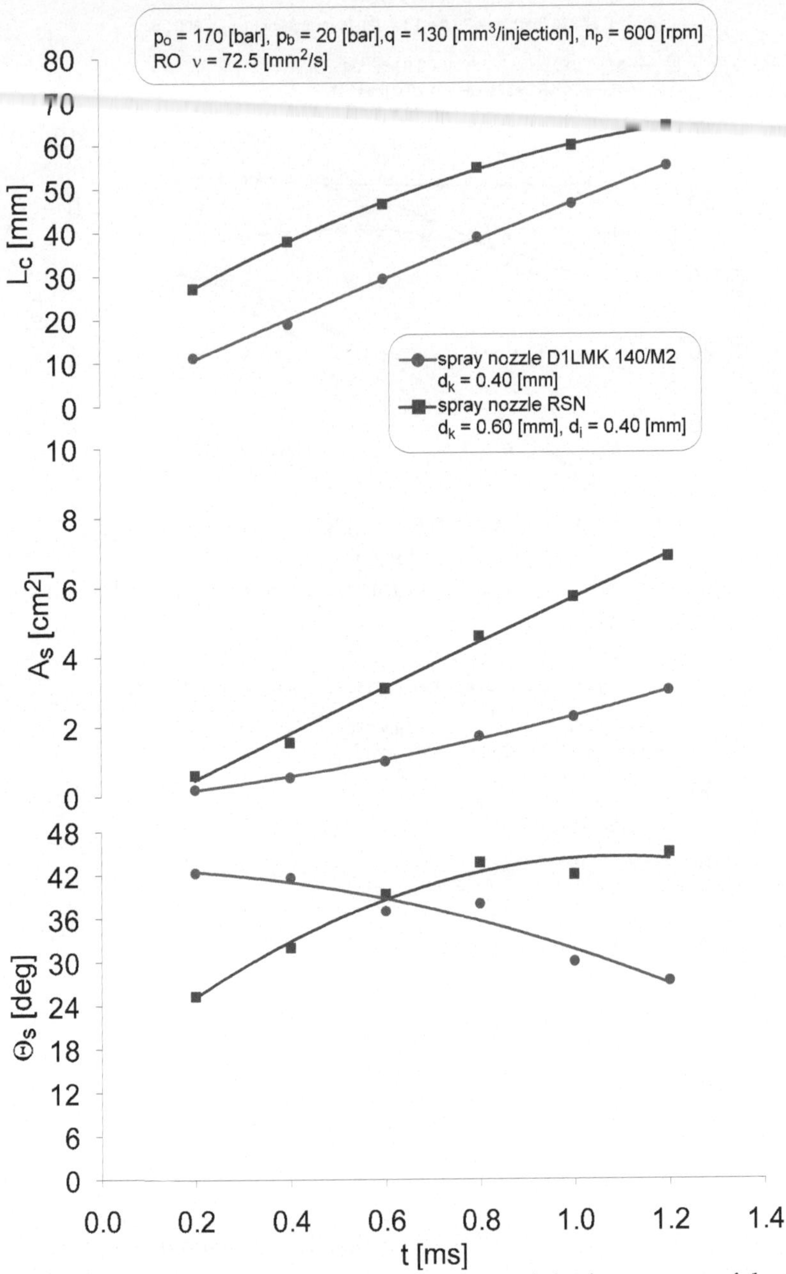

Fig. 15. The comparison of the apex angle, surface area, and the front-range of the spray generated by the classical injector and the RSN type when spraying RO

In Fig. 15, an additional comparison of the surface area, apex angle and range of the spray front for a spray of RO through a triple-hole standard injector and the RSN injector type, is depicted. The studies were carried out at p_b = 20 bar and a line pressure at injector opening p_o = 170 bar. The fuel dose was set at q = 130 mm³/injection, and the rotary velocity of the camshaft of the injection pump was n_p = 600 rpm. Despite the fact that smaller values of injection pressures were noted for the RSN injector (p_{wmax} = 300 bar, p_{wav} = 189 bar, and for the classical injector 376 bar and 236 bar, respectively), the surface area and range of the spray front were much greater in this case. Only the apex angle of the spray in the initial phase of the injection had a lower value for the spray generated by this injector (RSN type). Later in the cycle, however, this angle increased rapidly and at the end of the analysed period of spray development, the angle was greater by about 18 deg. Greater values of the parameters A_S, Θ_S, and L_C for the RSN injector probably resulted not only from the lack of throttling of the fuel flow in the needle seat, but also from the mechanical action of the outlet holes in the spray nozzle on the spray.

6. Radial distribution of fuel in spray drops generated by standard and RSN injectors

Investigations of fuel distribution were carried out using both injectors in a spray of droplets, at a constant injection pump speed of n_p = 600 rpm. The fuel dose was adjusted to 130 mm^3/injection and the line pressure at the injector was p_o = 170 bar. Fuel was injected into a background atmospheric of p_b = 1 bar; the fuel level H_p in the measuring vessels was read after each 1000-cycle period. The radial distribution of fuel in a spray was measured by directing the sprayed fuel into a series of standing measuring vessels. The inlet openings of the vessels were perpendicular to the axis of the spraying hole. Fuel distribution in a spray was investigated by placing the inlets of the measuring vessels at several distances from the edge of the inlet hole of the sprayer body – S_r. These were: 75, 150 and 210 mm. In addition, for each distance, the series of vessels was rotated by 45 deg, which enabled determination of the fuel distribution in four planes, mutually inclined at angles of 45 deg. Fig. 17 and 18 have the following legend: 'Position 90 deg', denoting the axis '–x + x' and the axis of a sprayer in one plane. 'Position 45 deg' denotes that the series of vessels had been turned through 45 deg in relation to position 90 deg.

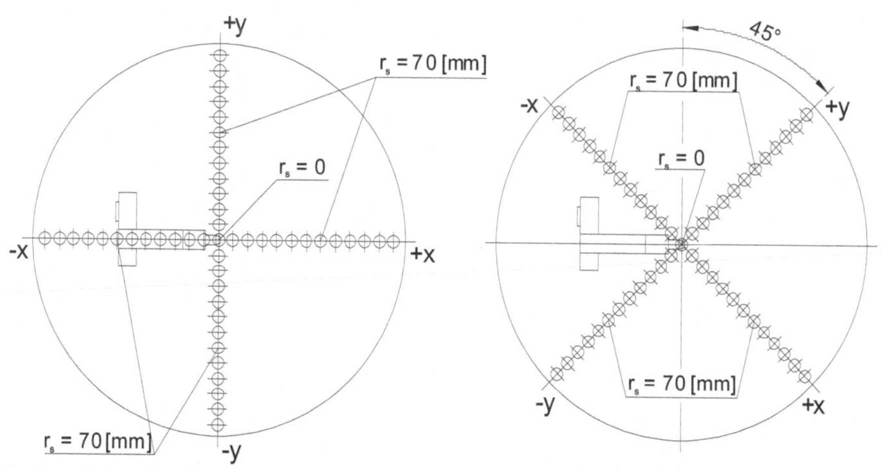

Fig. 16. A series of cylindrical measuring vessels used in determining fuel distribution in a spray of drops (top view)

The height of fuel in the measuring vessels was adopted (denoted by H_p) as a comparative measure to ascertain the fuel distribution in a spray of droplets. A radius at which a chosen fuel column was located, i.e., the radial distance from the theoretical axis of a spray, was denoted by r_s (Fig. 16). 'Direction x' and 'direction y' (legends on figures), denote vessels placed on the '–x + x' and '–y + y' axes, respectively, in Fig. 16.
Similar to the case of the direct observation studies – the standard injector with a D1LMK 140/M2 sprayer, and the new type injector – denoted as RSN, were studied.

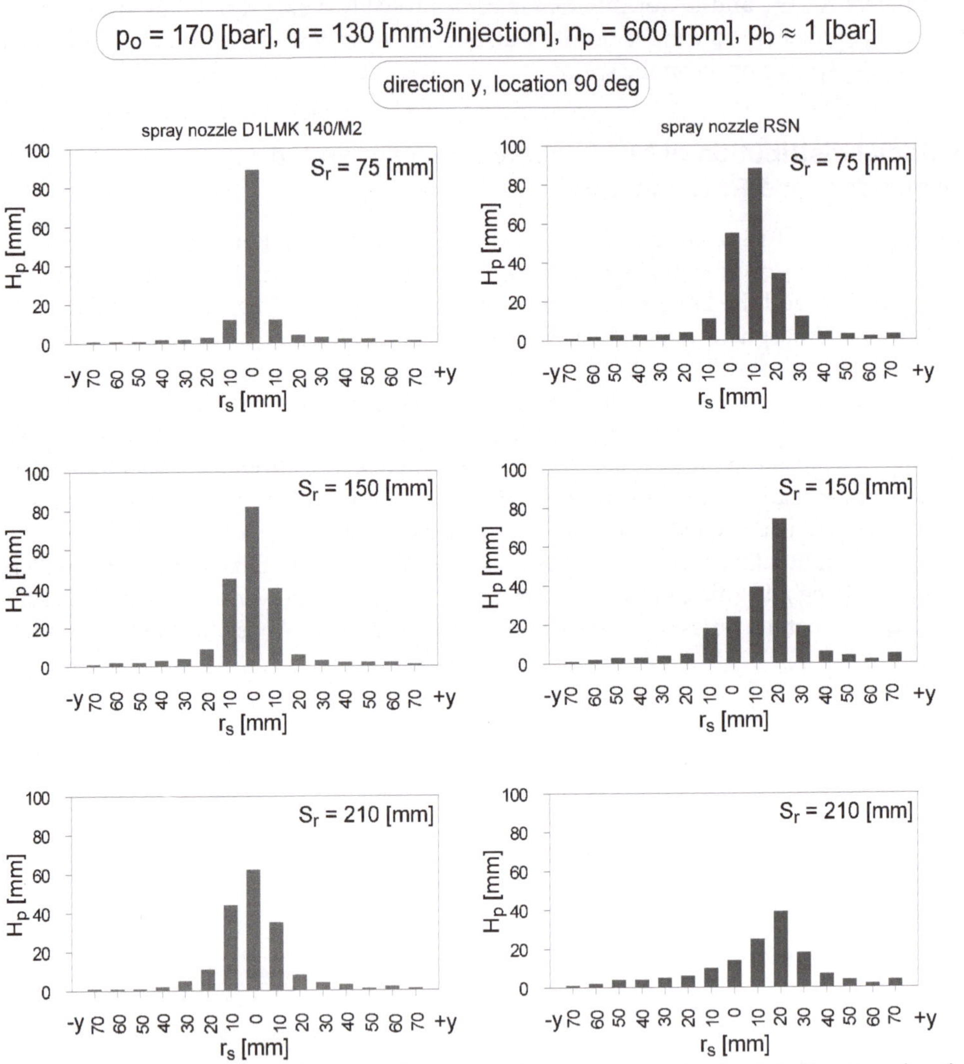

Fig. 17. Comparison of the radial distribution of fuel in a spray in the 'y' direction for the standard injector and the RSN type

Using histograms, Fig. 17 and 18 show the results of studies of the radial distribution of fuel in a spray of drops, formed by the standard injector (D1LMK 140/M2) and the RSN type. For simplicity, particular values of the radius r_s are plotted against the measured heights of fuel columns in the measuring vessels, H_p, rather than the related values of the spray density.

As seen in the standard injector, the usual situation prevailed, and the highest concentration of fuel lay at the core of the spray, i.e., the density of a unit spray has a maximum value at the spray axis, where large diameter droplets are most numerous, as stated earlier. A characteristic feature of fuel distribution in the standard spray is its symmetry around the spray axis (the axis in line with the axis of symmetry of the outlet hole), and the levelling off of the distribution as the distance from the sprayer increases (H_p values diminish in the centre and increase slightly towards the outside).

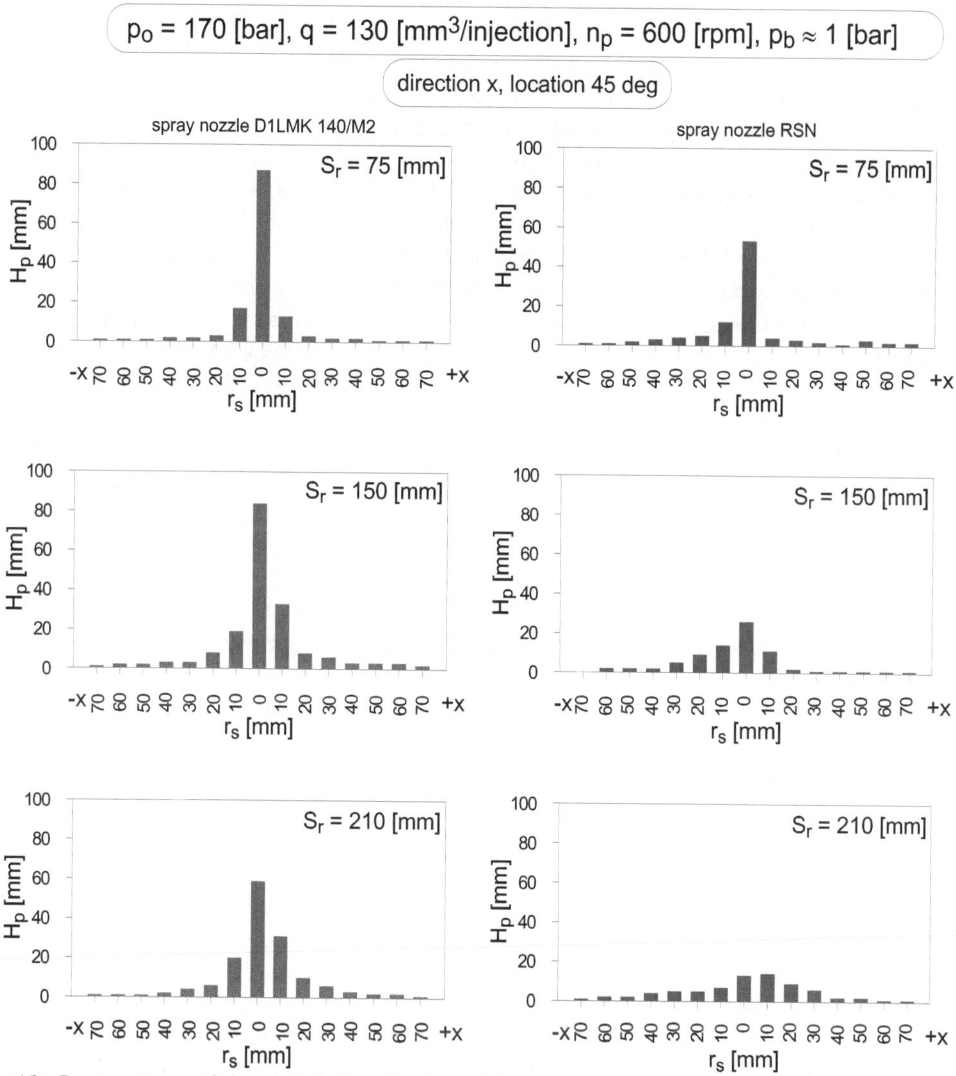

Fig. 18. Comparison of the radial distribution of fuel in a spray in the 'x' direction for the standard injector and the RSN type

The levelling off of the fuel distribution in a spray as the distance from the sprayer increases is caused by the size reduction of the droplets and the damping of their movement. Additionally, the turbulent movements in a spray tend to carry fuel towards the outer layers of the spray, and the distribution becomes more equal (Metz and Seika, 1998). This phenomenon is related to the fuel movement in the later phase of injection and it is also observed in the spray formed by the RSN-type injector. The levelling off of the fuel distribution with increased distance from the sprayer seems to be a phenomenon shared among sprays generated by both injector types.

A spray of fuel generated by the RSN sprayer shows asymmetry; the distribution in the 'x' direction differs from that in the 'y' direction. In the 'y' direction particularly, the concentration of fuel is considerably larger (also when the series of vessels is rotated through 45 deg). Moreover, in the 'y' direction a greater shift of the area of the maximum fuel concentration (core of a spray) may be observed in comparison to the 'x' direction. This leads to the conclusion that the fuel distribution in the spray formed by the RSN sprayer does not show any symmetry in relation to the theoretical axis of the spray.

The largest shift of the spray core from the theoretical axis for the RSN sprayer was observed in the 'y' direction. This effect appeared when the axis of the sprayer was in one plane with the axis at –x + x. In this position the axis of the needle rotation was perpendicular to the 'y' direction. The asymmetry of the core of the spray generated by the RSN sprayer may be explained by the change of the cross-sections of the outlet holes and the resulting mechanical action of the surface of the hole in the sprayer body on the fuel being discharged. The fuel, flowing through the spraying hole (particularly in the opening phase), hit the surface of the outlet hole. This changed the direction of the flow, which caused variations in the position of the core in the cross-section of the spray.

The spray generated by standard injector is axially symmetric. More fuel saturation in the spray core causes a different value of combustion air factor. This is unfavourable, because soot is usually produced in the rich mixture area (local deficiency of air) at a sufficiently high temperature (800–1400 K). This happens mainly in the core of the fuel spray and at its rear, where the concentration of fuel droplets is often higher.

Executed investigations of radial distribution of fuel in spray confirm that the spray generated by RSN injector is not symmetrical. The shift of the spray core outside (as effect of needle rotary) can be favourable on account of the possibly stronger impact of gas medium on spray zone, where the concentration of the fuel is higher. In this case, the secondary drop break-up will be more intensive. Smaller diameters of drops are obviously favourable with regard to soot and PM formation.

7. Conclusions

The parameters of the injection system have a decisive effect on the rate of combustion in the diesel engine, because of the influence on quality of formed air-fuel mixture. However, the optimal macrostructure of the spray, which is distributed in the cylinder volume, depends on the type and construction of the injector. On braking, the fuel stream in drops increases the area of contact between the fuel and air. It causes, first of all, fuel vaporisation and, then, its diffusion into air. The pressure energy generated by the injection system is consumed on spraying of the fuel stream which, together with the phenomena of physical and chemical parts of self-ignition delay, leads to fast increase in mixture entropy.

A better quality of fuel spraying guarantees RSN injector, which was confirmed by model investigations. The selected results have been presented in the paper.

The results of these investigations show that fuel sprays formed by using a RSN type injector differ from those generated by a standard injector. In particular, the parameters analysed, i.e., the range of the spray-front, the apex angle of the spray and its surface area, reach greater values for a spray formed by the new RSN type of sprayer; this may positively affect the ecological impact as well as the performance of engines fitted with injectors of this type.
Variation in the conditions of injection (pressure changes in the gaseous medium into which fuel is injected, change due to use of fuels of differing viscosity), affects the macrostructure of sprays generated differently by each type of injector. The best example may be the variance in the apex angle of the spray while spraying RO. In the standard injector, it was found that this angle diminished as the spray developed, while in the RSN injector the opposite tendency was observed.
The investigations of fuel distribution in a spray of droplets confirm that the spray generated by the RSN-type injector develops in a different way from that generated by the standard injector. In particular, the results of these studies show the asymmetry of the spray formed by the new type of injector.
More favourable parameters of the macrostructure of the spray generated by the RSN injector allow the air-fuel mixture to burn more completely. Next, it provides reducing of emission of toxic components from exhaust gases. However, for using a new type of injector, modification of the combustion chamber is needed. This modification has to consider higher values of spray macrostructure parameters. For example, a confirmed larger range of the spray formed by a new type of injector can be served. At injection into the combustion chamber without modification, the spray can settle on the walls of combustion chamber which can cause increase in PM emission. The authors conducted investigations in this range and intend to publish them in the subsequent papers.

8. Nomenclature

The Table 1 shows the parameters for the atomization of fuel, which were used in the study. Additionally, there are used description of parameters, if required.

Quantity	*Unit*	*Specification*
A_S	[cm^2]	Surface of view of fuel spray on perpendicular plane to spray nozzle axis
L_C	[mm]	Tip penetration of fuel spray
Θ_S	[deg]	Apex angle of fuel spray
H_p	[mm]	Fuel level (at measuring of the fuel radial distribution in a spray)
S_r	[mm]	Distance of an inlet area of the measuring vessel from the edge of outlet hole in spray nozzle body (measuring fuel radial distribution in a spray)
r_s	[mm]	Distance measuring point from the theoretical axis spray (measuring fuel radial distribution in a spray)
d_k	[mm]	Outlet hole diameter in a needle
d_i	[mm]	Outlet hole diameter in a spray nozzle body

Table 1. Description of parameters used in the study

The continuation of Table 1

Quantity	*Unit*	*Specification*
h_t	[mm]	Piston stroke of injector
α_i	[deg]	Angle of needle rotation
f_c	[mm²]	Geometrical flow area
q	[mm³/injection]	Fuel dose
t	[ms]	Time
n_p	[rpm]	Rotational speed of injection pump camshaft
p_o	[bar]	Static opening pressure of injector
p_b	[bar]	Background pressure
p_{wmax}	[bar]	Maximum fuel injection pressure
p_{wav}	[bar]	Average fuel injection pressure
v	[mm²/s]	Kinematic viscosity of fuel
DF	-	Diesel Fuel
RO	-	Rape Oil

9. References

Beck, N.J.; Uyehara, O.A. & Johnson, W.P. (1988) *Effects of Fuel Injection on Diesel Combustion*, SAE Transactions, Paper 880299.

Dürnholz, M. & Krüger, M. (1997) Hat der Dieselmotor als Fahrzeugantrieb eine zukunft?, *6. Aachener Kolloquium Fahrzeug - und Motorentechnik*, Akwizgran, Germany.

Hiroyasu, H. & Arai, M. (1990) *Structures of Fuel Sprays in Diesel Engines*, SAE Transactions, Paper 900475.

Kollmann, K. & Bargende, M. (1997) DI - Dieselmotor und DI - Ottomotor - Wohin geht die Pkw - Motorenentwicklung?, *Symposium Dieselmotorentechnik 98*, Technische Akademie Esslingen, Ostfildern, Germany.

Kuszewski, H. (2002) *Wpływ zmiennych przekrojów wylotowych wtryskiwacza z obrotową iglicą na rozpylanie oleju napędowego*, PhD Dissertation, Cracow University of Technology, Cracow, Poland.

Kuszewski, H. & Lejda, K. (2009) Experimental investigations of a new type of fueliing system for heavy-duty diesel engines, *International Journal of Heavy Vehicle Systems, Inderscience Enterprises Ltd*, Olney, UK.

Metz, N. & Seika, M. (1998) Die Luftqualität in Europa bis zum Jahre 2010 mit und ohne EURO IV Grenzwerte, *19. Internationales Wiener Motorensymposium*, Fortschrittberichte VDI Reihe 12, Nr 348, Wien, Austria.

Peake, S. (1997) *Vehicle and Fuel - Challenges Beyond 2000*, Automotive Publishing, London, UK.

Szlachta, Z. & Kuszewski, H. (2002) *Wpływ zmiennych przekrojów wylotowych wtryskiwacza z obrotową iglicą na rozpylanie oleju napędowego*, Rep. 5 T12D 026 22, Cracow, Poland.

Szymański, J. & Zabłocki, M. (1992) *Wtryskiwacz do silnika spalinowego*, Patent Application in Patent Department R.P, P-294889, Poland.

Varde, K.S. & Popa, D.M. (1983) Diesel Fuel Spray Penetration at High Injection Pressures, SAE Transactions, Paper 830448.

5

Effect of injector nozzle holes on diesel engine performance

1. Introduction

The four-stroke direct-injection diesel engine typical was measured and modeled by Bakar et al (2007) using GT-POWER computational model and has explored of diesel engine performance effect based on engine speeds. GT-POWER is the leading engine simulation tool used by engine and vehicle makers and suppliers and is suitable for analysis of a wide range of engine issues. The details of the diesel engine design vary significantly over the engine performance and size range. In particular, different combustion chamber geometries and fuel injection characteristics are required to deal effectively with major diesel engine design problem achieving sufficiently rapid fuel-air mixing rates to complete the fuel-burning process in the time available. According to Heywood (1988) and Ganesan (1999), a wide variety of inlet port geometries, cylinder head and piston shapes, and fuel-injection patterns are used to accomplish this over the diesel size range. The engine ratings usually indicate the highest power at which manufacturer expect their products to give satisfactory of power, economy, reliability and durability under service conditions. Maximum torque and the speed at which it is achieved, is usually given also by Heywood (1988). The importance of the diesel engine performance parameters are geometrical properties, the term of efficiency and other related engine performance parameters. The engine efficiencies are indicated thermal efficiency, brake thermal efficiency, mechanical efficiency, volumetric efficiency and relative efficiency (Ganesan, 1999). The other related engine performance parameters are mean effective pressure, mean piston speed, specific power output, specific fuel consumption, intake valve mach index, fuel-air or air-fuel ratio and calorific value of the fuel (Heywood, 1988; Ganesan, 1999; Semin et al., 2007). According to Heywood (1988) in the diesel engine geometries design written that diesel engine compression ratio is maximum cylinder volume or the displaced volume or swept and clearance volume divided by minimum cylinder volume. And the power delivered by the diesel engine and absorbed by the dynamometer is the product of torque and angular speed. The engine efficiencies, every its efficiencies defined by Ganesan (1999).

2. Important

In this chapter has investigated the effect of injector nozzle holes diameter geometries on the performance of diesel engine such as indicated power, indicated torque, fuel consumption and fuel in-engine cylinder. The investigation is using computational modelling based on variation engine speeds.

3. Engine Performance Review

In the diesel engine geometries design by Heywood (1988), the diesel engine compression ratio is maximum cylinder volume or the displaced volume or swept (V_d) and clearance volume (V_c) divided by minimum cylinder volume (V_c). The diesel engine compression ratio can be calculated as below:

$$r_c = \frac{V_d + V_c}{V_c} \tag{1}$$

and the power delivered by the diesel engine and absorbed by the dynamometer is the product of torque and angular speed. Diesel engine power definition as:

$$P = 2\pi NT \tag{2}$$

In the diesel engine efficiencies, every its efficiencies defined by Ganesan (1999). Indicated thermal efficiency (η_{ith}) is the ratio of energy (E) in the indicated power (ip) to the input fuel energy. Brake thermal efficiency (η_{bth}) is the ratio of energy in the brake power (bp), Mechanical efficiency (η_m) is defined as the ratio of brake power (bp) or delivered power to the indicated power (ip) or power provided to the piston and it can also be defined as the ratio of the brake thermal efficiency to the indicated thermal efficiency. Relative efficiency or efficiency ratio (η_{rel}) is the ratio of thermal efficiency of an actual cycle to that of the ideal cycle, the efficiency ratio is a very useful criteria which indicates the degree of development of the engine. Ganesan (1999) written that, the one of the very important parameters which decides the performance of four-stroke engines is volumetric efficiency (η_v), where four-stroke engines have distinct suction stroke and therefore the volumetric efficiency indicates the breathing ability of the engine. The volumetric efficiency is defined as the volume flow rate of air into the intake system divided by the rate at which the volume is displaced by the system. The normal range of volumetric efficiency at full throttle for SI engines is 80% to 85% and for CI engines is 85% to 90%.

$$\eta_{ith} = \frac{ip}{E} \tag{3}$$

$$\eta_{bth} = \frac{bp}{E} \tag{4}$$

$$\eta_m = \frac{bp}{ip} \tag{5}$$

$$\eta_v = \frac{\dot{m}_a}{\rho_a V_{disp} N/2} \tag{6}$$

$$\eta_{rel} = \frac{\text{Actual thermal efficiency}}{\text{Air - standard efficiency}} \tag{7}$$

The other related engine performance was defined by Heywood (1988), Kowalewicz (1984), Stone (1997) and Ganesan (1999). Mean effective pressure (*mep*), where n_R is the number of crank revolutions for each power stroke per cylinder (two for four-stroke, one for two-stroke cycles) as :

$$mep = \frac{Pn_R}{V_d N} \tag{8}$$

The measure of an engine's efficiency which will be called the fuel conversion efficiency is given by Heywood (1988):

$$nf = \frac{W_c}{m_f Q_{HV}} = \frac{(Pn_R / N)}{(m_f n_R / N) Q_{HV}} = \frac{P}{m_f Q_{HV}} \tag{9}$$

Specific fuel consumption as :

$$sfc = \frac{m_f}{P} \tag{10}$$

In the engine testing, both the air mass flow rate m_a and the fuel mass flow rate m_f are normally measured. The ratio of these flow rates is useful in defining engine operating conditions are air/fuel ratio (A/F) and fuel/air ratio (F/A).

The following relationships between diesel engine performance parameters can be developed. For power *P*:

$$P = \frac{\eta_f m_a N Q_{HV} (F/A)}{n_R} \tag{11}$$

$$P = \frac{\eta_f \eta_v N V_d Q_{HV} \rho_{a,i} (F/A)}{2} \tag{12}$$

For torque T :

$$T = \frac{\eta_f \eta_v V_d Q_{HV} \rho_{a,i} (F/A)}{4\pi} \tag{13}$$

For mean effective pressure :

$$mep = \eta_f \eta_v Q_{HV} \rho_{a,i} (F/A) \tag{14}$$

The specific power or the power per unit piston area is a measure of the engine designer's success in using the available piston area regardless of cylinder size. The specific power is :

$$\frac{P}{A_p} = \frac{\eta_f \eta_v N L Q_{HV} \rho_{a,i} (F/A)}{2} \tag{15}$$

Mean piston speed :

$$\frac{P}{A_p} = \frac{\eta_f \eta_v N \overline{S}_p Q_{HV} \rho_{a,i} (F/A)}{4} \tag{16}$$

Heywood (1988) written that, specific power is thus proportional to the product of mean effective pressure and mean piston speed. These relationship illustrated the direct importance to engine performance of high fuel conversion efficiency, high volumetric efficiency, increasing the output of a given displacement engine by increasing the inlet air density, maximum fuel/air ratio that can be useful burned in the engine and high mean piston speed.

4. Modelling of Injector Nozzle Holes

The four-stroke direct-injection (DI) diesel engine was presented in this chapter. The specification of the selected diesel engine was presented in Table 1. To develop the four-stroke direct-injection diesel engine modeling is step by step, the first step is open all of the selected diesel engine components to measure the engine components part size. Then, the engine components size data will be input to the software library of the all engine components data. To create the model, select window and then tile with template library from the menu. This will place the template library on the left hand side of the screen. The template library contains all of the available templates that can be used in computational modeling. Some of these templates those that will be needed in the project need to be copied into the project before they can be used to create objects and parts. For the purpose of this model, click on the icons listed and drag them from the template library into the project library. Some of these are templates and some are objects that have already been defined and included in the template library (Gamma Technologies, 2004). This chapter focused on fuel nozzle hole of fuel injector. The engine modeling is according to Semin et al. (2007) as shown in Fig. 1.

All of the parameters in the model will be listed automatically in the case setup and each one must be defined for first case of the simulation. The physically of the injector fuel nozzle hole material detailed were investigated in this research is shown in Fig. 2. In this figure was showed the detail of injection hole or fuel nozzle hole. The fuel nozzle holes would be changed in wide diameter of nozzle hole and in different number of nozzle hole.

Engine Parameters	Value	Engine Parameters	Value
Model	CF186F	Intake valve close (^{0}CA)	530
Bore (mm)	86.0	Exhaust valve open (^{0}CA)	147
Stroke (mm)	70.0	Exhaust valve close (^{0}CA)	282
Displacement (cc)	407.0	Max. intake valve open (mm)	7.095
Number of cylinder	1	Max. exhaust valve open (mm)	7.095
Connecting rod length (mm)	118.1	Valve lift periodicity (deg)	360
Piston pin offset (mm)	1.00	Fuel nozzle diameter (mm)	0.1
Intake valve open (^{0}CA)	395	Fuel nozzle hole number (pc)	4

Table 1. Specification of the selected diesel engine

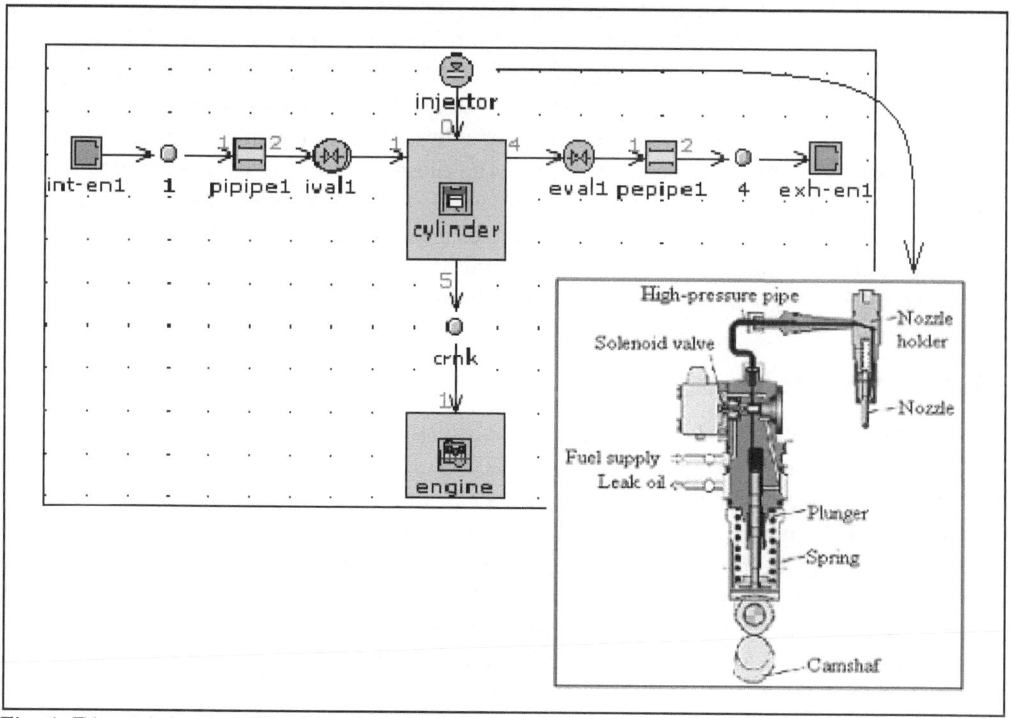

Fig. 1. Direct-injection diesel engine modeling

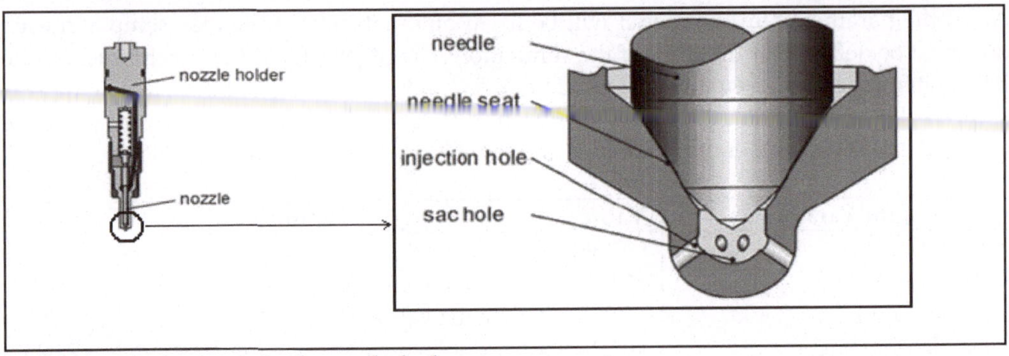

Fig. 2. Detail of injector fuel nozzle holes

Whenever the computational simulation is running, the computational model produces several output files that contain simulation results in various formats. Most of the output is available in the post-processing application. The software is powerful tool that can be used to view animation and order analysis output (Gamma Technologies, 2004). After the simulation was finished, report tables that summarize the simulations can be produced. These reports contain important information about the simulation and simulation result in a tabular form. The computational simulation of the engine model result is informed the engine performance. The running simulation result in this research is focused on the engine performance data based on variation of fuel nozzle material hole diameter size, diameter number and the different engine speed (rpm). The diesel engine model was running on any different engine speeds in rpm, there are 500, 1000, 1500, 2000, 2500, 3000 and 3500. The variations of injector fuel nozzle holes number are based on multi holes and multi diameter holes, the simulation model there are started from the injector fuel nozzle 1 hole, 2 holes, 3 holes, 4 holes, 5 holes, 6 holes, 7 holes, 8 holes, 9 holes and 10 holes.

5. Effect of Injector Nozzle Holes on Fuel in Engine Cylinder

The simulation results are shown in every cases, such as case 1 is on 500 rpm, case 2 is on 1000 rpm, case 3 is on 1500 rpm, case 4 is on 2000 rpm, case 5 is on 2500 rpm, case 6 is on 3000 rpm, case 7 is on 3500 rpm and case 8 on 4000 rpm. Numerous studies have suggested that decreasing the injector nozzle orifice diameter is an effective method on increasing fuel air mixing during injection (Baik, 2001). Smaller nozzle holes have found to be the most efficient at fuel/air mixing primarily because the fuel rich core of the jet is smaller. In addition, decreasing the nozzle hole orifice diameter, would reduce the length of the potential core region. Unfortunately, decreasing nozzle holes size causes a reduction in the turbulent energy generated by the jet.

Since fuel air mixing is controlled by turbulence generated at the jet boundary layer, this will offset the benefits of the reduced jet core size. Furthermore, jets emerging from smaller nozzle orifices were shown not to penetrate as far as those emerging from larger orifices. This decrease in penetration means that the fuel will not be exposed to all of the available air in the chamber. The effect of fuel nozzle holes number and geometries of in-cylinder engine liquid fuel are shown in Fig. 3 – Fig. 12,

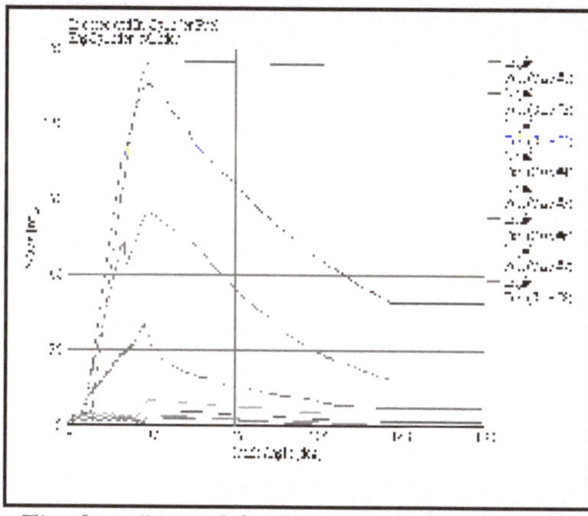

Fig. 3. Liquid fuel in cylinder of injector nozzle 1 holes

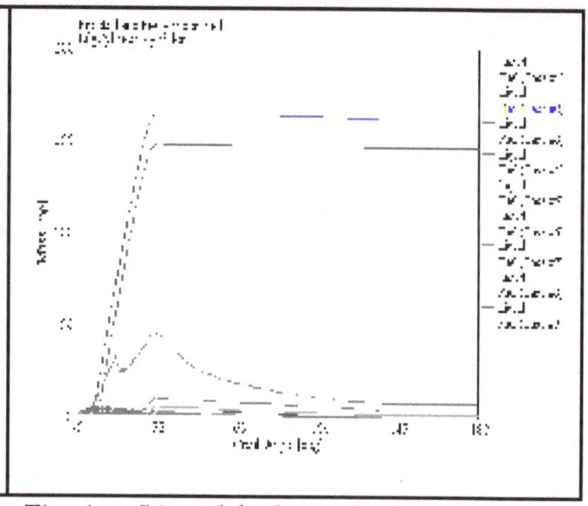

Fig. 4. Liquid fuel in cylinder of injector nozzle 2 holes

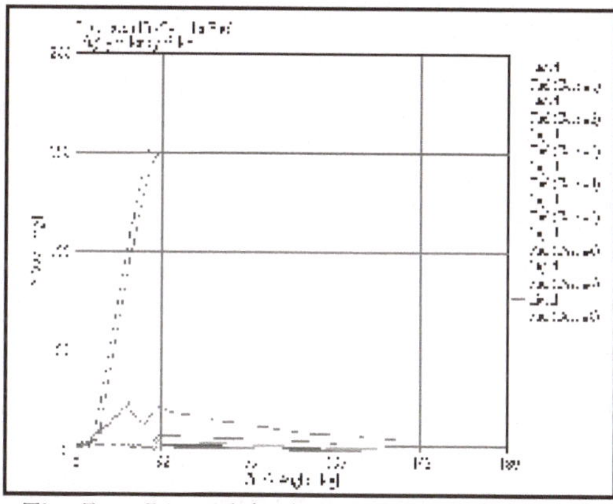

Fig. 5. Liquid fuel in cylinder of injector nozzle 3 holes

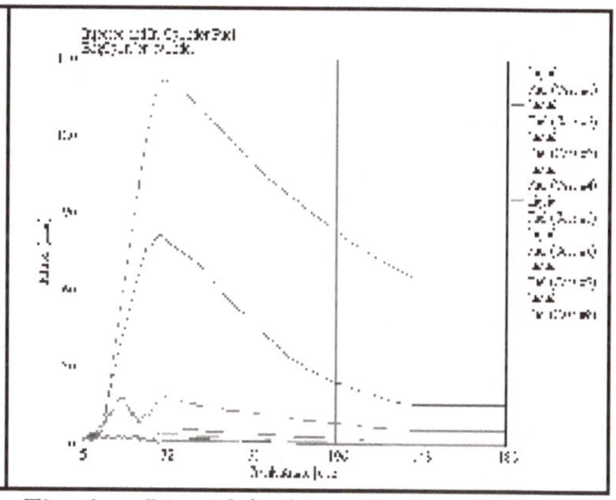

Fig. 6. Liquid fuel in cylinder of injector nozzle 4 holes

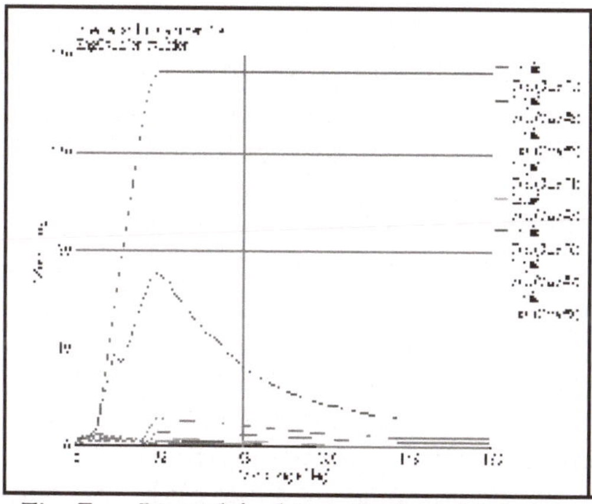

Fig. 7. Liquid fuel in cylinder of injector nozzle 5 holes

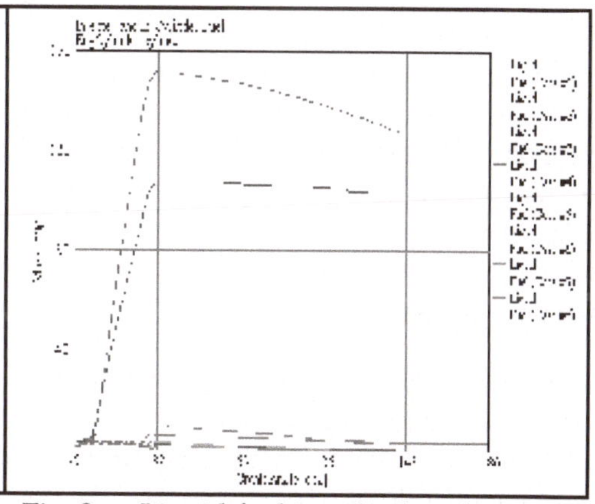

Fig. 8. Liquid fuel in cylinder of injector nozzle 6 holes

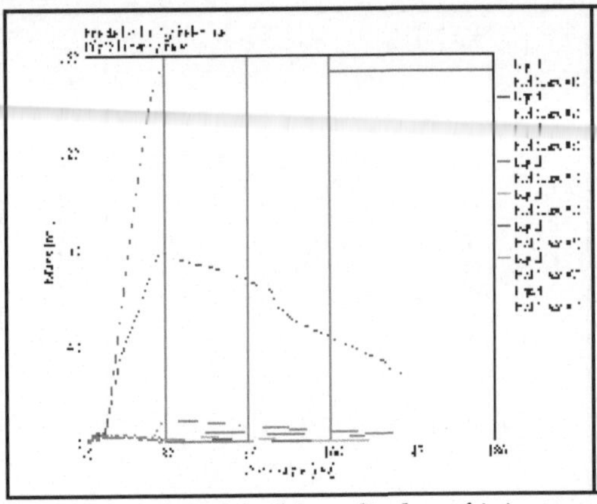

Fig. 9. Liquid fuel in cylinder of injector nozzle 7 holes

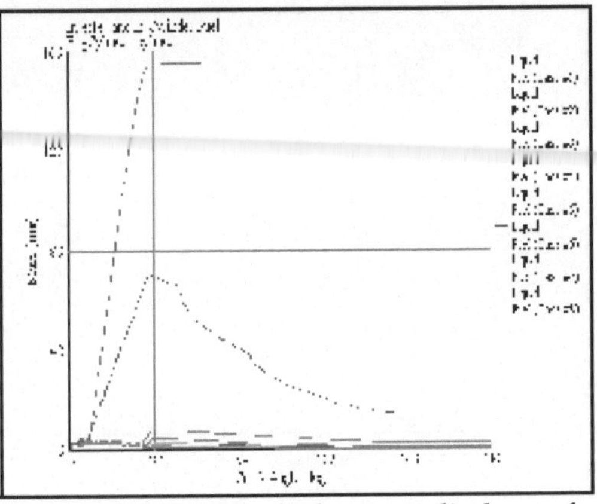

Fig. 10. Liquid fuel in cylinder of injector nozzle 8 holes

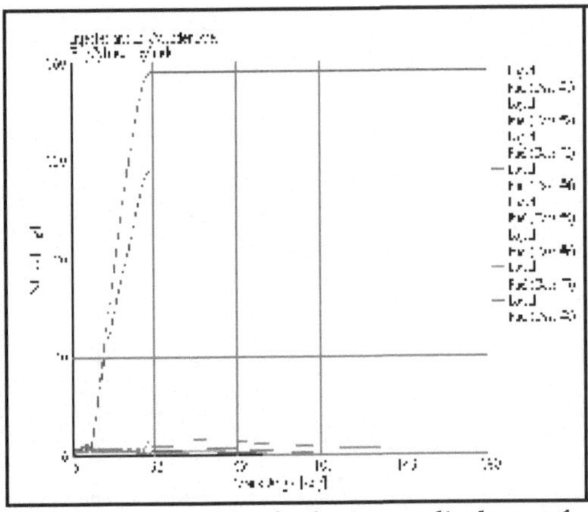

Fig. 11. Liquid fuel in cylinder of injector nozzle 9 holes

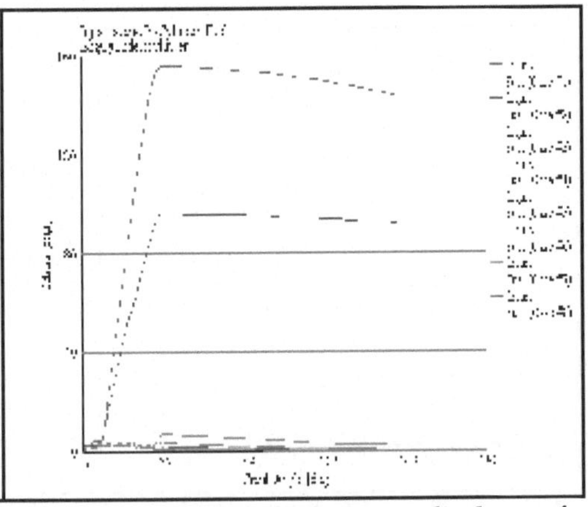

Fig. 12. Liquid fuel in cylinder of injector nozzle 10 holes

For excessively small nozzle size, the improvements in mixing related to decreased plume size may be negated by a reduction in radial penetration (Baumgarter, 2006). This behavior is undesirable because it restricts penetration to the chamber extremities where a large portion of the air mass resides. Furthermore, it hampers air entrainment from the head side of the plume because the exposed surface area of the plume is reduced. It has been suggested that a nozzle containing many small holes would provide better mixing than a nozzle consisting of a single large hole. The effect of injector nozzle multi holes in-cylinder engine unburned fuel are shown in Fig. 13 – Fig. 22.

The optimal nozzle design would be one that provided the maximum number of liquid fuel burn in combustion process and minimum number of liquid fuel unburned. Theoretically, a 10 holes nozzle satisfies this requirement. Unfortunately, jets emerging from a 10 holes nozzle tended to be very susceptible. All of the nozzles examined and the result shown that the seven holes nozzle provided the best results for any different engine speeds in simulation and the best performance shown on low speed engine.

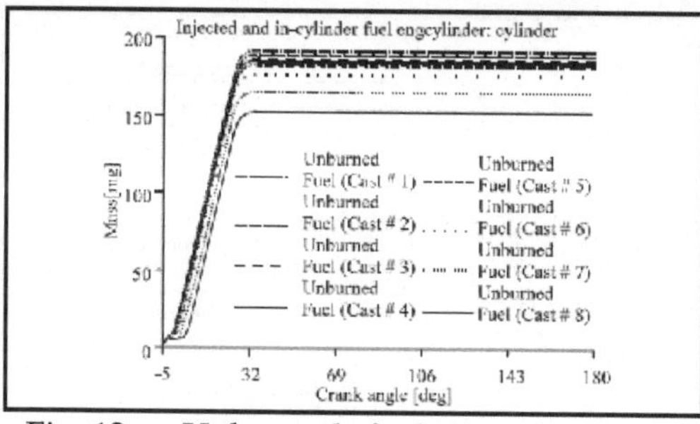

Fig. 13. Unburned fuel in cylinder of injector nozzle 1 holes

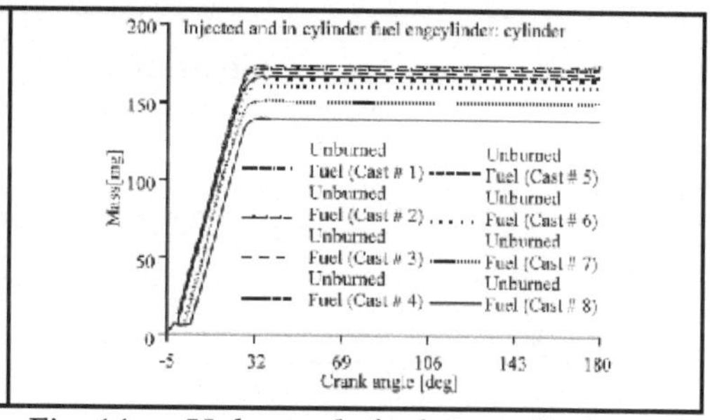

Fig. 14. Unburned fuel in cylinder of injector nozzle 2 holes

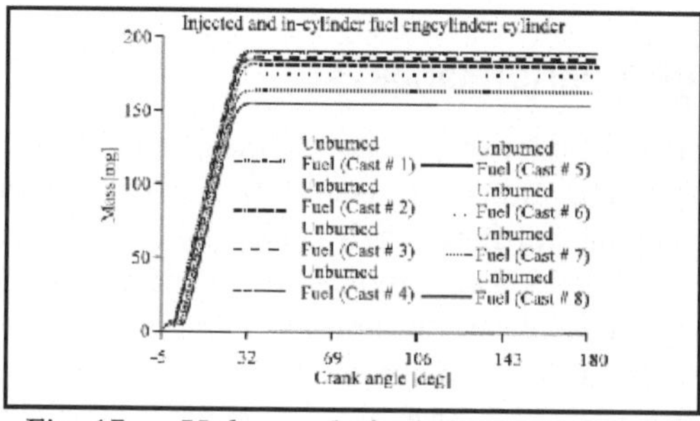

Fig. 15. Unburned fuel in cylinder of injector nozzle 3 holes

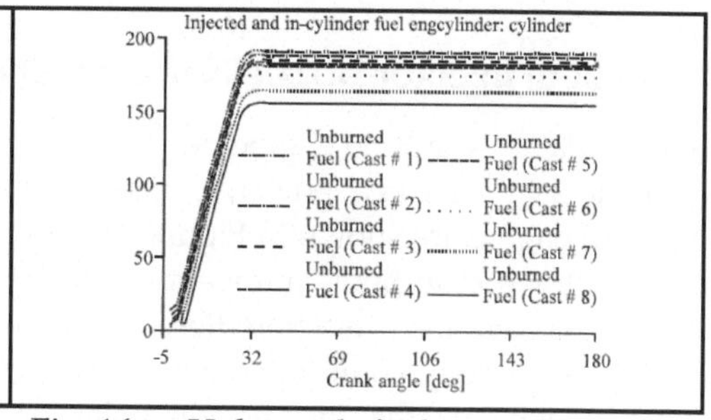

Fig. 16. Unburned fuel in cylinder of injector nozzle 4 holes

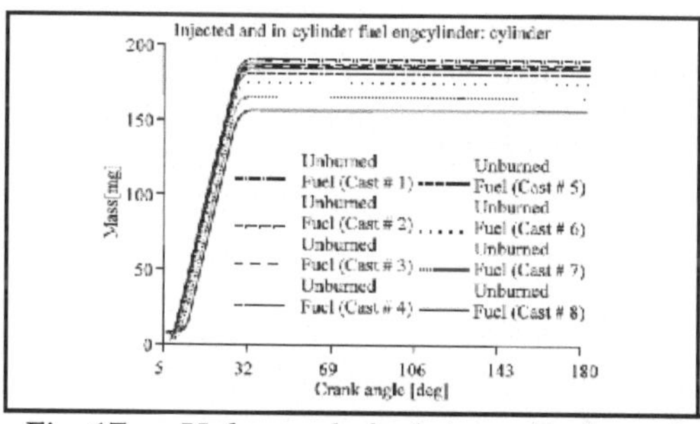

Fig. 17. Unburned fuel in cylinder of injector nozzle 5 holes

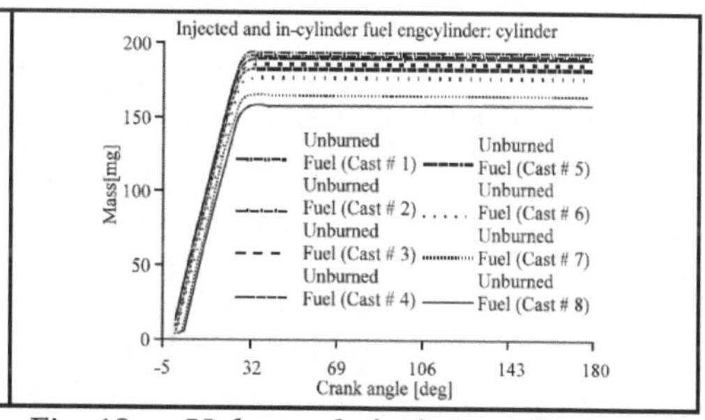

Fig. 18. Unburned fuel in cylinder of injector nozzle 6 holes

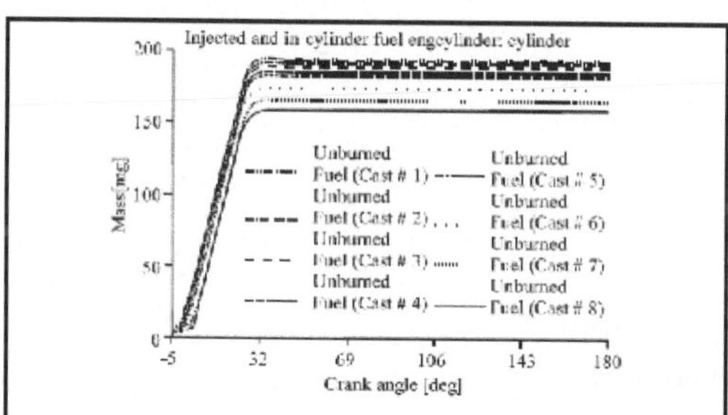

Fig. 19. Unburned fuel in cylinder of injector nozzle 7 holes

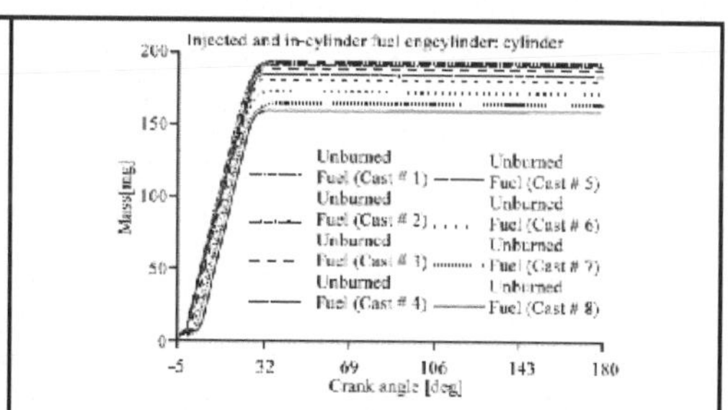

Fig. 20. Unburned fuel in cylinder of injector nozzle 8 holes

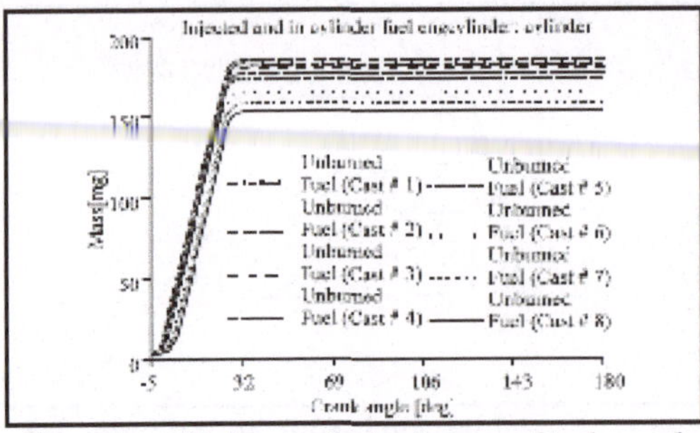

Fig. 21. Unburned fuel in cylinder of injector nozzle 9 holes

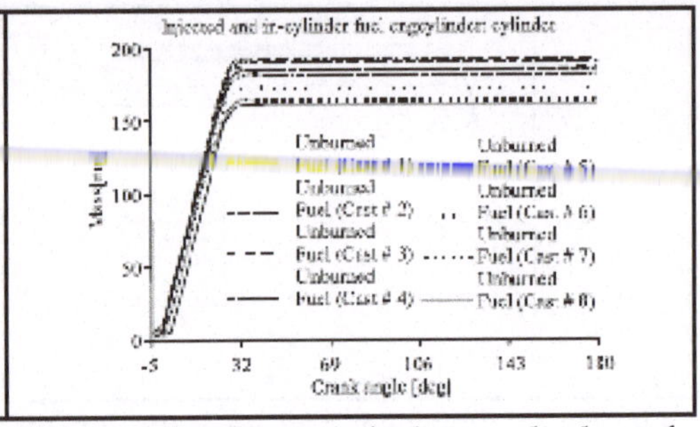

Fig. 22. Unburned fuel in cylinder of injector nozzle 10 holes

6. Effect of Injector Nozzle Holes on Engine Performance

The simulation result on engine performance effect of injector fuel nozzle holes number and geometries in indicated power, indicated torque and indicated specific fuel consumption (ISFC) of engine are shown in Figure 23 - 25. The injector fuel nozzle holes orifice diameter and injector nozzle holes numbers effect on indicated power, indicated torque and ISFC performance of direct-injection diesel engine was shown from the simulation model running output. An aerodynamic interaction and turbulence seem to have competing effects on spray breakup as the fuel nozzle holes orifice diameter decreases. The fuel drop size decreases if the fuel nozzle holes orifice diameter is decreases with a decreasing quantitative effect for a given set of jet conditions.

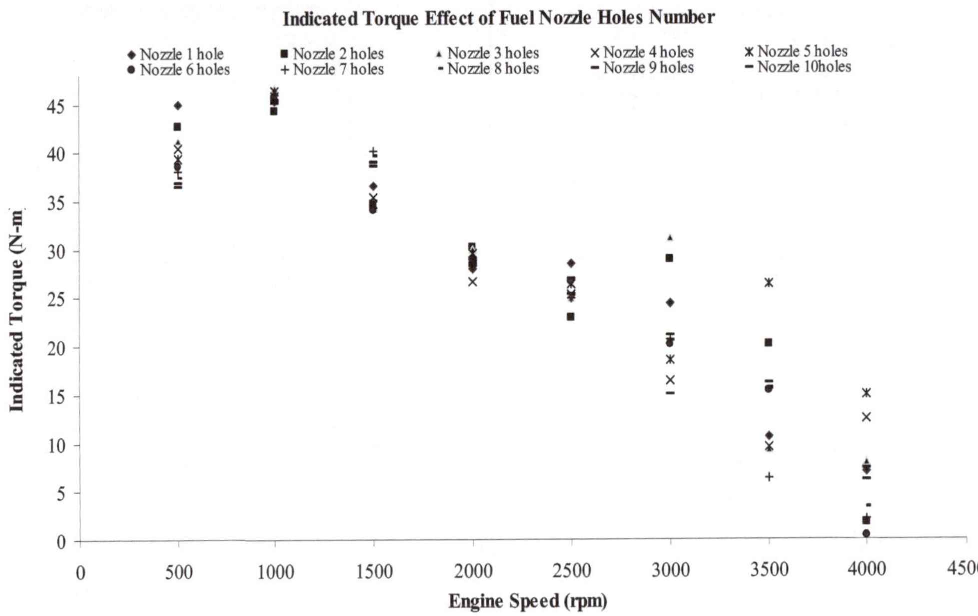

Fig. 23. Effect of fuel nozzle holes on indicated torque of diesel engine

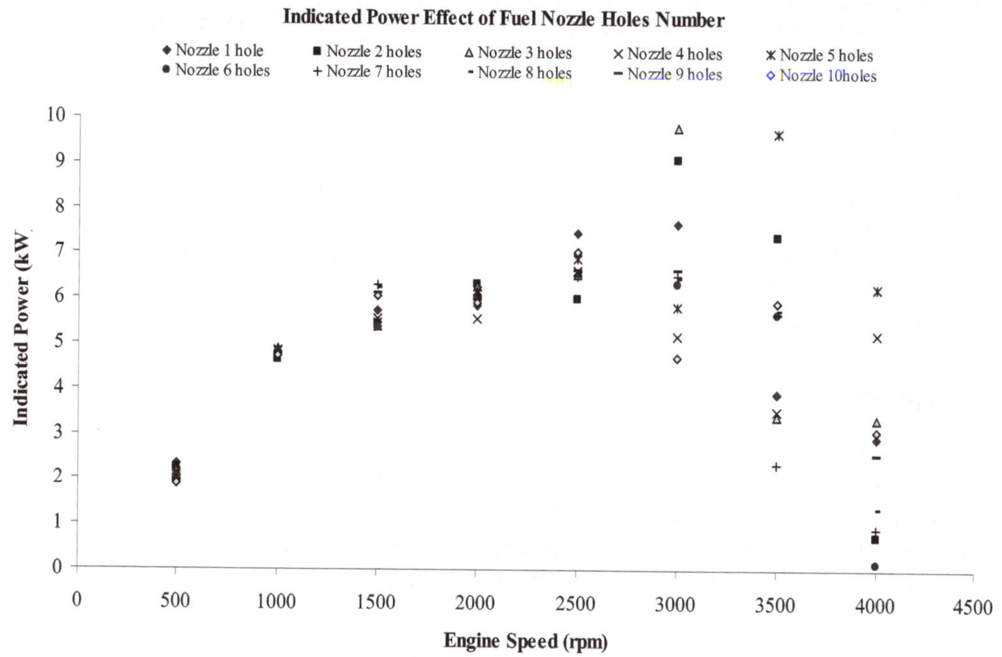

Fig. 24. Effect of fuel nozzle holes on indicated power of diesel engine

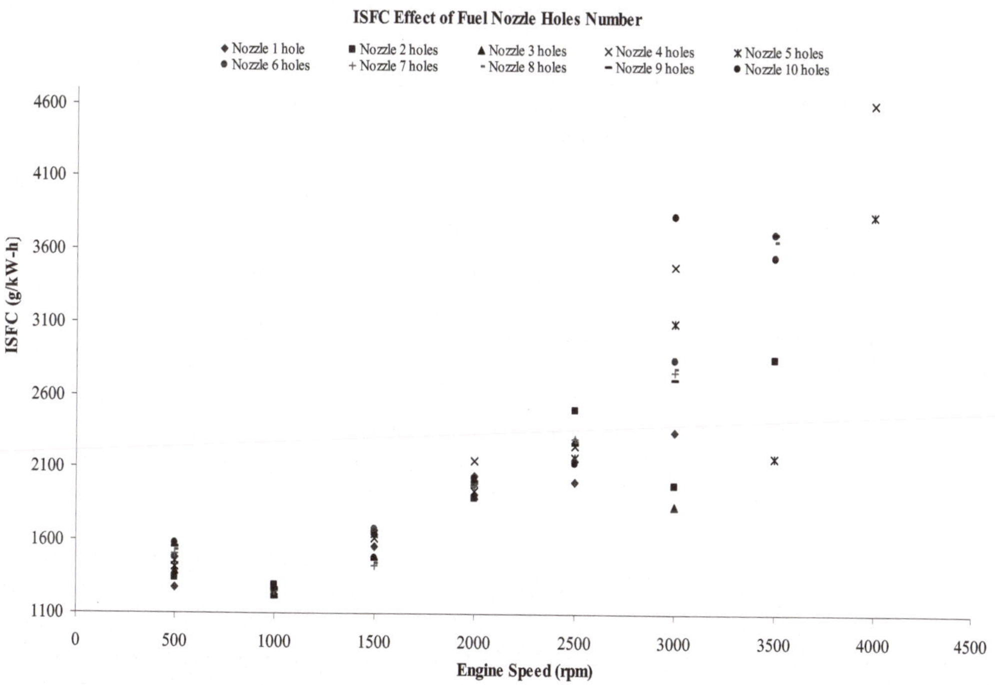

Fig. 25. Effect of fuel nozzle holes on ISFC of diesel engine

Fuel-air mixing increases as the fuel nozzle holes orifice diameter fuel nozzle holes decreases. Also soot incandescence is observed to decrease as the amount of fuel-air premixing upstream of the lift-off length increases. This can be a significant advantage for small orifice nozzles hole. However, multiple holes orifices diameter required to meet the desired mass flow rate as orifice diameter decreases. In this case, the orifices diameter need to placed with appropriate spacing and directions in order to avoid interference among adjacent sprays. The empirical correlations generally predict smaller drop size, slower penetrating speed and smaller spray cone angles as the orifice diameter decreases, however the predicted values were different for different relation. All of the nozzles have examined and the results are shown that the five holes nozzle provided the best results for indicted torque, indicated power and ISFC in any different engine speed in simulation.

7. Conclusion

All of the injector nozzle holes have examined and the results are shown that the seven holes nozzle have provided the best burning result for the fuel in-cylinder burned in any different engine speeds and the best burning is in low speed engine. In engine performance effect, all of the nozzles have examined and the five holes nozzle provided the best result in indicted power, indicated torque and ISFC in any different engine speeds.

8. References

Baik, Seunghyun. (2001). Development of Micro-Diesel Injector Nozzles Via MEMS Technology and Effects on Spray Characteristics, *PhD Dissertation*, University of Wisconsin-Madison, USA.

Bakar, R.A., Semin., Ismail, A.R. and Ali, Ismail., 2008. Computational Simulation of Fuel Nozzle Multi Holes Geometries Effect on Direct Injection Diesel Engine Performance Using GT-POWER. *American Journal of Applied Sciences 5 (2)*: 110-116.

Baumgarter, Carsten. (2006). *Mixture Formation in Internal Combustion Engines*, Spinger Berlin.

Gamma Technologies, (2004). *GT-POWER User's Manual 6.1*, Gamma Technologies Inc.

Ganesan, V. (1999). *Internal Combustion Engines 2nd Edition*, Tata McGraw-Hill, New Delhi, India.

Heywood, J.B. (1988). *Internal Combustion Engine Fundamentals - Second Edition*, McGraw-Hill, Singapore.

Kowalewicz, Andrzej., 1984. *Combustion System of High-Speed Piston I.C. Engines*, Wydawnictwa Komunikacji i Lacznosci, Warszawa.

Semin and Bakar, R.A. (2007). Nozzle Holes Effect on Unburned Fuel in Injected and In-Cylinder Fuel of Four Stroke Direct Injection Diesel Engine. *Presearch Journal of Applied Sciences 2 (11)*: 1165-1169.

Semin., Bakar, R.A. and Ismail, A.R. (2007). Effect Of Engine Performance For Four-Stroke Diesel Engine Using Simulation, *Proceeding The 5th International Conference On Numerical Analysis in Engineering*, Padang-West Sumatera, Indonesia.

Stone, Richard. (1997). *Introduction to Internal Combustion Engines-Second Edition*, SAE Inc, USA.

6

Accurate Modelling of an Injector for Common Rail Systems

1. Introduction

It is well known that the injection system plays a leading role in achieving high diesel engine performance; the introduction of the common rail fuel injection system (Boehner & Kumel, 1997; Schommers et al., 2000; Stumpp & Ricco, 1996) represented a major evolutionary step that allowed the diesel engine to reach high efficiency and low emissions in a wide range of load conditions.

Many experimental works show the positive effects of splitting the injection process in several pilot, main and post injections on the reduction of noise, soot and NOx emission (Badami et al., 2002; Brusca et al., 2002; Henelin et al., 2002; Park et al., 2004; Schmid et al., 2002). In addition, the success of engine downsizing (Beatrice et al., 2003) and homogeneous charge combustion engines (HCCI) (Canakci & Reitz, 2004; Yamane & Shimamoto, 2002) is deeply connected with the injection system performance and injection strategy.

However, the development of a high performance common rail injection system requires a considerable investment in terms of time, as well as money, due to the need of fine tuning the operation of its components and, in particular, of the electronic fuel injector. In this light, numerical simulation models represent a crucial tool for reducing the amount of experiments needed to reach the final product configuration.

Many common-rail injector models are reported in the literature. (Amoia et al., 1997; Bianchi et al., 2000; Brusca et al., 2002; Catalano et al., 2002; Ficarella et al., 1999; Payri et al., 2004). One of the older common-rail injector model was presented in (Amoia et al., 1997) and successively improved and employed for the analysis of the instability phenomena due to the control valve behaviour (Ficarella et al., 1999). An important input parameter in this model was the magnetic attraction force in the control valve dynamic model. This was calculated interpolating the experimental curve between driving current and magnetic force measured at fixed control valve positions. The discharge coefficient of the feeding and discharge control volume holes were determined and the authors asserted that the discharge hole operates, with the exception of short transients, under cavitating flow conditions at every working pressure, but this was not confirmed by (Coppo & Dongiovanni, 2007). Furthermore, the deformation of the stressed injector mechanical components was not taken into account. In (Bianchi et al., 2000) the electromagnetic attraction force was evaluated by means of a phenomenological model. The force was considered directly proportional to the square of the magnetic flux and the proportionality constant was experimentally determined under stationary conditions. The elastic deformation of the moving injector components were considered, but the injector body was treated as a rigid body. The models in (Brusca et al., 2002; Catalano et al., 2002) were very simple models. The aims in (Catalano et al., 2002) were to prove that pressure drops in an injection system are mainly caused by dynamic effects rather than friction losses and to analyse new common-rail injection system configurations in which the wave propagation phenomenon was used to increase the injection pressure. The model in (Brusca et al., 2002) was developed in the AMESim environment and its goal was to give the boundary conditions to a 3D-CFD code for spray simulation. Payri et al. (2004) report a model developed in the AMESim environment too, and suggest silicone moulds as an interesting tool for characterising valve and nozzle hole geometry.

A common-rail injector model employs three sub-models (electrical, hydraulic and mechanical) to describe all the phenomena that govern injector operation. Before one can use the model to estimate the effects of little adjustments or little geometrical modifications on the system performance, it is fundamental to validate the predictions of all the sub-models in the whole range of possible working conditions.

In the following sections of this chapter every sub-model will be thoroughly presented and it will be shown how its parameters can be evaluated by means of theoretical or experimental analysis. The focus will be placed on the electronic injector, as this component is the heart of any common rail system

2. Mathematical model

The injector considered in this investigation is a standard Bosch UNIJET unit (Fig. 1) of the common-rail type used in car engines, but the study methodology that will be discussed can be easily adapted to injectors manufactured by other companies.

The definition of a mathematical model always begins with a thorough analysis of the parts that make up the component to be modelled. Once geometrical details and functional relationships between parts are acquired and understood they can be described in terms of mathematical relationships. For the injector, this leads to the definition of hydraulic, mechanical, and electromagnetic models.

2.1 Hydraulic Model

Fig. 2 shows the equivalent hydraulic circuit of the injector, drawn following ISO 1219 standards. Continuous lines represent the main connecting ducts, while dashed lines represent pilot and vent connections. The hydraulic parts of the injector that have limited spatial extension are modelled with ideal components such as uniform pressure chambers and laminar or turbulent hydraulic resistances, according to a zero-dimensional approach. The internal hole connecting injector inlet with the nozzle delivery chamber (as well as the pipe connecting the injector to the rail or the rail to the high pressure pump) are modelled according to a one-dimensional approach because wave propagation phenomena in these parts play an important role in determining injector performance.

Fig. 3a shows the control valve and the relative equivalent hydraulic circuit. R_A and R_Z are the hydraulic resistances used for modelling flow through control-volume orifices A (dis-

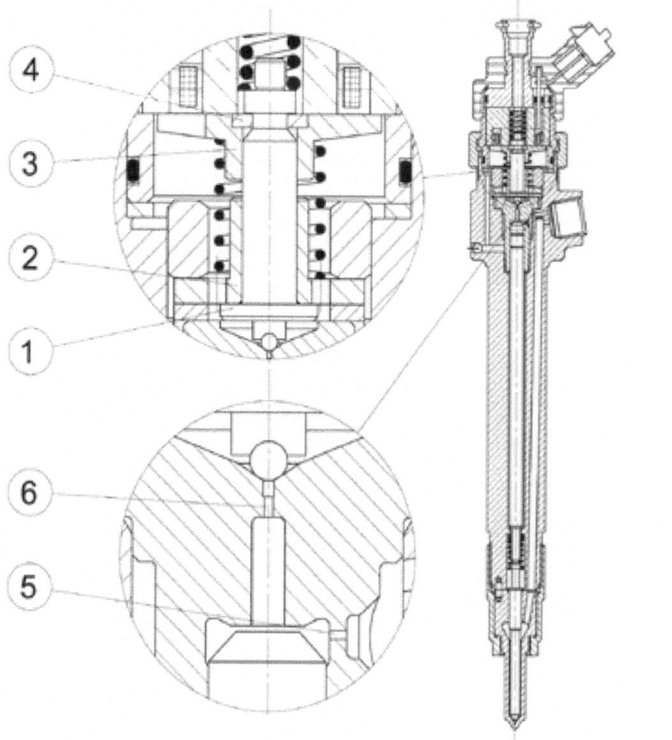

1. Control valve pin	4. C-shaped connecting pin and anchor
2. Pin guide and upper stop	5. Control volume feeding (Z) hole
3. Control valve anchor	6. Control volume discharge (A) hole

Fig. 1. Standard Bosch UNIJET injector

charge) and Z (feeding), respectively. The variable resistance R_{AZ} models the flow between chambers C_{dZ} and C_{uA}, taking into account the effect of the control piston position on the actual flow area between the aforementioned chambers. The solenoid control valve V_c is represented using its standard symbol, which shows the forces that act in the opening (one generated by the current I flowing through the solenoid, the other by the pressure in the chamber C_{dA}) and closing direction (spring force).

Fig. 3b illustrates the control piston and nozzle along with the relative equivalent hydraulic circuit. The needle valve V_n is represented with all the actions governing the needle motion, such as pressures acting on different surface areas, force applied by the control piston and spring force. The chamber C_D models the nozzle delivery volume, C_S is the sac volume, whereas the hydraulic resistance R_{hi} represents the i-th nozzle hole through which fuel is injected in the combustion chamber C_e. The control piston model considers two different surface areas on one side, so as to take into account the different contribution of pressure in the chambers C_{uA} and C_{dZ} to the total force applied in the needle valve closing direction.

Leakages both between control valve and piston and between needle and its liner are modelled by means of the resistances R_P and R_n respectively, and the resulting flow, which is collected in chamber C_T (the annular chamber around the control piston), is then returned to

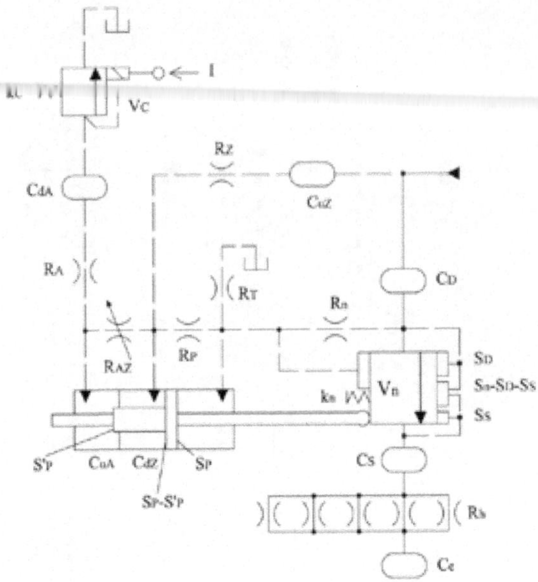

Fig. 2. Injection equivalent hydraulic circuit

tank after passing through a small opening, modelled with the resistance R_T, between control valve and injector body.

2.1.1 Zero-dimensional hydraulic model

The continuity and compressibility equation is written for every chamber in the model

$$\sum Q = \frac{V}{E_l}\frac{dp}{dt} + \frac{dV}{dt} \tag{1}$$

where $\sum Q$ is the net flow-rate coming into the chamber, $(V/E_l)(dp/dt)$ the rate of increase of the fluid volume in the chamber due to the fluid compressibility and (dV/dt) the deformation rate of the chamber volume.

Fluid leakages occurring between coupled mechanical elements in relative motion (e.g. needle and its liner, or control piston and control valve body) are modelled using laminar flow hydraulic resistances, characterized by a flow rate proportional to the pressure drop Δp across the element

$$Q = K_L \Delta p \tag{2}$$

where the theoretical value of K_L for an annulus shaped cross-section flow area can be obtained by

$$K_L = \frac{\pi d_m g^3}{12 l \rho \nu} \tag{3}$$

In case of eccentric annulus shaped cross-section flow area, Eq. 3 gives an underestimation of the leakage flow rate that can be as low as one third of the real one (White, 1991).

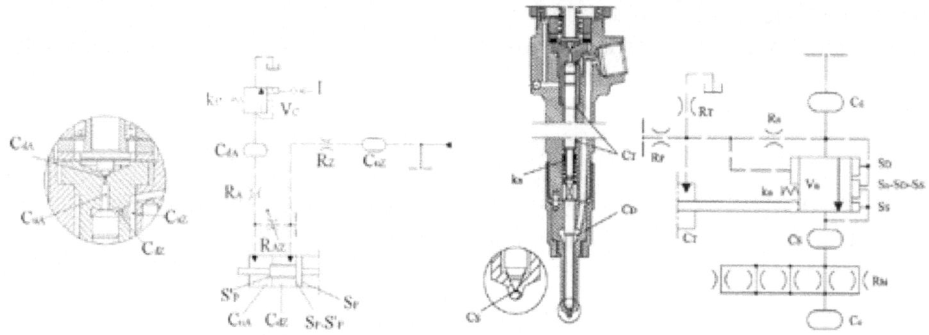

(a) Control valve (b) Needle and control piston

Fig. 3. Injection equivalent hydraulic circuit

Furthermore, the leakage flow rate, Equations 2 and 3, depends on the third power of the radial gap g. At high pressure the material deformation strongly affects the gap entity and its value is not constant along the gap length l because pressure decreases in the gap when approaching the low pressure side (Ganser, 2000). In order to take into account these effects on the leakage flow rate, the value of K_L has to be experimentally evaluated in the real injector working conditions.

Turbulent flow is assumed to occur in control volume feeding and discharge holes, in nozzle holes and in the needle-seat opening passage. As a result, according to Bernoulli's law, the flow rate through these orifices is proportional to the square root of the pressure drop, Δp, across the orifice, namely,

$$Q = \mu A \sqrt{\frac{2\Delta p}{\rho}} \tag{4}$$

The flow model through these orifices plays a fundamental role in the simulation of the injector behavior in its whole operation field, so the evaluation of the μ factor is extremely important.

2.1.2 Hole A and Z discharge coefficient

The discharge coefficient of control volume orifices A and Z is evaluated according to the model proposed in (Von Kuensberg Sarre et al., 1999). This considers four flow regimes inside the hole: laminar, turbulent, reattaching and fully cavitating.

Neglecting cavitation occurrence, a preliminary estimation of the hole discharge coefficient can be obtained as follows

$$\frac{1}{\mu} = \sqrt{K_I + f\frac{l}{d} + 1} \tag{5}$$

where K_I is the inlet loss coefficient, which is a function of the hole inlet geometry (Munson et al., 1990), l is the hole axial length, d is the hole diameter, and f is the wall friction coefficient, evaluated as

$$f = MAX\left(\frac{64}{Re}, 0.316\, Re^{0.25}\right) \tag{6}$$

where Re stands for the Reynolds number.
The ratio between the cross section area of the vena contracta and the geometrical hole area, μ_{vc}, can be evaluated with the relation:

$$\frac{1}{\mu_{vc}^2} = \frac{1}{\mu_{vc_0}^2} - 11.4\frac{r}{d} \tag{7}$$

where $\mu_{vc_0} = 0.61$ (Munson et al., 1990) and r is the fillet radius of the hole inlet.
It follows that the pressure in the vena contracta can be estimated as

$$p_{vc} = p_u - \frac{\rho_l}{2}\left(\frac{Q}{A\mu_{vc}}\right)^2 \tag{8}$$

If the pressure in the vena contracta (p_{vc}) is higher then the oil vapor pressure (p_v), cavitation does not occur and the value of the hole discharge coefficient is given by Equation 5. Otherwise, cavitation occurs and the discharge coefficient is evaluated according to

$$\mu = \mu_{vc}\sqrt{\frac{p_u - p_v}{p_u - p_d}} \tag{9}$$

The geometrical profile of the hole inlet plays a crucial role in determining, or avoiding, the onset of cavitation in the flow. In turn, the occurrence of cavitation strongly affects the flow rate through the orifice, as can be seen in Figure 4, which shows two trends of predicted flow rate (Q/Q_0) in function of pressure drop ($\Delta p = p_u - p_d$) through holes with the same diameter and length, but characterized by two different values of the r/d ratio (0.2 and 0.02), when p_u is kept constant and p_d is progressively decreased. In absence of cavitation, ($r/d = 0.2$), the relation between flow rate and pressure drop is monotonic while, if cavitation occurs ($r/d = 0.02$), the hole experiences a decrease in flow rate as pressure drop is further increased. This behavior agrees with experimental data reported in the literature (Lefebvre, 1989).

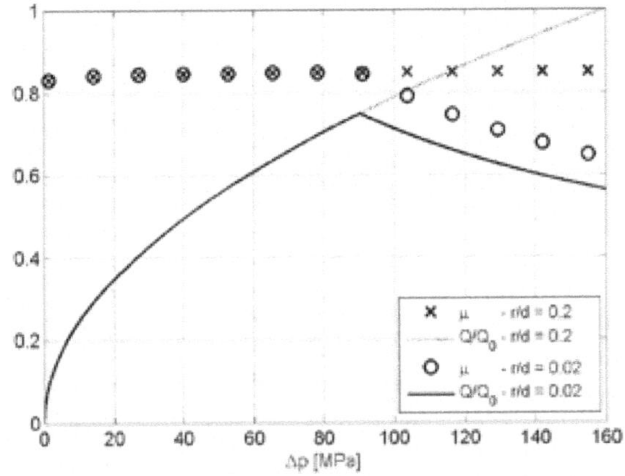

Fig. 4. Predicted flow through an orifice in presence/absence of cavitation

Obviously, such behavior would reflect strongly on the injector performance if the control volume holes happened to cavitate in some working conditions. Therefore, in order to accurately model the injector operation, it is necessary to accurately measure the geometrical profile of the control volume holes A and Z; by means of silicone moulds, as proposed by (Payri et al., 2004), it is possible to acquire an image of the hole shape details, as shown in Figure 5.

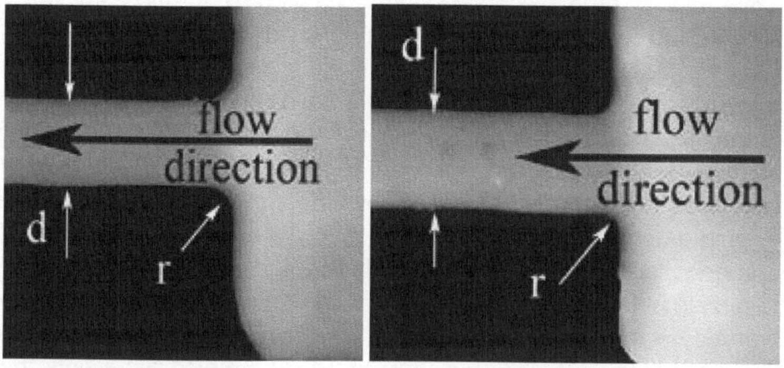

(a) A hole (b) Z hole

Fig. 5. Moulds of the control valve holes

By means of imaging techniques it is possible to measure the r/d ratio of the hole under investigation. Table 1 reports the results obtained for the injector under investigation. The value of K_I, in Equation 5, is a function of r/d only (Von Kuensberg Sarre et al., 1999) and, hence, easily obtainable.
Knowing that during production a hydro-erosion process is applied to make sure that, under steady flow conditions, all the holes yield the same flow rate, it is possible to define an iterative procedure to calculate the hole diameter using the discharge coefficient model presented above and the the steady flow rate value. This approach is preferrable to the estimation of the hole diameter with imaging techniques because it yields a result that is consistent with the discharge coefficient model used.

	r/d	K_I	d [μm]
Hole A	0.23±5%	0.033	280±2%
Hole Z	0.22±5%	0.034	249±2%

Table 1. Characteristics of control volume holes

In the control valve used in our experiments, under a pressure drop of 10 MPa, with a back pressure of 4 MPa, the holes A and Z yielded $6.5 \pm 0.2\, cm^3/s$ and $5.3 \pm 0.2\, cm^3/s$, respectively. With these values it is possible to calculate the most probable diameter of the control volume holes, as reported in Table 1. It is worth noting that the precision with which the diameters were evaluated was higher than that of the optical technique used for evaluating the shape of the control volume holes. This resulted from the fact that K_I shows little dependence on r/d when the latter assumes values as high as those measured. As a consequence, the experimental uncertainty in the diameter estimation is mainly originated from the uncertainty given on the stationary flow rate through the orifices.

2.1.3 Discharge coefficient of the nozzle holes

The model of the discharge coefficient of the nozzle holes is designed on the base of the unsteady coefficients reported in (Catania et al., 1994; 1997). These coefficients were experimentally evaluated for minisac and VCO nozzles in the real working conditions of a distributor pump-valve-pipe-injector type injection system. The pattern of this coefficient versus needle lift evidences three different phases. In the first phase, during injector opening, the moving needle tip strongly influences the efflux through the nozzle holes. In this phase, the discharge coefficient progressively increases with the needle lift. In the second phase, when the needle is at its maximum stroke, the discharge coefficient increases in time, independently from the pressure level at the injector inlet. In the last phase, during the needle closing stroke, the discharge coefficient remains almost constant. These three phases above mentioned describe a hysteresis-like phenomenon. In order to build a model suitable for a common rail injector in its whole operation field these three phases need to be considered.

Therefore, the nozzle hole discharge coefficient is modeled as needle lift dependent by considering two limit curves: a lower limit trend (μ_h^d), which models the discharge coefficient in transient efflux conditions, and an upper limit trend (μ_h^s), which represents the steady-state value of the discharge coefficient for a given needle lift. The evolution from transient to stationary values is modeled with a first order system dynamics.

It was experimentally observed (Catania et al., 1994; 1997) that the transient trend presents a first region in which the discharge coefficient increases rapidly with needle lift, following a sinusoidal-like pattern, and a second region, characterized by a linear dependence between discharge coefficient and needle lift. Thus, the following model is adopted:

$$\mu_h^d(\xi) = \begin{cases} \mu_h^d(\xi_0)\sin(\frac{\pi}{2\xi_0}\xi) & 0 \le \xi < \xi_0 \\ \frac{\mu_h^d(\xi_M)-\mu_h^d(\xi_0)}{\xi_M-\xi_0}(\xi-\xi_0)+\mu_h^d(\xi_0) & \xi \ge \xi_0 \end{cases} \tag{10}$$

where ξ is the needle-seat relative displacement, and ξ_0 is the transition value of ξ between the sinusoidal and the linear trend.

The use of the variable ξ, rather than the needle lift, x_n, emphasizes the fact that all the mechanical elements subject to fuel pressure, including nozzle and needle, deform, thus the real variable controlling the discharge coefficient is not the position of the needle, but rather the effective clearance between the latter and the nozzle.

The maximum needle lift, ξ_M, varies with rail pressure due to the different level of deformation that this parameter induces on the mechanical components of the injector. The relation between ξ_M and the reference rail pressure p_{r0} is assumed to be linear as

$$\xi_M = K_1\, p_{r0} + K_2 \tag{11}$$

where K_1 and K_2 are constants that are evaluated as explained in the section 2.3.3.

Similarly, the value of ξ_0 in Equation 10 is modeled as a function of the operating pressure p_{r0} in order to better match the experimental behavior of the injection system. Thus, the following fit is used

$$\xi_0 = K_3\, p_{r0} + K_4 \tag{12}$$

and K_3 and K_4 are obtained at the end of the model tuning phase (table 4).

In order to define the relation between the steady state value of the nozzle-hole discharge coefficient (μ_h^s) and the needle-seat relative displacement (ξ) the device in Figure 6 was designed. It contains a camshaft that can impose to the needle a continuously variable lift up to

1 mm. Then, a modified injector equipped with this device was connected to the common rail injection system and installed in a Bosch measuring tube, in order to control the nozzle hole downstream pressure. The steady flow rate was measured by means of a set of graduated burettes.

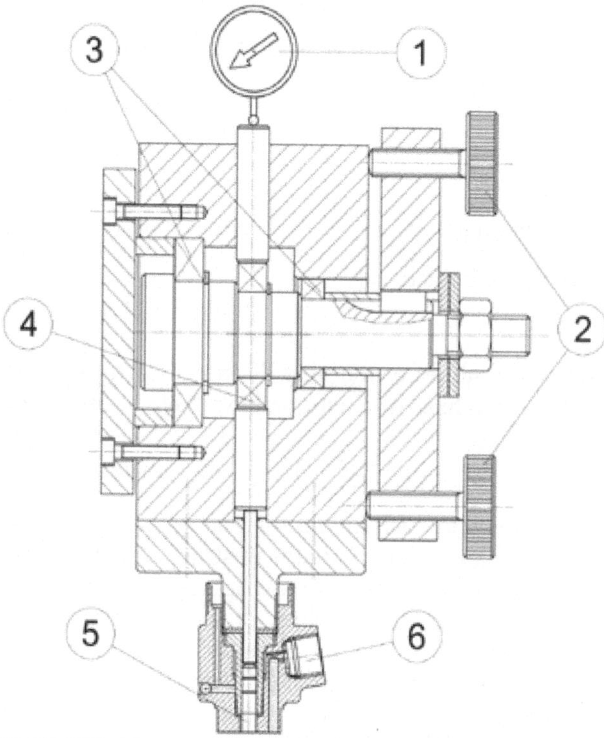

1. Dial indicator	4. Eccentric ball bearing (e = 1mm)
2. Handing for varying needle lift	5. Injector control piston
3. Axis support bearing	6. Injector inlet

Fig. 6. Device for fixed needle-seat displacement imposition

Figure 7a shows the trends of steady-state flow rate versus needle lift at rail pressures of 10 and 20 MPa, while the back pressure in the Bosch measuring tube was kept to either ambient pressure or 4 MPa; whereas Figure 7b shows the resulting stationary hole discharge coefficient, evaluated for the nozzle under investigation.

Taking advantage of the reduced variation of μ_h^s with operation pressure, it is possible to use the measured values to extrapolate the trends of steady-state discharge coefficient for higher pressures, thus defining the upper boundary of variation of the nozzle hole discharge coefficient values.

During the injector opening phase the unsteady effects are predominant and the sinusoidal-linear trend of the hole discharge coefficient, Equation 10, was considered; when the needle-seat relative displacement approaches its relative maximum value ξ_M^r, the discharge coefficient increases in time, which means that the efflux through the nozzle holes is moving to the stationary conditions. In order to describe this behavior, a transition phase between the

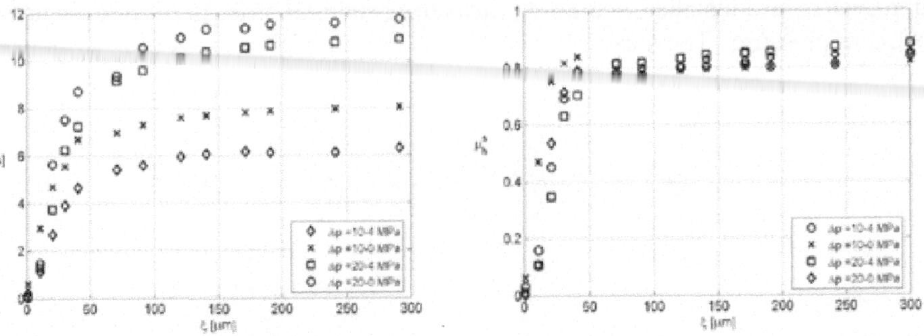

(a) Steady flow rate

(b) Stationary discharge coefficient

Fig. 7. Stationary efflux through the nozzle

unsteady and the stationary values of the hole discharge coefficient at this needle lift was considered. This phase was modeled as a temporal exponential curve, namely,

$$\mu_h = \mu_h^d(\xi_M^r) + [\mu_h^s(\xi_M^r) - \mu_h^d(\xi_M^r)]\ [1 - exp\,(-\frac{t - t_0}{\tau})] \tag{13}$$

where t_0 is the instant in time at which the needle-seat relative displacement approaches its maximum value ξ_M^r, $\mu_h^d\ (\xi_M^r)$ and $\mu_h^s\ (\xi_M^r)$ are the unsteady and the stationary hole discharge coefficients evaluated at this needle-seat relative displacement, and τ is the time constant of this phenomenon, which have to be defined during the model tuning phase.

Figure 8 shows the computed nozzle hole discharge coefficient, μ_h, dependence upon needle-seat relative displacement, ξ, in accordance to the proposed model, in a wide range of operating conditions (which are showed by rail pressure p_{r0} and energisation time ET_0 in the legend).

Examining the discharge coefficient, μ_h, trends for the three main injections (ET_0 = 780 μs, 700 μs and 670 μs) during the opening phase, it is interesting to note that for a given value of the needle lift, lower discharge coefficients are to be expected at higher operating pressures. This can be explained considering that the flow takes longer to develop if the pressure differential, and thus the steady state velocity to reach is higher.

The main injection trends also show the transition from the sinusoidal to the linear dependence of the transient discharge coefficient on needle lift.

The phase in which the needle has reached the maximum value and the discharge coefficient increases in time from unsteady to stationary values is not very evident in main injections, because the former increases enough during the opening phase to approach the latter. This happens because the needle reaches sufficiently high lifts as to have reduced effect on the flow in the nozzle holes, and the longer injection allows time for complete flow development.

Conversely, during pilot injections (ET_0=300 μs), the needle reaches lower maximum lifts, hence lower values of the unsteady discharge coefficient, so that the phase of transition to the stationary value lasts longer. The beginning of this transition can be easily identified by analyzing the curves marked with dots and crosses in Figure 8. The point at which they depart from their main injection counterpart (same line style but without markers) marks the beginning of the exponential evolution in time to stationary value of discharge coefficient.

For both pilot and main injections, the nozzle hole discharge coefficient remains constant, and equal to the stationary value, during the injector closing phase, as shown by the horizontal profile of the trends in Figure 8.
The needle-seat discharge coefficient μ_s has to be modeled too. It is assumed as needle lift dependent according to (Xu et al., 1992) where this coefficient was experimentally evaluated after removing the nozzle tip. A three segment trend is considered, as shown in Fig. 8, but it is worth to point out that it plays a marginal role in the injection system simulation because its values are higher than 0.8 for most needle lift values.

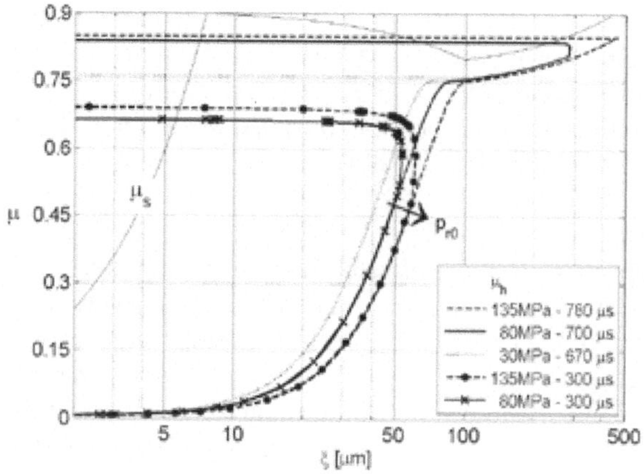

Fig. 8. Needle-seat and holes discharge coefficient

2.1.4 One-dimensional model: pipe flow model

A one-dimensional modelling approach is followed in order to model the fluid flow in the pipe connecting injector and rail and in the injector internal duct that carries the fluid from the inlet to the delivery chamber. This is necessary to correctly take into account pressure wave propagation that occurs in those elements. The pipe flow conservation equations are written for a single-phase fluid because in the common-rail injection system cavitation does not appear in the connecting pipe. An isothermal flow is assumed and only the momentum and mass conservation equations need to be solved

$$\frac{\partial \mathbf{w}}{\partial t} + A\frac{\partial \mathbf{w}}{\partial x} = \mathbf{b} \tag{14}$$

where $\mathbf{w} = \left\{ \begin{array}{c} u \\ p \end{array} \right\}$, $A = \left\{ \begin{array}{cc} u & 1/\rho \\ \rho c^2 & u \end{array} \right\}$, $\mathbf{b} = \left\{ \begin{array}{c} -4\tau/\rho d \\ 0 \end{array} \right\}$
and τ is the wall shear stress that is evaluated under the assumption of steady-state friction (Streeter et al., 1998).
The eigenvalues of the hyperbolic system of partial differential Equations 14 are $\lambda = u \pm c$, real and distinct. The celerity c of the wave propagation can be evaluated as

$$c = \sqrt{\frac{c_l}{\left(1 + K_p \frac{E_l}{E_p} \frac{d_p}{t_p}\right)}} \tag{15}$$

where the second term within brackets takes into account the effect of the pipe elasticity; K_p is the pipe constraint factor, depending on pipe support layout, E_p the Young's modulus of elasticity of the pipe material, d_p the pipe diameter and t_p the pipe wall thickness (Streeter et al., 1998). Being the pipe ends rigidly constrained, the pipe constraint factor K_p can be evaluated as

$$K_p = 1 - \nu_p^2 \tag{16}$$

where ν_p is the Poisson's modulus of the pipe material.
Pipe junctions are treated as minor losses and only the continuity equation is locally written. As mentioned before, this simple pipe flow model is not suitable when cavitation occurs. This is not a limitation when common-rail injection system are modelled because of the high pressure level at which these systems always work. In order to model conventional injection systems, as pump-pipe-nozzle systems, it is necessary to employ a pipe flow model able to simulate the cavitation occurrence. For this purpose the authors developed an appropriate second order model (Dongiovanni et al., 2003).

2.1.5 Fluid properties

Thermodynamic properties of oil are affected by temperature and pressure that remarkably vary in the common rail injection system operation field. Density, wave propagation speed and kinematic viscosity of the ISO4113 air-free test oil had been evaluated as function of pressure and temperature (Dongiovanni, 1997). These oil properties were approximated with analytic functions of the exponential type in the range of pressures from 0.1 to 200 MPa and temperatures from 10 °C to 120 °C. These analytic relations were derived from the actual property values supplied by the oil maker, by using the least-square method for non-linear approximation functions with two independent variables. The adopted formulae are:

$$\rho_l(p,T) = K_{\rho 1} + \left[1 - exp\left(-\frac{p}{K_{\rho 2}}\right)\right] K_{\rho 3}\, p^{K_{\rho 4}} \tag{17}$$

$$E_l(p,T) = K_{E1} + \left[1 - exp\left(-\frac{p}{K_{E2}}\right)\right] K_{E3}\, p^{K_{E4}} \tag{18}$$

$$\nu_l(p,T) = K_{\nu 1} + K_{\nu 2}\, p^{K_{\nu 3}} \tag{19}$$

The K_{Ei}, $K_{\rho i}$ and $K_{\nu i}$ are polynomial functions of temperature T

$$K_i = \sum_{j=0}^{l_i} K_{i,j} T^j \qquad i = 1,2,3,4 \tag{20}$$

and the numerical coefficients that appear in them are reported in Table 2 according with SI units: pressure $[p] = bar$, temperature $[T] =^\circ C$, density $[\rho_l] = kg/m^3$, bulk modulus $[E_l] = MPa$ and kinematic viscosity $[\nu_l] = mm^2/s$
Finally, the celerity of the air free oil is evaluate in accordance with $c_l = \sqrt{E_l/\rho_l}$.
By using these approximation functions, the maximum deviation between experimental and analytical values in the examined range of pressure and temperature has been estimated as being lower than ±0.2% for density, ±1.2% for bulk modulus, ±0.6% for celerity and ±18% for kinematic viscosity.

K_ρ	j= 0	j= 1	j= 2
$K_{\rho 1,j}$	8.3636e2	-6.7753e-1	-
$K_{\rho 2,j}$	1.5063e2	-2.4202e-1	-
$K_{\rho 3,j}$	1.7784e-1	1.4640e-3	1.5402e-5
$K_{\rho 4,j}$	7.8109e-1	-8.1893e-4	-

K_E	j= 0	j= 1	j= 2
$K_{E1,j}$	1.7356e3	-1.0908e1	2.2976e-2
$K_{E2,j}$	7.5540e1	-	-
$K_{E3,j}$	1.5050	-3.7603e-3	-
$K_{E4,j}$	9.4448e-1	3.9441e-4	-

K_ν	j=0	j=1	j=2	j=3
$K_{\nu 1,j}$	6.4862	-1.5847e-1	1.6342e-3	-6.0334e-6
$K_{\nu 2,j}$	4.0435e-4	-2.3118e-6	-	-
$K_{\nu 3,j}$	1.4346	-6.2288e-3	3.3500e-5	-

Table 2. Polynomial coefficients for ISO4113 oil

2.2 Electromagnetic model

A model of the electromechanical actuator that drives the control valve must be realized in order to work out the net mechanical force applied by the solenoid on its armature, for a given current flowing in the solenoid. The magnetic force applied by the solenoid on the armature F_{Ea} can be obtained by applying the principle of energy conservation to the armature-coil system (Chai, 1998; Nasar, 1995). In the general form it can be written as follows:

$$V\, I\, dt = F_{Ea} dx_a + dW_m \tag{21}$$

where $V\, I\, dt$ represents the electric energy input to the system, $F_{Ea}\, dx_a$ is the mechanical work done on the armature and dW_m is the change in the magnetic energy.
From Faraday's law, voltage V may be expressed in terms of flux linkage ($N\frac{d\Phi}{dt}$) and Equation 21 becomes

$$N\, I\, d\Phi = F_{Ea} dx_a + dW_m \tag{22}$$

as shown in (Chai, 1998; Nasar, 1995); by considering Φ and x_a as independent variables, Equation 22 can be reduced to

$$F_{Ea} = -\left.\frac{\partial W_m}{\partial x_a}\right|_{\Phi} \tag{23}$$

The magnetic circuit geometry of the control valve needs to be thoroughly analyzed in order to evaluate the magnetic energy stored in the gap. Fig. 9a shows the path of the significant magnetic fluxes, having neglected secondary leakage fluxes and flux fringing.
Exploiting the analogy between Ohm's and Hopkinson's law, it is possible to obtain the magnetic equivalent circuit of Fig. 9b where NI is the ampere-turns in the exciting coil and $\Re_j$ ($j = 1,..,5$) are the magnetic reluctances. When the magnetic flux flows across a cross-section area A_a constant along the path length l, the value of the j-th reluctance can be obtained by:

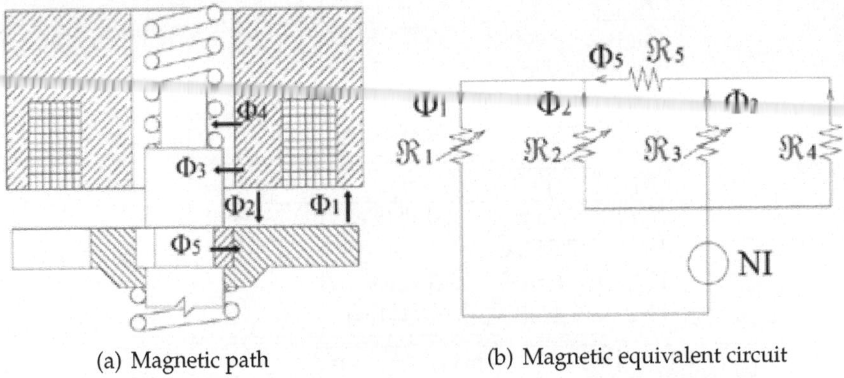

Fig. 9. Magnetic model sketch

$$\Re_j = \frac{l_j}{\mu_0 A_{aj}} \quad (j = 1,2) \tag{24}$$

When the flux flows across a radial path, the reluctance can be evaluated as

$$\Re_j = \frac{1}{2\pi\mu_0 t_j} ln \left(\frac{d_e}{d_i} \right)_j \quad (j = 3,4,5) \tag{25}$$

being t the radial thickness, d_e and d_i the external and internal diameter of the gap volume. Reluctance of the ferromagnetic components was neglected because it is several order of magnitude lower than the corresponding gap reluctance.

Circuit of Fig. 9b is solved using Thevenin's theorem, and the equivalent circuit reluctance connected to the magnetomotive force generator is determined as

$$\Re = \Re_1 + \frac{\Re_2\Re_3\Re_4 + \Re_2\Re_5(\Re_3 + \Re_4)}{\Re_3\Re_4 + (\Re_2 + \Re_5)(\Re_3 + \Re_4)} \tag{26}$$

The magnetic energy W_m is stored in the volume of the electromechanical actuator, but only the portion of energy stored in the gap between control-valve body and magnetic core depends on the armature lift x_a. Consequently, being the magnetization curve of non-ferromagneti materials (oil in the gaps) linear, Equation 23 can be written as

$$F_{Ea} = -\frac{1}{2}\Phi^2 \frac{d\Re}{dx_a} = -\frac{1}{2}\left(\frac{NI}{\Re}\right)^2 \frac{d\Re}{dx_a} \tag{27}$$

To complete the model, it was necessary to take into account the saturation phenomenon that occurs to every ferromagnetic material. That is, a magnetic flux cannot increase indefinitely, as the material presents a maximum magnetic flux density after which the curve $B - H$ is almost flat. In this model we assumed a simplified magnetization curve, given by :

$$B = \begin{cases} \mu H & H < H^* \\ \mu H^* + \mu_0 (H - H^*) & H \geq H^* \end{cases} \tag{28}$$

thus neglecting material hysteresis and non-linearity.

As a result of the saturation phenomenon, the maximum force of attraction is limited because the maximum magnetic flux which can be obtained in the j-th branch of the circuit is approximately

$$\Phi_{Mj} \approx \mu H_j^* A_j \tag{29}$$

being μ_0 negligible with respect μ.
The most important parameters in the electromagnetic model are set as reported in Table 3.

N	$B^* = \mu H^*$ [T]	t_3 [mm]	t_4 [mm]	t_5 [mm]
32	2.5	0.65	1.5	0.05

Table 3. Most important electromagnetic model parameters

The model was employed to evaluate the inductance of the solenoid when mounted on the injector body. In this case, with the valve actuator in the closed position, an inductance of 134 μH was evaluated. Employing a sinusoidal wave generator at a frequency of 5 kHz, which is high enough to make negligible the mechanical system movements, an inductance of 137 μH was measured. The accordance between experimental and theoretical inductance value indirectly validates the electromagnetic model and the parameters value.
Fig. 10a shows the theoretical (solid line) driving actuator force when the actual energizing current (dashed line) is used to feeding the injector solenoid.

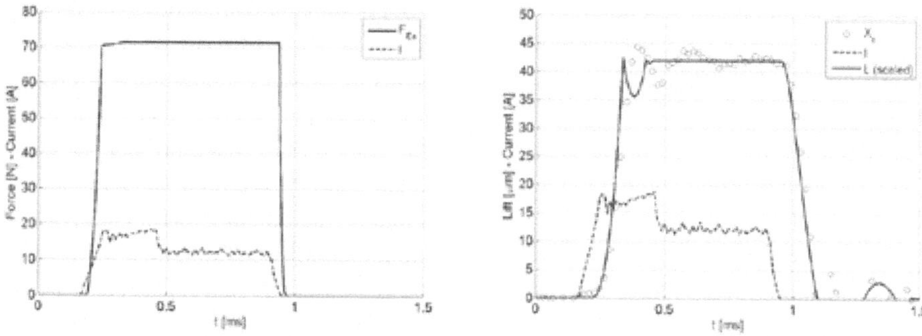

(a) Magnetic force and feeding current (b) Inductance and armature lift

Fig. 10. Magnetic model results

Furthermore, we point out that the measure of the injector coil inductance $L = N/\Re^2$ could be used to indirectly evaluate the control valve lift, due to the dependence of reluctance upon armature distance from the solenoid (Equation 24).
Bearing in mind that, by applying Ohm's law to the solenoid coil, the inductance L could be evaluated as:

$$L = \frac{\int (V - RI)\, dt}{I} \tag{30}$$

hence only the measurement of solenoid current I and voltage V would be required to calculate L.

Fig. 10b draws the theoretical inductance L, which was calculated according to Equation 24 and opportunely scaled, compared to the experimental valve lift x_c, showing a good agreement between the two trends, and hence the potential of this non-invasive measurement technique. However, Equation 30 is only applicable when electric current is flowing in the solenoid coil so, for example, it is not possible to use this method to record the the control valve closure trend because, as Fig. 10b shows, this usually begins when the solenoid current is null.
A possible way to solve this problem would be to inject an additional, small amplitude, high frequency (around 1 MHz) current into the coil, but this technique has not yet been tested by the authors.

2.3 Mechanical model

All mechanical devices that can move during injector functioning (i.e. needle, control piston and control valve) are modelled using the conventional mass-spring-damper scheme, governed by a mechanical equilibrium equation, in which the dynamic parameters are function of element position.

$$m_j \frac{d^2 x_j}{dt^2} + \overline{\beta}_j \frac{dx_j}{dt} + \overline{k}_j x_j + \overline{F}_{0j} = F_j \tag{31}$$

where m_j is the displacing mass, $\overline{\beta_j}$ the damping coefficient, $\overline{k}_j$ the spring stiffness and $\overline{F}_{0_j}$ the spring preload; the bar above the symbols indicates that these coefficients are evaluated according to the relative position of the moving elements.

2.3.1 Control piston, needle and nozzle model

The high working pressures in the common-rail injection system stress its components and cause appreciable deformation of them. In order to take into account the effects of the axial deformation of nozzle and injector body, the nozzle is modelled by means of a conventional mass-spring-damper scheme as well as the needle and the control piston, while the injector body is modelled by means of a simple spring having appropriate stiffness.
Injector needle, control piston and nozzle form a three degrees of freedom mechanical system, which can be modelled as shown in Figure 11a. Three equilibrium equations are needed to describe the system motion, one for each element. With reference to Fig. 11a, the dynamic equilibrium Equation 31 is written using the following definition for the control piston ($j = P$), the needle ($j = n$) and the nozzle ($j = N$):
external force F_j:

$$\begin{aligned} F_P &= -F_c + p_T S_P + F_{R(Pn)} - F_{R(Pb)} \\ F_n &= -p_T S_n - F_{R(Pn)} + F_S + F_{R(nN)} \\ F_N &= -F_S + p_e S_n - F_{R(nN)} \end{aligned} \tag{32}$$

where $F_{R(ij)}$ is the force that i-th and j-th element apply to each other when they are in contact,

$$F_c = p_{uA} S_P' + p_{dZ}(S_P - S_P') \tag{33}$$

and

$$F_S = p_S S_S + p_D S_D + [\gamma\, p_S + (1-\gamma)\, p_D)]\,(S_n - S_D - S_S) \tag{34}$$

where $\gamma = 0$ indicates that the nozzle is closed while $\gamma = 1$ indicates open nozzle conditions. Damping coefficient $\overline{\beta_j}$, stiffness $\overline{k_j}$ and preload $\overline{F_{0j}}$ are evaluated as follows for:
control piston

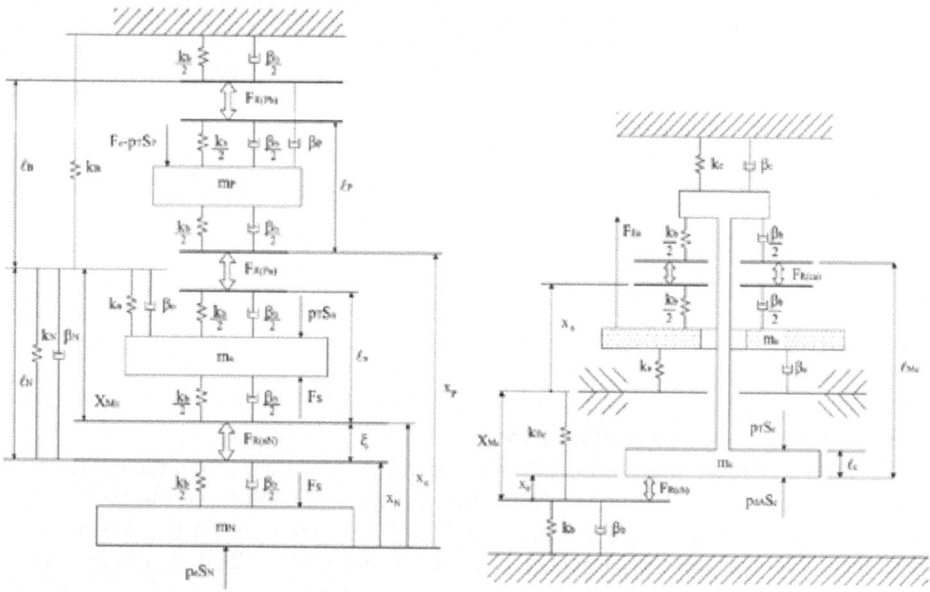

(a) Control piston and needle (b) Control valve

Fig. 11. Dynamic models

$$\begin{array}{llll} x_P < X_{MP} - l_P & \overline{\beta_P} = \beta_P & \overline{k_P} = 0 & \overline{F_{0P}} = 0 \\ X_{MP} - l_P \leq x_P & \overline{\beta_P} = \beta_b + \beta_P & \overline{k_P} = k_b & \overline{F_{0P}} = -k_b(X_{MP} - l_P) \end{array} \tag{35}$$

needle

$$\begin{array}{llll} x_n - x_N < 0 & \overline{\beta_n} = \beta_b + \beta_n & \overline{k_n} = k_b + k_n & \overline{F_{0n}} = F_{0n} \\ 0 \leq x_n - x_N < X_{Mn} - l_n & \overline{\beta_n} = \beta_n & \overline{k_n} = k_n & \overline{F_{0n}} = F_{0n} \\ X_{Mn} - l_n \leq x_n - x_N & \overline{\beta_n} = \beta_b + \beta_n & \overline{k_n} = k_b + k_n & \overline{F_{0n}} = F_{0n} - k_b X_{Mn} \end{array} \tag{36}$$

nozzle

$$\begin{array}{llll} x_n - x_N < 0 & \overline{\beta_N} = \beta_b + \beta_N & \overline{k_N} = k_b + k_N & \overline{F_{0N}} = 0 \\ 0 \leq x_n - x_N & \overline{\beta_N} = \beta_N & \overline{k_N} = k_N & \overline{F_{0N}} = 0 \end{array} \tag{37}$$

2.3.2 Control valve model

The control valve contains two mobile parts: the pin element having mass m_c and the armature element of mass m_a; they can be modelled with the two degrees of freedom scheme shown in Fig. 11b. The two dynamic equilibrium equations are written in a similar fashion as Equation 31 where $j = a$ indicates the armature and $j = c$ the control-pin. The external forces F_j can be evaluated as

$$\begin{array}{l} F_a = F_{Ea} - F_{R(ca)} \\ F_c = (p_{dA} - p_T)S_c + F_{R(ca)} + F_{R(cb)} \end{array} \tag{38}$$

where F_{Ea} represents the electromagnetic action that the current generates when it flows in the solenoid coil determined as shown in section 2.2.

Damping coefficient $\overline{\beta_j}$, stiffness $\overline{k_j}$ and preload $\overline{F_{0j}}$ are evaluated as follows:
pin element

$$
\begin{array}{llll}
x_c < 0 & \overline{\beta_c} = \beta_b + \beta_c & \overline{k_c} = k_b + k_c & \overline{F_{0c}} = F_{0c} \\
0 \le x_c < X_{Mc} - l_c & \overline{\beta_c} = \beta_c & \overline{k_c} = k_c & \overline{F_{0c}} = F_{0c} \\
X_{Mc} - l_c \le x_c & \overline{\beta_c} = \beta_b + \beta_c & \overline{k_c} = k_b + k_c & \overline{F_{0c}} = F_{0c} - k_b(X_{Mc} - l_c)
\end{array}
\tag{39}
$$

armature

$$
\begin{array}{llll}
l_{Mc} - X_{Mc} + x_c \ge x_a & \overline{\beta_a} = \beta_a & \overline{k_a} = k_a & \overline{F_{0a}} = F_{0a} \\
x_a > l_{Mc} - X_{Mc} + x_c & \overline{\beta_a} = \beta_b + \beta_a & \overline{k_a} = k_b + k_a & \overline{F_{0a}} = F_{0a} - k_b(l_{Mc} - X_{Mc} + x_c)
\end{array}
\tag{40}
$$

2.3.3 Mechanical components deformation

The axial deformation of needle, nozzle and control piston have to be taken into account. These elements are considered only axially stressed, while the effects of the radial stress are neglected. For the sake of simplicity, the axial length of control piston (l_P), needle (l_n), and nozzle (l_N) can be evaluated as function of the axial compressive load (F_C) in each element. Therefore, the deformed length l of these elements, which are considered formed by m parts having cross section A_j and initial length l_{0_j}, is evaluated as follows

$$
l = \sum_{j}^{m} l_{0_j} \left(1 - \frac{F_{C_j}}{EA_j}\right) \tag{41}
$$

where E is Young's modulus of the considered material.
The axial deformation of the injector body is taken into account by introducing in the model the elastic elements indicated as k_B and k_{Bc} in Figure 11.
The injector body deformation cannot be theoretically calculated very easily, because one should need to take into account the effect and the deformation of the constraints that fix the injector on the test rig. For this reason, in order to evaluate the elasticity coefficient of k_B and k_{Bc}, an empirical approach is followed, which consists in obtaining a relation between the axial length of these elements and the fluid pressure inside the injector body. As direct consequence, the maximum stroke of the needle-control piston (ξ_M) and of the control-valve (X_{Mc}) can be expressed as a function of the injector structural stress.

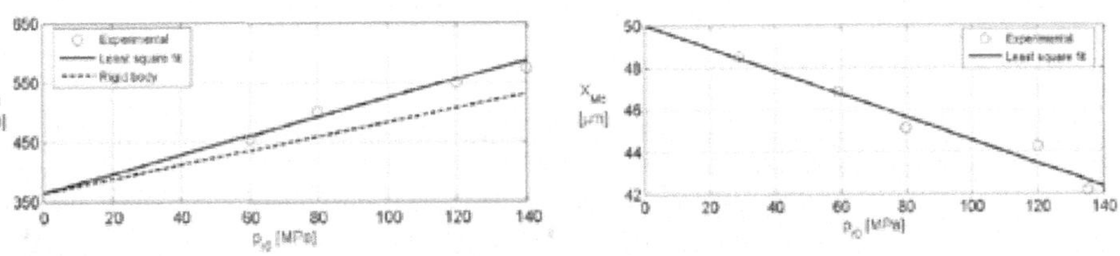

(a) Needle (b) Control valve

Fig. 12. Effect of pressure on the maximum moving element lift

Figure 12 reports the actual maximum needle-control piston lift (circular symbols) as a function of rail pressure. At the rail pressure of 30 MPa the maximum needle-control piston lift was

not reached, so no value is reported at this rail pressure. The continuous line represents the least-square fit interpolating the experimental data and the dashed line shows the maximum needle-control piston lift calculated by considering only nozzle, needle and control-piston axial deformation. The difference between the two lines represents the effect of the injector body deformation on the maximum needle-control piston lift. This can be expressed as a function of rail pressure and, for the considered injector, can be estimated in 0.41 $\mu m/MPa$. By means of the linear fit (continuous line) reported in Figure 12 it is possible to evaluate the parameters $K_1 = 1.59\ \mu m/MPa$ and $K_2 = 364\ \mu m$ that appear in Eq. 11.
In order to evaluate the elasticity coefficient k_{Bc}, an analogous procedure can be followed by analyzing the maximum control-valve lift dependence upon fuel pressure, as shown in Figure 12. It was found that the effect of injector body deformation was that of reducing the maximum control valve stroke of 0.06 $\mu m/MPa$.

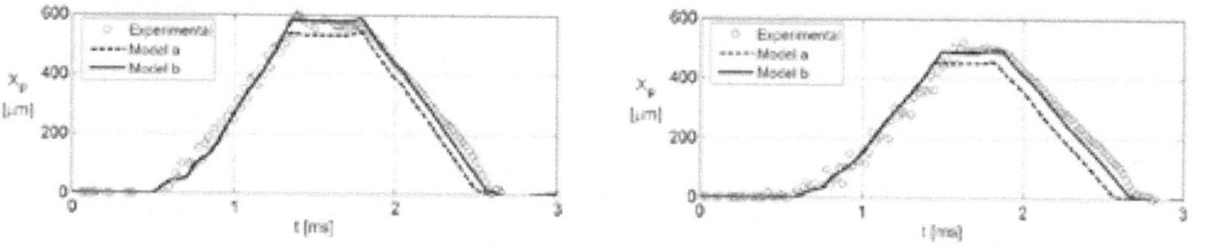

(a) p_{r0}=140 MPa, ET_0= 1230 μs (b) p_{r0}=80 MPa, ET_0= 1230 μs

Fig. 13. Deformation effects on needle lift

The relevance of the deformation effects on the injector predicted performances is shown in Fig. 13. The left graph shows the control piston lift at a rail pressure of 140 MPa generated with an energizing time ET_0 of 1230μs, while the right graph shows the same trend at a rail pressure of 80 MPa, and generated with the same value of ET_0. The experimental results are drawn by circular symbols, while lines refer to theoretical results. The dashed lines (Model a) show the theoretical control piston lift evaluated by only taking in to account the axial deformation of the moving elements and nozzle, while the continuous lines (Model b) show the theoretical results evaluated by taking into account the injector body deformation too. The difference between the two models is significant, and so is the underestimation of the volume of fluid injected per stroke (4.3% with p_{r0}=140MPa and ET_0 of 1230μs, 3.6% with p_{r0}=80MPa, ET_0 of 1230μs). This highlights the necessity of accounting for deformation of the entire injector body, if accurate predictions are sought.
Indeed, the maximum needle lift evaluation plays an important role in the simulation of the injector behaviour in its whole operation field because it influences both the calculation of the injected flow rate (as the discharge coefficients of needle-seat and nozzle holes depend also on needle lift) and of the injector closing time, thus strongly affecting the predicted volume of fuel injected per cycle.
The deformation of the injector body also affects the maximum control valve stroke, and a similar analysis can be performed to evaluate its effects on injector performance. Our study showed that this parameter does not play as important a role as the maximum needle stroke, because the effective flow area of the A hole is smaller than the one generated by the displacement of the control valve pin, and thus it is the A hole that controls the efflux from the control volume to the tank.

2.3.4 Masses, spring stiffness and damping factors

Components mass and springs stiffness k_j can be easily estimated. Whenever a spring is in contact to a moving element, the moving mass m_j value used in the model is the sum of the element mass and a third of the spring mass. In this way it is possible to correctly account for the effect of spring inertia too.

The evaluation of the damping factors β_j in Equation 31 is considerably more difficult. Considering the element moving in its liner, like needle and control piston, the damping factor takes into account the damping effects due to the oil that moves in the clearance and the friction between moving element and liner. The oil flow effect can be modelled as a combined Couette-Poiseuille flow (White, 1991) and the wall shear stress on the moving element surface can be theoretically evaluated. Experimental evidences show that friction effects are more relevant than the fluid-dynamics effects previously mentioned. Unfortunately, these can not be theoretically evaluated because their intensity is linked to manufacturing tolerances (both geometrical and dimensional). Therefore, damping factors must be estimated during the model tuning phase.

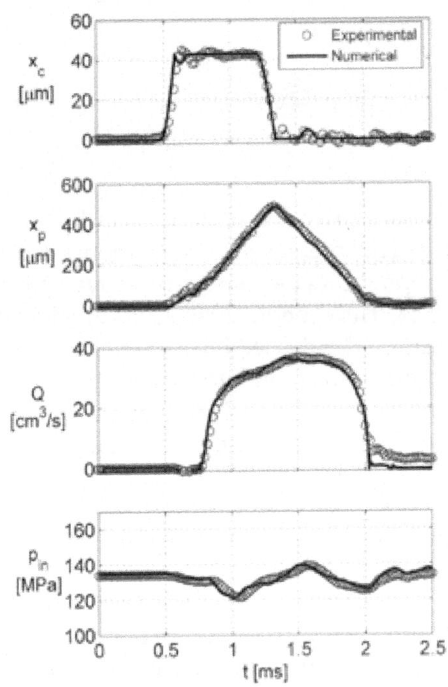

(a) Main injection: ET_0=780μs, p_{r0}=135 MPa

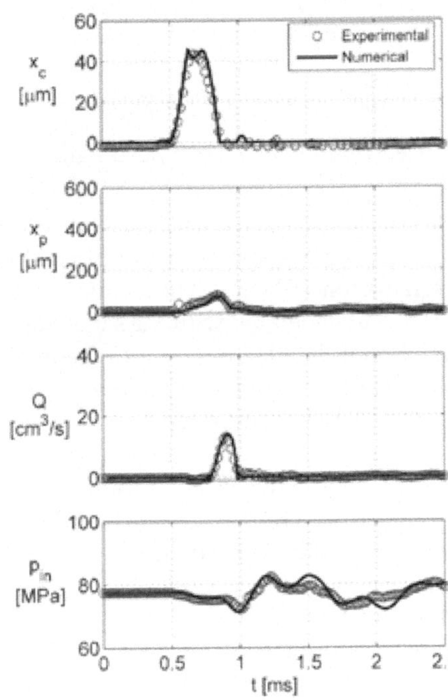

(b) Pilot injection: ET_0=300μs, p_{r0}=80 MPa

Fig. 14. Comparison between numerical and theoretical results

3. Model tuning and results

Any mathematical model requires to be validated by comparing its results with the experimental ones. During the validation phase some model parameters, which cannot be experimentally or theoretically evaluated, have to be carefully adjusted.
The model here presented was tested comparing numerical and experimental control valve lift x_c, control piston lift x_P, injected flow rate Q and injector inlet pressure p_{in} in several operating conditions. Figure 15 shows two of these validation tests and the good accordance between experimental and numerical results is evident.
Table 4 shows the value of the parameters that were adjusted during the tuning phase. These values can be used as starting points for the development of new injector models, but their exact value will have to be defined during model tuning for the reasons explained above.
After the tuning phase the model can be used to reproduce the injection system performance in its whole operation field. By way of example, Fig. 15 shows the experimental and numerical volume injected per stroke V_f and the percentage error of the numerical estimation.

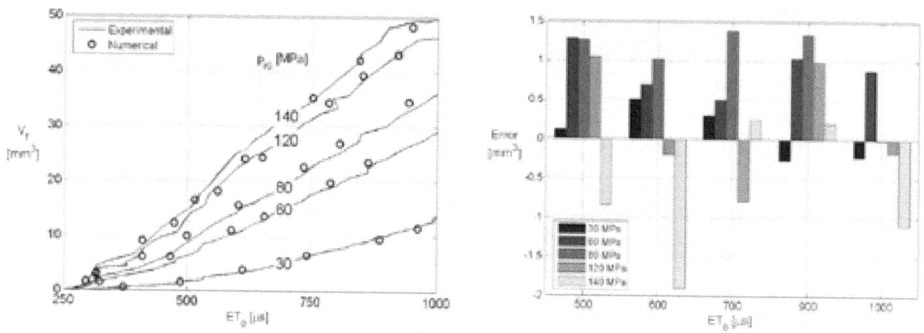

(a) Injected fluid volume per stroke (b) Model error

Fig. 15. Model validation

Eq. 10		Eq. 12		Eq. 13	Eq. 31					
$\mu_h^d(\xi_0)$	$\mu_h^d(\xi_M)$	K_3	K_4	τ	β_n	β_N	β_P	β_c	β_a	
0.75	0.85	0.28 $\mu m/MPa$	63 μm	25 μs	6.1	6310	6.5	28	5.1	[kg/s]

Table 4. Tuning defined parameters

4. Nomenclature

Symbol	Definition	Unit
A	Geometrical area	m^2
C	Uniform pressure chamber	
c	Wave propagation speed	m/s
d	Hole || Pipe diameter	m
e	Eccentricity	m
E	Young's modulus	Pa
ET	Injector solenoid energisation time	s
F	Force	N
f	Friction factor	
I	Electric current	A
K	Coefficient	
k	Spring stiffness	N/m
l	Length	m
m	Mass	kg
N	Number of coil turns	
p	Pressure	Pa
Q	Flow rate	m^3/s
r	Rail || Fillet radius	m
R	Hydraulic resistance	
Re	Reynolds number	
S	Surface area	m^2
t	Time	s
u	Average cross-sectional velocity of the fluid	m/s
V	Valve || Volume	m^3
W	Energy	J
X	Distance	m
x	Displacement || Axial coordinate	m
β	Damping factor	kg/s
γ	switch (0=nozzle closed,1=nozzle open)	
Δ	Increment || Drop	
Φ	Magnetic flux	Wb
ξ	Needle-seat relative displacement	m
μ	Contraction || Discharge coefficient	
ρ	Density	kg/m^3
τ	Wall shear stress || Time constant	Pa || s
$\Re$	Reluctance	H^{-1}

Subscript	Definition
A	Control-volume discharge hole
a	Armature
B	Injector body
b	Seat
C	Compression
c	Control valve
D	Delivery

Symbol	Definition	Unit
d	Downstream	
E	Electromechanical	
e	Injection environment || External	
f	Fuel	
h	Hole	
l	Inlet loss || Liquid phase	
in	Injector inlet	
M	Maximum value	
m	Magnetic	
N	Nozzle	
n	Needle	
P	Piston	
R	Reaction Force	
r	Rail	
S	Sac	
s	Needle–seat	
T	Tank	
u	Upstream	
v	Vapour	
vc	Vena contracta	
Z	Control-volume feeding hole	
0	Reference value	

Superscripts	Definition
d	Dynamic
r	Relative
s	Steady-state

5. References

Amoia, V., Ficarella, A., Laforgia, D., De Matthaeis, S. & Genco, C. (1997). A theoretical code to simulate the behavior of an electro-injector for diesel engines and parametric analysis, *SAE Transactions* **970349**.

Badami, M., Mallamo, F., Millo, F. & Rossi, E. E. (2002). Influence of multiple injection strategies on emissions, combustion noise and bsfc of a di common rail diesel engines, *SAE paper* **2002-01-0503**.

Beatrice, C., Belardini, P., Bertoli, C., Del Giacomo, N. & Migliaccio, M. (2003). Downsizing of common rail d.i. engines: Influence of different injection strategies on combustion evolution, *SAE paper* **2003-01-1784**.

Bianchi, G. M., Pelloni, P. & Corcione, E. (2000). Numerical analysis of passenger car hsdi diesel engines with the 2nd generation of common rail injection systems: The effect of multiple injections on emissions, *SAE paper* **2001-01-1068**.

Boehner, W. & Kumel, K. (1997). Common rail injection system for commercial diesel vehicles, *SAE Transactions* **970345**.

Brusca, S., Giuffrida, A., Lanzafame, R. & Corcione, G. E. (2002). Theoretical and experimental analysis of diesel sprays behavior from multiple injections common rail systems, *SAE paper* **2002-01-2777**.

Canakci, M. & Reitz, R. D. (2004). Effect of optimization criteria on direct-injection homogeneous charge compression ignition gasoline engine performance and emissions using fully automated experiments and microgenetic algorithms, *J. of Engineering for Gas Turbines and Power* **126**: 167–177.

Catalano, L. A., Tondolo, V. A. & Dadone, A. (2002). Dynamic rise of pressure in the common rail fuel injection system, *SAE paper* **2002-01-0210**.

Catania, A., Dongiovanni, C., Mittica, A., Badami, M. & Lovisolo, F. (1994). Numerical analysis vs. experimental investigation of a distribution type diesel fuel injection system, *J. of Engineering for Gas Turbines and Power* **116**: 814–830.

Catania, A. E., Dongiovanni, C., Mittica, A., Negri, C. & Spessa, E. (1997). Experimental evaluation of injector-nozzle-hole unsteady flow-coefficients in light duty diesel injection systems, *Proceedings of the Ninth Internal Pacific Conference on Automotive Engineering*, Bali, Indonesia.

Chai, H. (1998). *Electromechanical Motion Devices*, Pearson Professional Education.

Coppo, M. & Dongiovanni, C. (2007). Experimental validation of a common-rail injector model in the whole operation field, *J. of Engineering for Gas Turbines and Power* **129**(2): 596–608.

Dongiovanni, C. (1997). Influence of oil thermodynamic properties on the simulation of a high pressure injection system by means of a refined second order accurate implicit algorithm, *ATA Automotive Engineering* pp. 530–541.

Dongiovanni, C., Negri, C. & Roberto, R. (2003). A fluid model for simulation of diesel injection systems in cavitating and non-cavitating conditions, *Proceedings of the ASME ICED Spring Technical Conference*, Salzburg, Austria.

Ficarella, A., Laforgia, D. & Landriscina, V. (1999). Evaluation of instability phenomena in a common rail injection system for high speed diesel engines, *SAE paper* **1999-01-0192**.

Ganser, M. A. (2000). Common rail injectors for 2000 bar and beyond, *SAE paper* **2000-01-0706**.

Henelin, N. A., Lai, M.-C., Singh, I. P., Zhong, L. & Han, J. (2002). Characteristics of a common rail diesel injection system under pilot and post injection modes, *SAE paper* **2002-010218**.

Lefebvre, A. (1989). *Atomization and Sprays*, Hemisphere Publishing Company.

Munson, B. R., Young, D. F. & Okiishi, T. H. (1990). *Fundamentals of Fluid Mechanics*, Wiley.

Nasar, S. (1995). *Electric machines and power systems : Vol. 1. Electric Machines*, McGraw-Hill.

Park, C., Kook, S. & Bae, C. (2004). Effects of multiple injections in a hsdi diesel engine equipped with common rail injection system, *SAE paper* **2004-01-0127**.

Payri, R., Climent, H., Salvador, F. J. & Favennec, A. G. (2004). Diesel injection system modelling. methodology and application for a first-generation common rail system, *Proceedings of the Institution of Mechanical Engineering* **Vol. 218 Part D**.

Schmid, M., Leipertz, A. & Fettes, C. (2002). Influence of nozzle hole geometry, rail pressure and pre-injection on injection, vaporization and combustion in a single-cylinder transparent passenger car common rail engine, *SAE paper* **2002-01-2665**.

Schommers, J., Duvinage, F., Stotz, M., Peters, A., Ellwanger, S., Koyanagi, K. & Gildein, H. (2000). Potential of common rail injection system passenger car di diesel engines, *SAE paper* **2000-01-0944**.

Streeter, V. L., White, E. B. & Bedford, K. W. (1998). *Fluid Mechanics*, McGraw-Hill.

Stumpp, G. & Ricco, M. (1996). Common rail - an attractive fuel injection system for passenger car di diesel engines, *SAE Transactions* **960870**.

Von Kuensberg Sarre, C., Kong, S.-C. & Reitz, R. D. (1999). Modeling the effects of injector nozzle geometry on diesel sprays, *SAE paper* **1999-01-0912**.

White, F. M. (1991). *Viscous Fluid Flow*, McGraw-Hill.

Xu, M., Nishida, K. & Hiroyasu, H. (1992). A practical calculation method for injection pressure and spray penetration in diesel engines, *SAE Transactions* **920624**.

Yamane, K. & Shimamoto, Y. (2002). Combustion and emission characteristics of direct-injection compression ignition engines by means of two-stage split and early fuel injection, *J. of Engineering for Gas Turbines and Power* **124**: 660–667.

The investigation of the mixture formation upon fuel injection into high-temperature gas flows

1. Introduction

Combustion of a fuel in the combustion chambers of a gas-turbine engine and a gas-turbine plant is closely connected with the processes of mixing (Lefebvre, 1985). Investigations of these processes carried out by both experimental and computational methods have recently become especially crucial because of the necessity of solving ecological problems.

One of the most pressing problems at present is account for the influence of droplets on an air flow. In some of the regimes of chamber operation this may lead to a substantial, almost twofold, change in the long range of a fuel spray and, consequently, to corresponding changes in the distributions of the concentrations of fuel phases.

In this chapter physical models of the processes of interphase heat and mass transfer and computational techniques based on them are suggested. The present work is a continuation of research by Maiorova & Tretyakov, 2008. We set out to calculate the fields of air velocity and temperature as well as of the distribution of a liquid fuel in module combustion chambers with account for the processes of heating and evaporation of droplets in those regimes typical of combustion chambers in which there is a substantial interphase exchange. It is clear that when a "cold" fuel is supplied into a "hot" air flow, the droplets are heated and the air surrounding them is cooled. It is evident that at small flow rates of the fuel this cooling can be neglected. The aim of this work is to answer two questions: how much the air flow is cooled by fuel in the range of parameters typical of real combustion chambers, and how far the region of flow cooling extends. Moreover, the dependence of the flow characteristics on the means of fuel spraying (pressure atomizer, jetty or pneumatic) and also on the spraying air temperature is investigated.

2. Statement of the Problem

Schemes of calculated areas are presented on fig. 1. Calculations were carried out for the velocity and temperature of the main air flow U_0 = 20 m/s and T_0 = 900 K, fuel velocity V_f = 8 m/s, fuel temperature T_f = 300 K. The gas pressure at the channel inlet was equal to 100 kPa.

The first model selected for investigation (fig. 1-a) is a straight channel of rectangular cross section 150 mm long into which air is supplied at a velocity U_0 and temperature T_0. It was assumed that the stalling air flow at the inlet had a developed turbulent profile and that the spraying air had a uniform profile. Injection of a fuel with a temperature T_f into the channel at a velocity V_f is made through a hole in the upper wall of the channel with the aid of an injector installed along the normal to the longitudinal axis of the channel halfway between the side walls. In modeling the pneumatic injector it is considered that, coaxially with the fuel supply, the spraying air is fed at a velocity U_1 and temperature T_1 into the channel through a rectangular hole of size 4.5 ×3.75 mm. In modeling a jetty injector, we assume that the spraying air is absent.

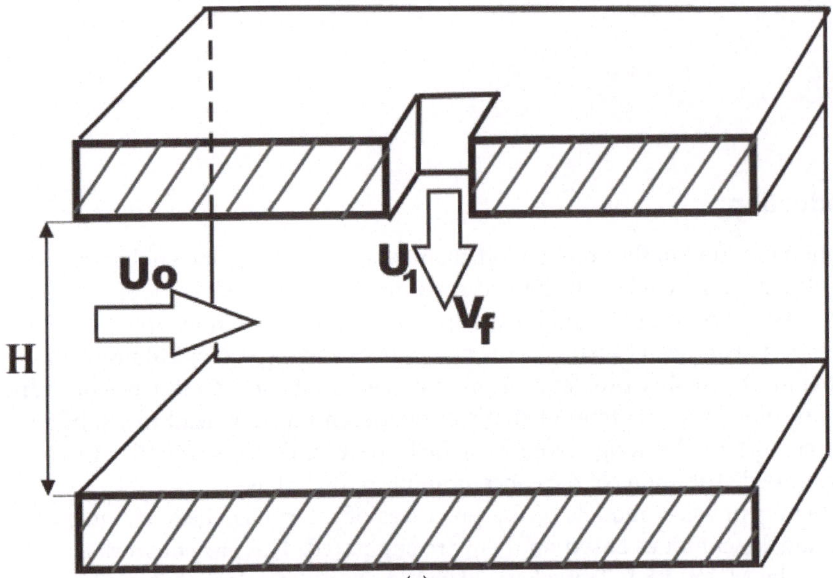

(a)

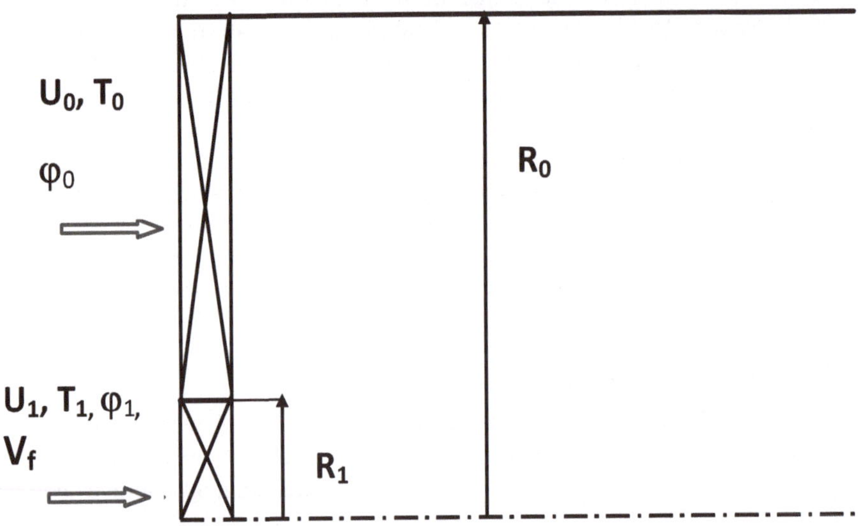

(b)

Fig. 1. Schemes of calculated areas

The variable parameters of the calculation were the velocity and temperature of the spraying air: U_1 = 0–20 m/s and T_1 = 300–900 K, as well as the summed coefficient of air excess through the module α = 1.35–5.4.
The values of the regime parameters are presented in Table 1. Regime 1 corresponds to jet spraying of a fuel, regime 2 – to pneumatic spraying of a fuel by a cold air jet; and regime 3 – to pneumatic spraying by a hot air jet in the limiting case of equality between the temperatures of the spraying air and main flow.

Variant	α	Regime 1	Regime 2		Regime 3	
		U_1, m/s	U_1, m/s	T_1, K	U_1, m/s	T_1, K
1	5.4	0	20	300	20	900
2	2.7	0	20	300	20	900
3	1.35	0	20	300	20	900

Table 1. Operating Parameters for the flow in a straight channel.

The second model (fig. 1-b) is the flow behind two coaxial tubes in radius of 5 and 40 mm, tube length is 240 mm. Heat-mass transfer of drop-forming fuel with the co-swirling two-phase turbulent gas flows is calculated. In this case injection of a fuel is made through a pressure or pneumatic atomizer along the longitudinal axis. Regime parameters corresponds regimes 2 and 3 from table 1 and α = 3.3. Inlet conditions were constant axial velocity, turbulent intensity and length. Axial swirlers are set in inlet sections. The tangential velocity set constant in the outer channel. The flow in the central tube exit section corresponded to solid body rotation law. The wane angles in inner and outer channels (φ_1 and φ_0) varied from 0 to 65°.

3. Calculation Technique

Calculations of the flow of a gas phase are based on numerical integration of the full system of stationary Reynolds equations and total enthalpy conservation equations written in Euler variables. The technique of allowing for the influence of droplets on a gas flow is based on the assumption that such an allowance can be made by introducing additional summands into the source terms of the mass, momentum, and energy conservation equations. The transfer equations were written in the following conservative form:

$$\mathrm{div}\left(\rho_g \vec{U}_g \Phi - \Gamma_\Phi \mathrm{grad}\Phi\right) = S_\Phi + S_\Phi^{int} \tag{1}$$

Here S_Φ^{int} is the interphase source term that describes the influence of droplets on the corresponding characteristics of flow. The density and pressure are ensemble-averaged (according to Reynolds) and all the remaining dependent variables – according to Favre, i.e., with the use of density as a weight coefficient.
Written in the form of Eq. (1), the system of equations of continuity ($\Phi \equiv 1, \Gamma_\Phi \equiv 0, S_\Phi \equiv 0$), motion ($\Phi = U_{gi}$, i = 1, 2, 3), and of total enthalpy conservation h ($S_h \equiv 0$) is solved by the Simple finite-difference iteration method (Patankar, 1980). The walls were considered

thermally insulated. To find the coefficients of turbulent diffusion, use is made of the Boussinesq hypothesis on the linear dependence of the components of the tensor of turbulent stresses on the components of the tensor of deformation rates of average motion and two equations of transfer of turbulence characteristics (k–ε) in the modification that takes into account the influence of flow turbulence Reynolds numbers on the turbulent characteristics of flow (Chien, 1982). Here, the boundary conditions of zero velocity are imposed on the solid walls. For swirl flows calculations the model was modernized to take into account the swirl effect in turbulence structure (Koosinlin at al., 1974).

In the absence of chemical reactions the gas mixture is considered to consist of two components: kerosene vapors (with a molecular weight of 0.168 kg /mole) and air (with a conventional molecular weight of 0.029 kg /mole). For the mass fraction of kerosene vapors m_f the equation of transfer of the type (1) is solved, and the mass fraction of air is determined from the condition under which the sums of the mass fractions of all the components are equal to unity.

The calculations of the distribution of fuel are based on the solution of a system of equations of motion, heating, and evaporation of individual droplets written in the Lagrange variables. The influence of turbulent pulsations onto the motion of droplets and on the change in their shape in the process of their motion is considered to be negligibly small. Then the equations that describe the processes of motion, heating, and evaporation have the following form:

$$m_d \frac{d\vec{V}_d}{dt} = \vec{R} \tag{2}$$

$$\frac{dT_d}{dt} = 6\lambda_g (1 + 0.6Re^{1/2}\, Pr^{1/3}) \frac{T_g - T_d}{\rho_d D_d^2 c_f} \tag{3}$$

$$\frac{dD_d}{dt} = \frac{8\lambda_g}{2D\rho_d c_{pg}} \ln\left(1 + \frac{c_{pg}}{L}(T_g - T_d)\right)\left(1 + 0.23Re^{1/2}\right) \tag{4}$$

We consider that the law of the resistance of droplets is the same as that of the resistance of solid spherical particles of diameter D_d

$$\vec{R} = 0.5 C_R S \rho_g W\vec{W}, \quad C_R = 24Re^{-1} + 4.4Re^{-0.5} + 0.32, \quad S = D_d^2/4 \tag{5}$$

In modeling a fuel spray it was assumed that it had a polydisperse structure with the size distribution of droplets obeying the Rosin–Rammler law (Dityakin at al., 1977) with exponent 3 and mean-median diameter 50 μm. The range of the sizes of droplets was divided into 14 intervals. The angle distribution of droplets was taken to be uniform, and the working fuel was TS-1 kerosene.

The interphase source terms are calculated together with the distribution of the liquid fuel from the conditions of the fulfillment of the laws of conservation of momentum, mass, and heat of the gas–droplet system. It is considered that the corresponding terms in the equations for the turbulence characteristics can be neglected.

Since physically the source term S_1^{int} in the continuity equation, just as the source term in the equation of transfer of m_f, $S_{m_f}^{int}$, is the increase in the concentration of the fuel vapor per unit time equal to the rate of liquid evaporation, then

$$S_1^{int} = S_{m_f}^{int} = \dot{C}_v = -\frac{dC_f}{dt} \quad (6)$$

where $\dot{C}_v$ is the rate of change of C_v due to the interphase exchange.

The interphase source terms in the equations of conservation of momentum components are the components of the vector of the rate of change in the gas momentum due to the exchange with droplets in a unit volume $\vec{S}_g^{int}$. These quantities are determined from the equation of conservation of momentum for the gas–droplet system:

$$\Delta(m_d\vec{V}_d)+\Delta(m_g\vec{U}_g)= 0 \quad (7)$$

where m_g is the mass of the isolated element of the gas volume Δv. Here and below, it is assumed that the volume of fuel droplets is negligibly small as compared to the volume occupied by the gas.

Assuming Δt_d (the residence time of a droplet in the volume element Δv) to be small enough, we may replace the second term in (7) by $\vec{S}_g^{int}\Delta v\Delta t_d$. This gives us an approximate expression to determine $\vec{S}_g^{int}$:

$$\vec{S}_{g,i}^{int} = -\frac{\Delta(m_g\vec{U}_g)}{\Delta v\Delta t_{d,i}} = -\frac{m_{d,i}}{\Delta v}\frac{\Delta\vec{V}_{d,i}}{\Delta t_{d,i}} - \vec{V}_{d,i}\frac{\Delta C_{f,i}}{\Delta t_{d,i}} \quad (8)$$

where ΔV_d is a change in the droplet velocity during its residence in the elementary volume, i means individual droplet.

The last term in relation (8) describes the gas momentum increment at the expense of the vapor fuel phase momentum related to the elementary volume Δv and the time of droplet evaporation in this volume, since $\Delta C_v = -\Delta C_f$. It is assumed that the fuel vapor and air in the volume Δv mix instantaneously. When $\Delta v \rightarrow 0$, $\Delta t_d \rightarrow 0$, we obtain an exact expression for $\vec{S}_g^{int}$ in a differential form:

$$\vec{S}_{g,i}^{int} = -C_{f,i}\frac{d\vec{V}_{d,i}}{dt} - \vec{V}_{d,i}\frac{dC_{f,i}}{dt} \quad (9)$$

Here $C_{f,i}$ denotes a fraction of the ith droplet in the volumetric concentration of liquid.

The summed value of the rate of change in the momentum of a unit volume of gas is equal to

$$\vec{S}_g^{int} = \sum_i \vec{S}_{g,i}^{int} \quad (10)$$

where summation is carried out over all the droplets.

The interphase source term in the transfer equation of the variable $U_\varphi r$, S_φ^{int} is determined from the equation of conservation of angular momentum for the gas–droplet system. This team has the folowing form:

$$S_\varphi^{int} = -r\left(C_{f,i}\frac{dV_{d\varphi,i}}{dt} - V_{d\varphi,i}\frac{dC_{f,i}}{dt}\right) \quad (11)$$

The interphase source term in the equation for enthalpy S_h^{int} that describes heat exchange between droplets and the gas flow is determined from the equation of conservation of the total enthalpy of the gas–droplet system, which has the form

$$\Delta(m_d h_d) + \Delta(m_g h_g) = -L\Delta m_g \quad (12)$$

The expression on the right-hand side of equality (12) determines the energy spent on the transition of the droplet liquid of mass $\Delta m_d = -\Delta m_g$ into the gaseous state, and Δh_d and Δm_d are changes in the enthalpy and mass of the droplet during its residence in the volume Δv.
Assuming the time Δt_d to be small enough, we replace the second term in expression (12) by $S_h^{int}\Delta v \Delta t_d$. Then the approximate expression for determining $S_{h,i}^{int}$

$$S_{h,i}^{int} = \frac{\Delta(m_g h_g)}{\Delta v \Delta t_{d,i}} = -\frac{\Delta(m_{d,i} h_{d,i})}{\Delta v \Delta t_{d,i}} + L\frac{\Delta m_{d,i}}{\Delta v \Delta t_{d,i}} \quad (13)$$

Using the definition of the enthalpy $h_d = c_f T$, we will rewrite (13) in the form

$$S_{h,i}^{int} = L\frac{\Delta C_{f,i}}{\Delta t_{d,i}} - C_{f,i}\left(c_f \frac{\Delta T_{d,i}}{\Delta t_{d,i}} + T_{d,i}\frac{\Delta c_{f,i}}{\Delta t_{d,i}}\right) \quad (14)$$

When $\Delta v \to 0$, $\Delta t_d \to 0$, we obtain an expression for $S_{h,i}^{int}$ in a differential form:

$$S_{h,i}^{int} = L\frac{dC_{f,i}}{dt} - C_{f,i}\left(c_f \frac{dT_{d,i}}{dt} + T_{d,i}\frac{dc_{f,i}}{dt}\right) \quad (15)$$

The summed value of S_h^{int} (inflow of heat from the liquid phase to the unit volume of gas) is equal to

$$S_h^{int} = \sum_i S_{h,i}^{int} \quad (16)$$

where summation is carried out over all the droplets.
The values $\Delta\vec{V}_{d,i}/\Delta t_{d,i}$, $\Delta T_{d,i}/\Delta t_{d,i}$ and $\Delta C_{f,i}/\Delta t_{di}$ or $d\vec{V}_{d,i}/dt$,, $dT_{d,i}/dt$ and $dC_{f,i}/dt$ are taken from the solution of the equation of motion and heating of an individual droplet.
The technique of calculation of a two-phase flow is based on the solution of a conjugate problem of flow of the gas and liquid media and heat exchange between them. First the problem of the motion of a gas is solved without account for the influence of the motion of droplets on the flow and then, based on the velocity and temperature fields obtained, the distribution of the liquid fuel is calculated as well as the interphase source terms. At the second stage, the gasdynamic and temperature fields are recalculated with account for the interphase sources (the results of the first stage are used as the initial conditions). When needed, the process is repeated several times. The convergence criteria of the iteration process are considered to be the absence of changes in the velocity and temperature fields from iteration to iteration for the gas flow and stabilization over the iterations of the coordinate of the maximum value of the concentration of droplets at the outlet of the model within the limits of one mesh of the finite-difference grid.

4. Testing of the Calculation Technique

The first model (fig. 1-a) was originally used to test the calculation method. As a result of methodical calculations a finite-difference grid uniform in the x and z directions was selected. The grid along the y axis was made finer toward the channel walls according to the exponential law with exponent 0.91. The total number of nodes in the grid was 111·111·41 = 505,161.

In experiences of authors it was spent laser visualization of a stream and postprocessing of photos by a method of the gradient analysis. The comparison on Fig.2 shows that the computational technique describes well the experimental data on the configuration of the fuel spray.

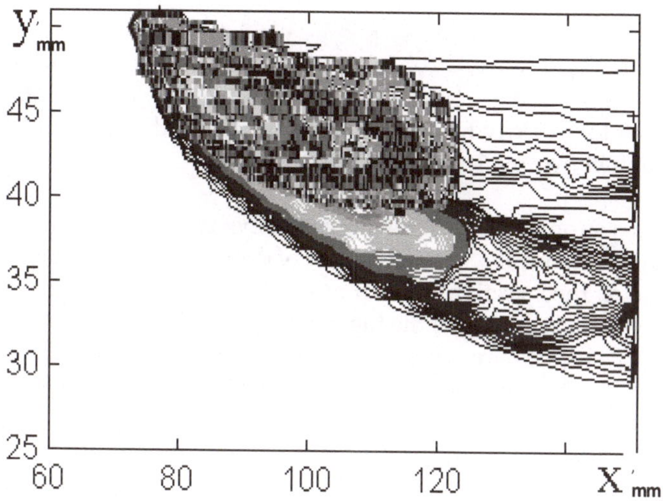

Fig. 2. Isolines of volumetric concentrations of fuel droplets behind the pressure atomizer in the central longitudinal section of the rectangular mixer; gray lines - calculation, color lines - experimental data (gradient analysis); T_0 = 300 K

5. Results of Calculations

The results of calculation for the straight channel of rectangular cross section are presented in fig. 3 - 11. The velocity field in the vicinity of the place of fuel injection for the jetty (U_1 = 0) and pneumatic (U_1 = 20 m/s) sprayings are given in Figs. 3 and 4, respectively. Here and below, the results were made nondimensional through division by the characteristic dimension H = 50 mm, which is the height of the channel (R_0 for the axisymmetric mixer), and by the characteristic velocity U_0 = 20 m/s.

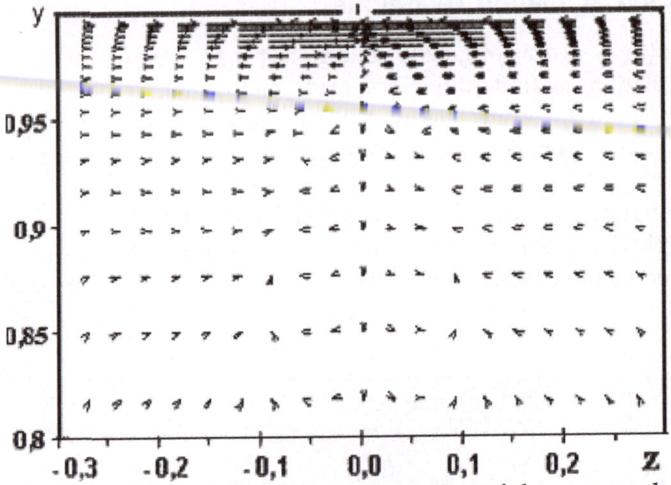

Fig. 3. Calculated vector velocity field in the cross section of the rectangular mixer x = 0.28 with jetty supply of fuel (regime 1, $U_1 = 0$); $\alpha = 1.35$

In the absence of fuel supply at $U_1 = 0$ the flow is homogeneous and isothermal. In the case of the jetty spraying, as a result of the interaction of droplets with the main air flow, on both sides of the center of the injection hole, zones of reverse flow initiated by droplets are observed (Fig. 3), which increase with the fuel flow rate. At the same time, the very values of the secondary flow velocities are almost an order of magnitude smaller than the characteristic flow velocity. In the longitudinal section the shape of the velocity profiles preserves its inlet configuration.

In supplying spraying air (Fig. 4) the main role in the formation of the gas velocity fields is played by the interaction of air streams of the main and spraying air. Thus, behind the injected jet a secondary flow is formed in the form of a three-dimensional zone of reverse flows. The influence of the process of interaction of droplets with air on the flow structure is practically unnoticeable for the cases considered (the patterns of flow for all the regimes are practically identical). Moreover, the depth of penetration of fuel-air jets into the stalling air flow decreases with increase in the temperature of the spraying air due to the decrease in the injected gas momentum.

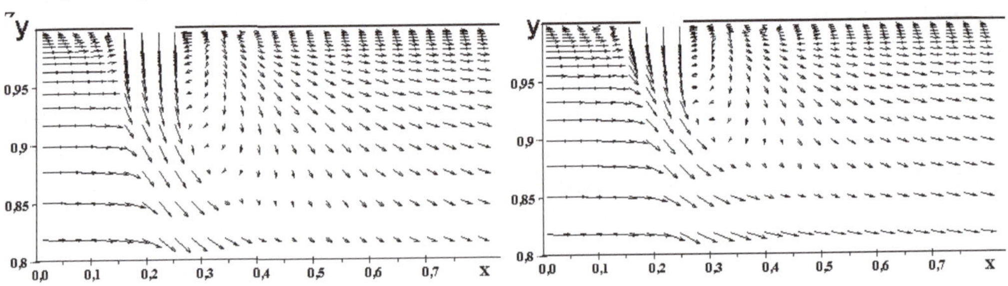

(a) (b)

Fig. 4. Calculated vector velocity field in the central longitudinal section of the rectangular mixer with pneumatic supply of a fuel; a) spraying by a cold air jet (regime 2, $U_1 = 20$ m/s, $T_1 = 300$ K); b) spraying by a hot air jet (regime 3, $U_1 = 20$ m/s, $T_1 = 900$ K)

Figures 5 - 8 present the distributions of dimensionless volumetric concentrations of a liquid fuel c_f. The results were made nondimensional through division by the value of the main air flow density at the inlet.

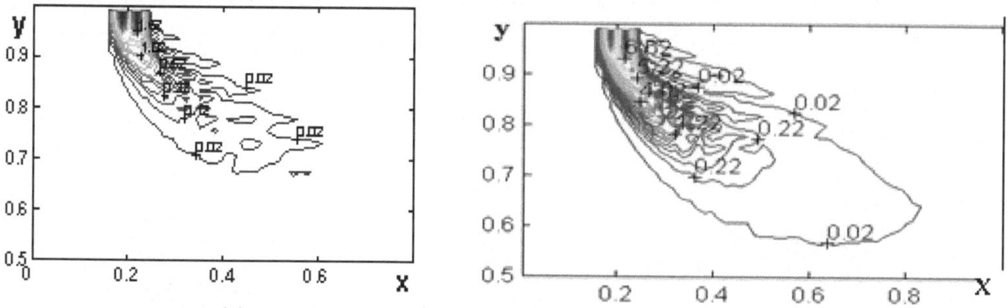

(a) (b)

Fig. 5. Isolines of volumetric concentrations of fuel droplets in the central longitudinal section of the rectangular mixer with jetty supply of fuel (regime 1, $U_1 = 0$); a) $\alpha = 5.4$; b) $\alpha = 1.35$

A comparative analysis of the concentration fields in jetty spraying for various values of α (fig. 5-6) shows that for the higher values of the fuel flow rate there corresponds a wider fuel spray. This spray is more extended, its inner region is characterized by higher values of concentrations, and it occupies a greater volume. Moreover, the patterns of the distributions of concentrations of droplets are identical, indicating the insignificant influence of droplets on the gas flow velocity fields.

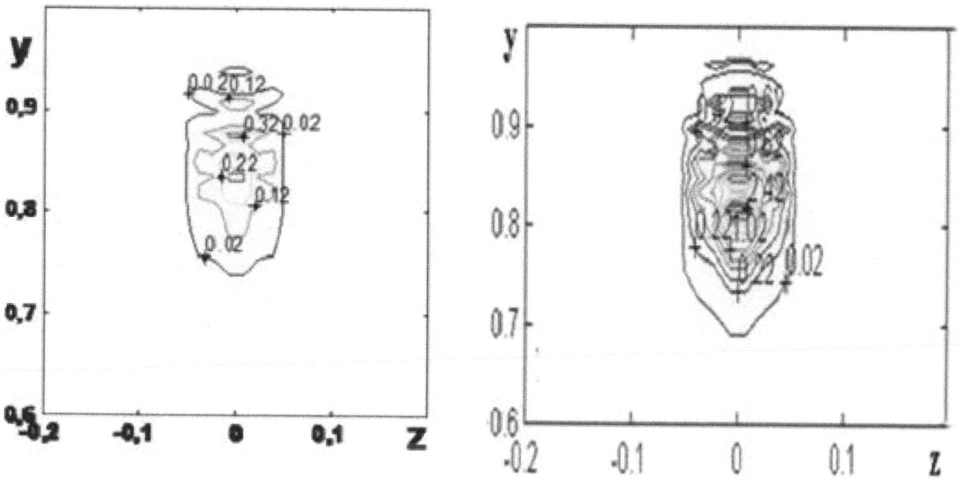

(a) (b)

Fig. 6. Isolines of volumetric concentrations of fuel droplets in the transverse section $x = 0.28$ of the rectangular mixer with jetty supply of fuel (regime 1, $U_1 = 0$); a) $\alpha = 5.4$; b) $\alpha = 1.35$

Figure 7 presents the calculated distributions of dimensionless volumetric concentrations of a liquid fuel c_f for cold spraying in the characteristic sections of a mixer: in the longitudinal section that passes through the center of the injection hole (z = 0) and in the transverse section immediately behind the injection hole at the distance x = 0.28 from the inlet section. A comparison with Fig. 5-6 shows that in the case of pneumatic spraying the patterns of fuel distribution change appreciably. However, in this case too the influence of exchange by momentum between the air and droplets on the distribution of concentrations is hardly noticeable.

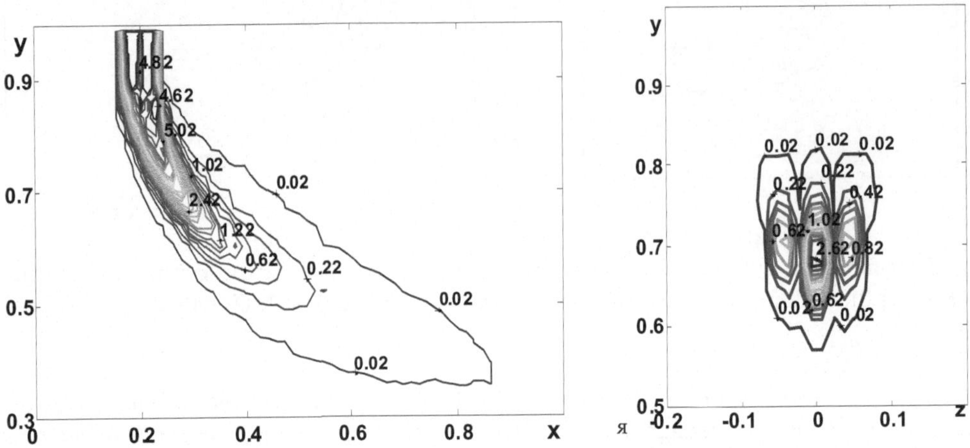

(a) (b)

Fig. 7. Isolines of volumetric concentrations of fuel droplets in the central longitudinal (a) and transverse x = 0.28 (b) sections of the rectangular mixer with pneumatic supply of fuel; spraying by a cold air jet (regime 2, U_1 = 20 m/s, T_1 = 300 K); α = 1.35

From the graphs of the distributions of the volumetric concentrations of fuel droplets it is seen that on the whole the latter follow the air flow. The splitting of the fuel jet in the transverse direction in pneumatic spraying is associated with the appearance of intense circulation flows in the wake of the spraying air jet. The absence of such splitting in jetty spraying indicates that the secondary flows induced by droplets are insufficiently intense. We note that when a high-temperature air jet is injected into a stalling flow, the depth of fuel penetration into a mixer is smaller than in the case of spraying by a cold jet (see Fig. 8). This effect is due, first of all, to the lessening of the penetrating power of an air jet (injection of a gas of a smaller density) and, second, to the enhancement of the processes of heating and evaporation of droplets in a high-temperature air flow of the injected jet.

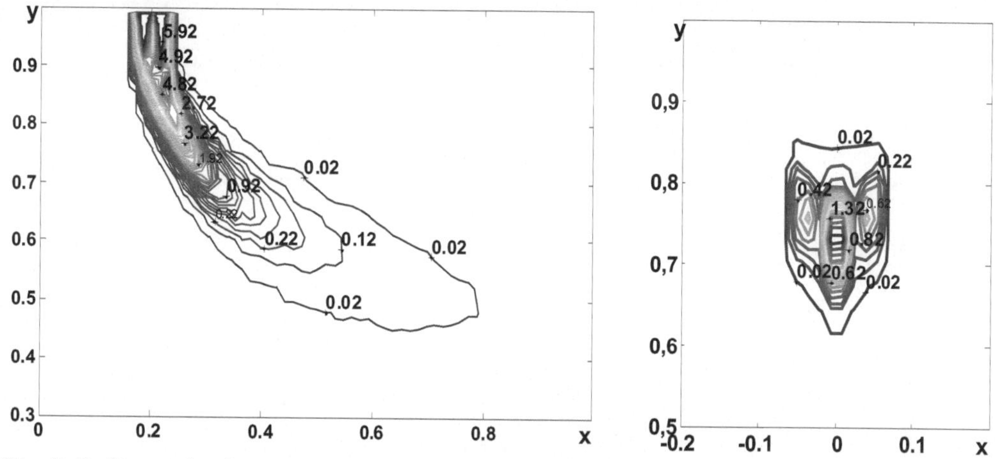

Fig. 8. Isolines of volumetric concentrations of fuel droplets in the central longitudinal (a) and transverse x = 0.28 (b) sections of the rectangular mixer with pneumatic supply of fuel; spraying by a hot air jet (regime 3, U_1 = 20 m/s, T_1 = 900 K); α = 1.35

Thus, in both jetty and pneumatic spraying of a fuel for the regimes considered it is possible to neglect the exchange of momentum between the gas and droplets and judge the interaction of droplets with the air flow from temperature fields. Quantitatively the intensity of heat transfer is characterized, firstly, by the dimensions of the region in which the gas temperature is smaller than that of the surrounding flow (in this case the boundary of this region is T = 900 K) and, secondly, by the minimum gas temperature in the computational domain. The former quantity indicates the part of the space where the air temperature underwent a change and the latter – the quantity of heat taken by droplets from the gas. The values of the minimum gas temperatures are given in Table 2 for all the operating conditions considered.

Regimes	Variant 1	Variant 2	Variant 3
1	638	539	447
2	300	300	300
3	724	612	502

Table 2. Minimum Gas Temperature, K, in the rectangular mixer

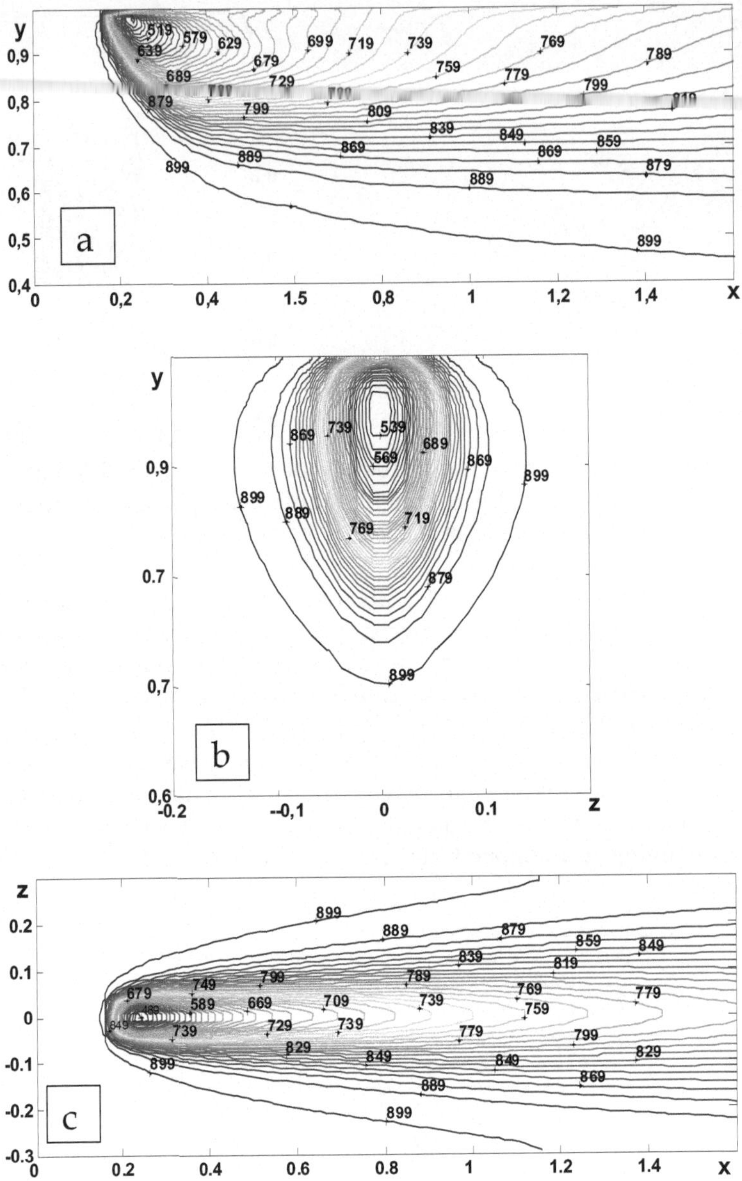

Fig. 9. Isolines of air temperatures in the central longitudinal (a), transverse x = 0.28 (b) and cross y = 0.95 (c) sections of the rectangular mixer of the rectangular mixer with jetty supply of fuel (regime 1, $U_1 = 0$); $\alpha = 1.35$

The calculations have shown that even in the absence of supply of the spraying air the gas temperature depends substantially on the values of operating conditions. The distributions of air temperatures in the absence and in the presence of a spraying air are presented in Figs. 9 and 10 - 11 respectively. Figure 9 characterizes the direct influence of heat exchange between the gas and droplets on temperature fields, since in the absence of this exchange air has the same initial temperature over the entire region of flow. From the distributions of temperatures in the longitudinal sections of the model it is seen that at $\alpha = 1.35$ the region of heat transfer at $x = 1.6$ extends in the direction of the y axis to the distance $\Delta y = 0.55$. As calculations showed, at $\alpha = 5.4$ this distance is equal to $\Delta y = 0.42$. The minimum temperatures that correspond to these variants are equal to 447 and 683 K (Table 2). For the variant $\alpha = 2.7$ this quantity is equal to 539 K. Thus, on increase in the fuel flow rate through a jet injector the influence of droplets on temperature fields becomes more and more appreciable.

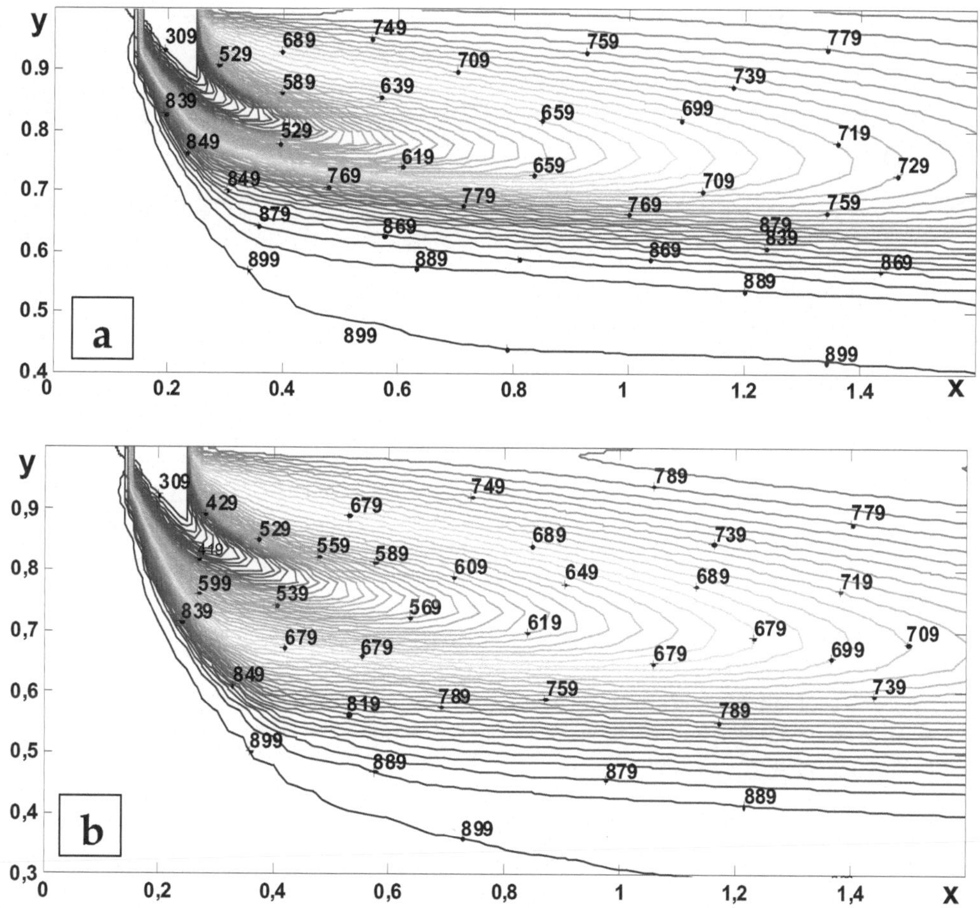

Fig. 10. Isolines of air temperatures in the central longitudinal section of the rectangular mixer with pneumatic supply of fuel; spraying by a cold air jet (regime 2, $U_1 = 20$ m/s, $T_1 = 300$ K); a) $\alpha = 5.4$; b) $\alpha = 1.35$

As calculations show, on injection of a cold spraying air (Fig. 10), when heat transfer is mainly determined by the interaction of the main and spraying flows, this effect is virtually unnoticeable. When a hot spraying air is injected ($T_1 = 900$ K), heat transfer will again be determined by the interaction of air flows with droplets, and therefore the influence of the fuel flow rate on the formation of temperature fields becomes appreciable (Table 2). The corresponding graphs are presented in Fig. 11. It is seen that in these cases the influence of droplets manifests itself virtually in the entire flow region.

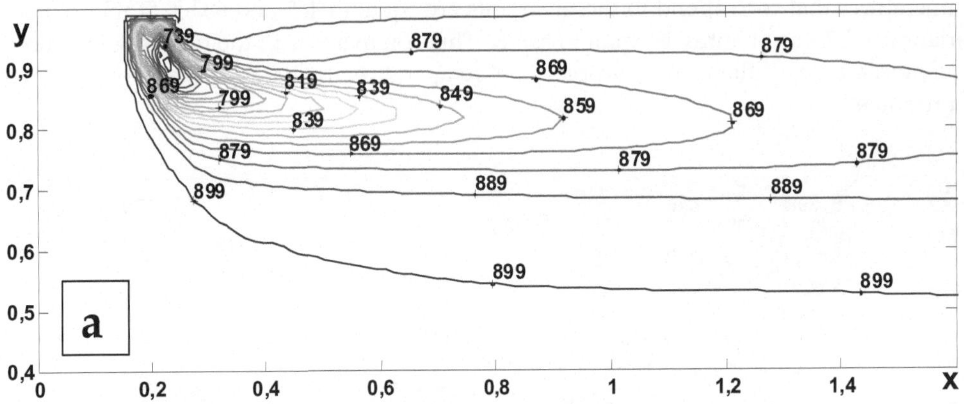

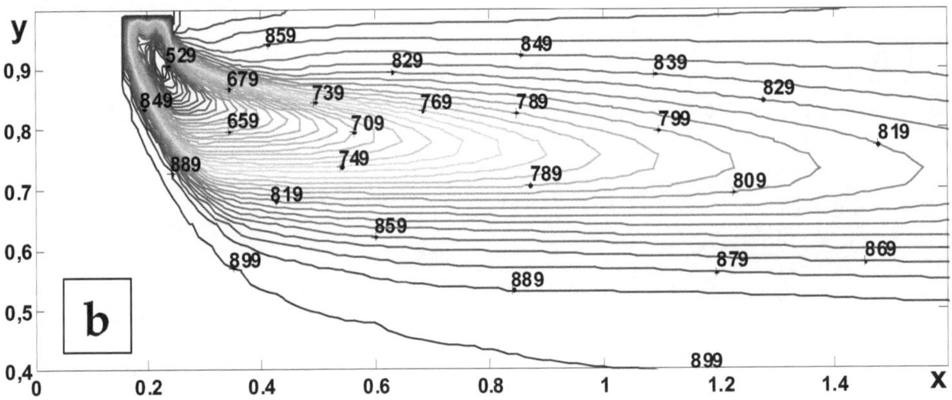

Fig. 11. Isolines of air temperatures in the central longitudinal section of the rectangular mixer with pneumatic supply of fuel; spraying by a hot air jet (regime 3, U_1 = 20 m /s, T_1 = 900 K); a) α = 5.4; b) α = 1.35

Considering the model of heat transfer suggested in the present work, two moments must be noted. The first is that the change in the gas temperature occurs owing to the transfer of heat from the gas to droplets and is spent to heat and evaporate them. As calculations show, both latter processes are essential despite the fact that the basic fraction of droplets (D_d < 100 μm) evaporates rather rapidly in the high-temperature air flow (T_1 = 900 K). The second moment is that heating and evaporation are the mechanisms that underlie heat transfer in the very gas phase and they are also two in number. The first is the conventional diffusion transfer of heat and the second – its convective transfer due to secondary flows which are either initiated by droplets or result from the flow of the stalling stream around the spraying air jets.

In the case of jetty supply of fuel the incipient secondary flows are of low intensity, and droplets are weakly entrained by such flows. This is expressed as the absence of individual vortex structures in the distributions of both concentrations and temperatures in the transverse sections of the module. The lowering of the gas temperature occurs exclusively at the expense of interphase exchange. Vortex structures are clearly seen in transverse sections with pneumatic spraying on the graphs of the distribution of fuel concentrations. A comparison between the distributions of temperatures and concentrations in these cases shows that the concentration profiles are much narrower than the corresponding temperature profiles in both longitudinal and transverse directions. This is associated with the intense diffusion heat fluxes, with the droplets mainly following the air flow. Attention is also drawn to the fact that the penetrating ability of a "cold" fuel-air jet is higher than that of a "hot" one due to the following two reasons: the great energy of the "cold" jet and the more intense process of heating and evaporation of droplets in the "hot" jet.

A comparison of gas cooling in spraying of a fuel by a hot air jet and in jetty spraying shows that although the fuel is injected into flows with identical temperatures, in the second case the lowering of the gas temperature is more appreciable. This seems to be due to the fact that on injection of droplets into a stalling air flow the velocity of droplets relative to the gas is higher than in the case of injection into a cocurrent flow. The rate of the evaporation of droplets is also higher and, consequently, the complete evaporation of droplets occurs over smaller distances and in smaller volumes, thus leading to the effect noted. The total quantity of heat transferred from air to droplets is the same in both cases, but the differences observed allow one to make different fuel-air mixtures by supplying a fuel either into a cocurrent air flow or into a stalling one.

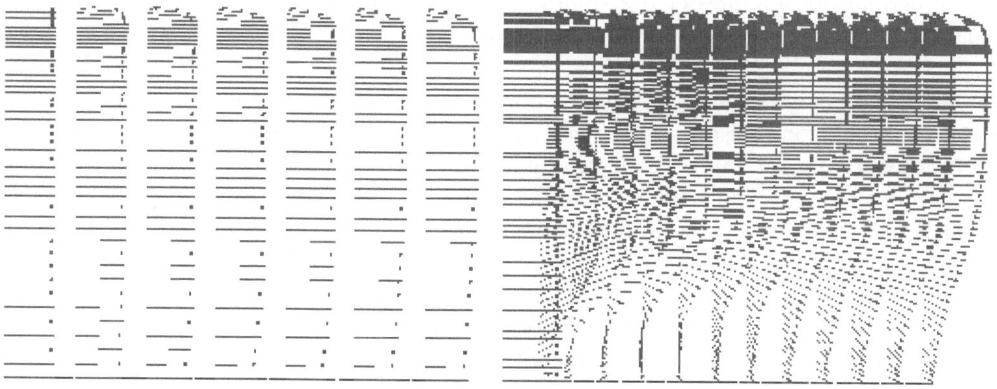

(a) (b)

Fig. 12. Calculated vector velocity field in the longitudinal section of the axisymmetric mixer; a) $-\varphi_1 = \varphi_0 = 30°$, b) $\varphi_1 = \varphi_0 = 60°$

The results of calculation for the axisymmetric mixer (fig. 1-b) are presented in fig. 12 - 18. The above-stated conclusions are applicable and to a flow beyond the coaxial tubes. However in the case of the swirl the region of flow cooling significantly depends on the operating conditions. This effect is connected with the absence or presence of paraxial reverse zone. The velocity field in the vicinity of the place of fuel injection are given in fig. 12. As calculations have shown, the basic role in formation of velocity fields is played by a swirl. In swirling flows with $\varphi_1 > 45°$ there occurs flow separated zone. Flow patterns at mixture of streams with identical ($T_1 = T_0$ =900 K) and various (T_1 = 300 K, T_0 =900 K) temperature are almost the

same. The influence of the mean of spraying and the process of interaction of droplets with air on the flow structure is practically unnoticeable for the cases considered.
In fig. 13 - 14 pictures of trajectories of the droplets projected on longitudinal section of the mixer are resulted.

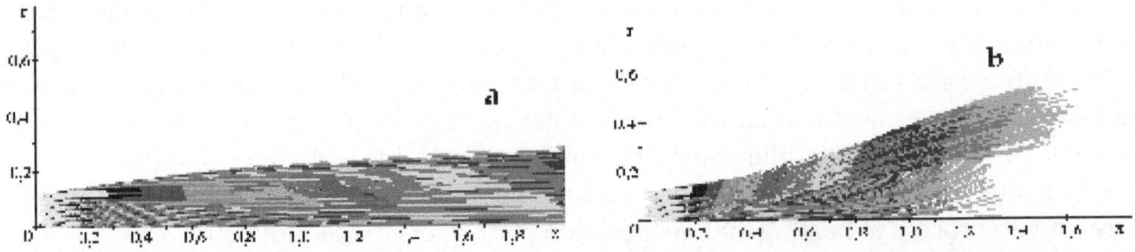

Fig. 13. Trajectories of the droplets in the axisymmetric mixer upon fuel injection into isothermal swirling flows (spraying by pneumatic atomizer with spray angle 40°); $T_0 = T_1 = 900$ K; a) $\varphi_1 = \varphi_0 = 30°$, b) $\varphi_1 = \varphi_0 = 60°$

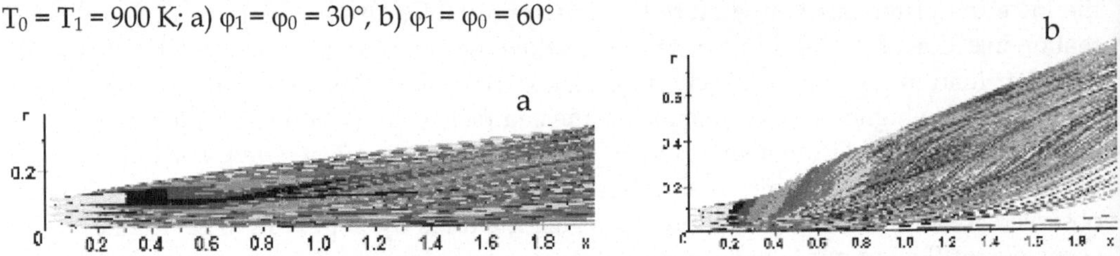

Fig. 14. Trajectories of the droplets in the axisymmetric mixer upon fuel injection into nonisothermal swirling flows (spraying by pneumatic atomizer with spray angle 40°); $T_0 =$ 900 K; $T_1 = 300$ K; a) $\varphi_1 = \varphi_0 = 30°$, b) $\varphi_1 = \varphi_0 = 60$

To various colors in drawing there correspond trajectories with various initial diameters of droplets. From comparison of the presented pictures of trajectories it is visible, that distinctions in interaction of a fuel spray with an air flow lead to significant differences in distributions of drops in a working volume. In the case of reverse zone (fig. 13 b and 14 b) droplets are shifted to the wall. The temperature mode also plays the important role in formation of a fuel spray. It is visible, that at $T_1 = T_0 = 900$ K, owing to evaporation of drops, their trajectories appear more shortly, than at motion in a flow with $T_1 = 300$ K. As calculations have shown the influence of interphase exchange on trajectories and the distribution of concentrations is insignificant.

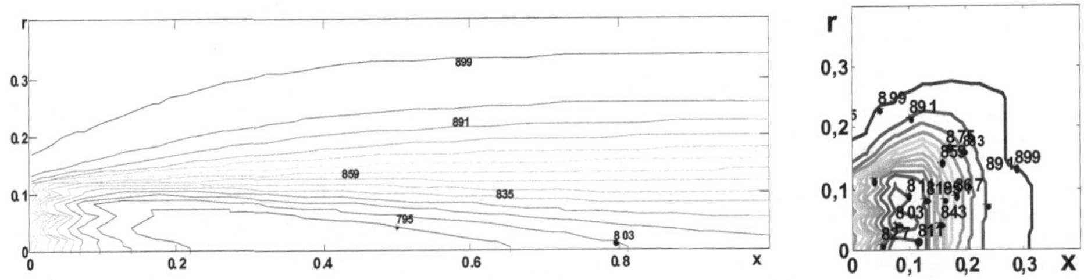

(a) (b)

Fig. 15. Isolines of air temperatures in the longitudinal section of the axisymmetric mixer upon fuel injection into isothermal swirling flows (spraying by pneumatic atomizer); $T_0 = T_1 = 900$ K; - a) $-\varphi_1 = \varphi_0 = 30°$, b) $\varphi_1 = \varphi_0 = 60°$

So just as in the case of rectangular mixer it is possible to neglect the exchange of momentum between the gas and droplets and to judge the interaction of droplets with an air flow from temperature fields. It's clear that the greatest cooling of a gas flow by droplets occurs on the maximum gas temperature. The distributions of air temperatures on injection of a hot spraying air are given in Fig. 15. That temperature fields to the full are determined by the interaction of air flows with droplets. From comparison of drawings in fig 15 a) and b) it is visible, that areas of influence of droplets on a gas flow are various also they are determined in the core by flow hydrodynamics. In a case $\varphi_1 = \varphi_0 = 30°$, the flow is no separated and the area of cooling of gas is stretched along an axis. In a case $\varphi_1 = \varphi_0 = 60°$ there exists the paraxial reverse zone. As result the last droplets are shifted to the wall together with cooled gas. Analogous isothermals of gas at fuel spraying from one source (supply by pressure atomizer) are resulted in fig. 16 a) and `16 b).

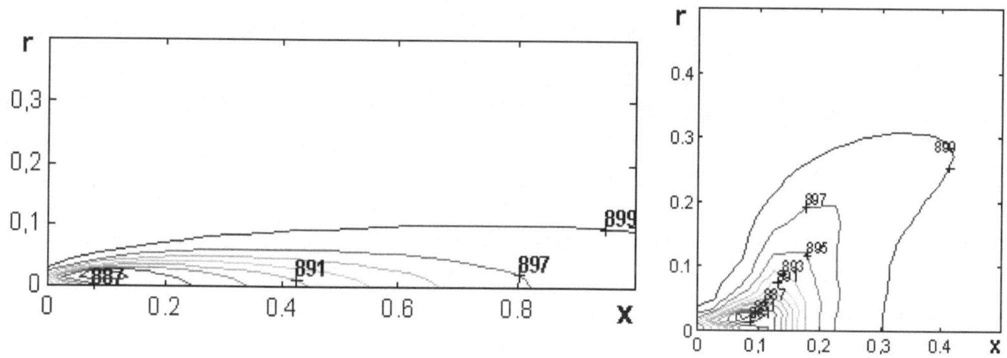

(a) (b)

Fig. 16. Isolines of air temperatures in the longitudinal section of the axisymmetric mixer upon fuel injection into isothermal swirling flows (spraying by pressure atomizer); $T_0 = T_1 = 900$ K; - a) $\varphi_1 = \varphi_0 = 30°$, b) $\varphi_1 = \varphi_0 = 60°$

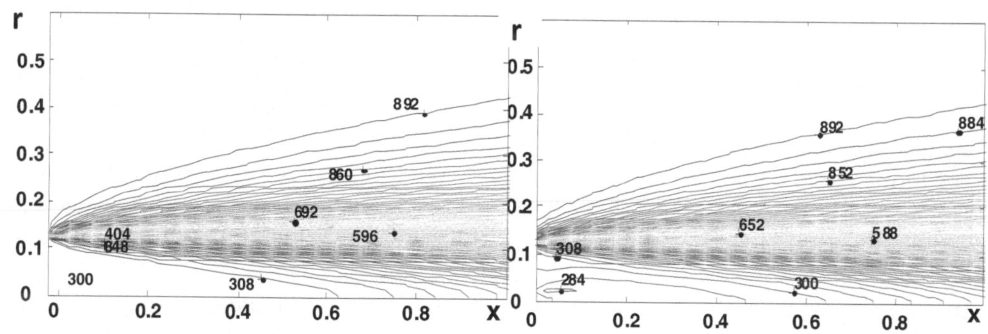

a) (b)

Fig. 17. Isolines of air temperatures in the longitudinal section of the axisymmetric mixer upon fuel injection into nonisothermal swirling flows (spraying by pneumatic atomizer); $T_0 = 900$ K; $T_1 = 300$ K; $\varphi_1 = \varphi_0 = 30°$; a) - without an interphase exchange; b) - taking into account an interphase exchange

During injection of a cold spraying air the heat transfer is determined both the interaction of the main and spraying flows and the interaction of air flows with droplets. Gas isotherms in this case are resulted on fig. 17 and 18, accordingly for $\varphi_1 = \varphi_0 = 30°$ and $\varphi_1 = \varphi_0 = 60°$.

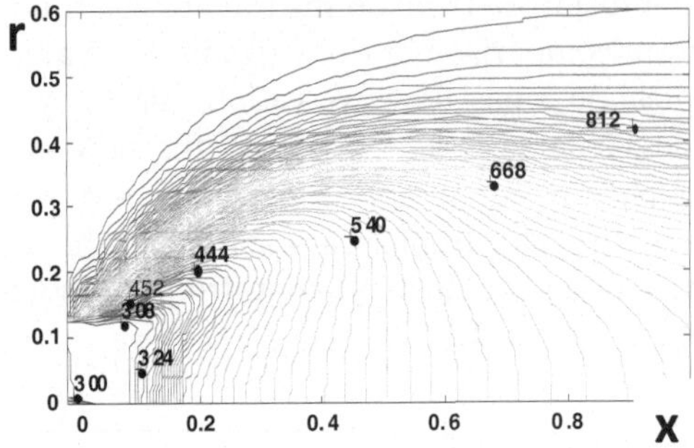

(a)

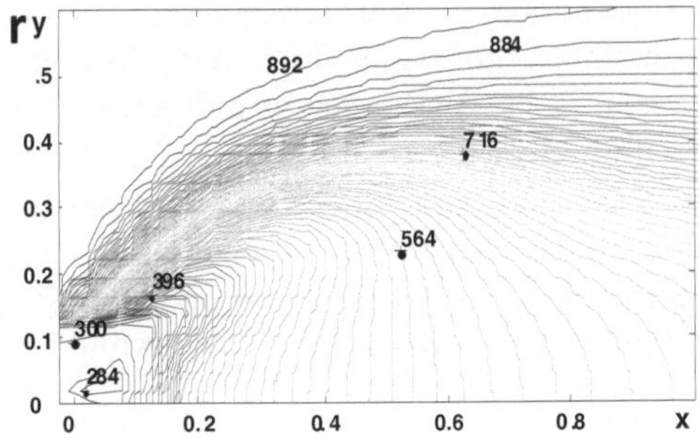

(b)

Fig. 18. Isolines of air temperatures in the longitudinal section of the axisymmetric mixer upon fuel injection into nonisothermal swirling flows (spraying by pneumatic atomizer); T_0 = 900 K; T_1 = 300 K; $\varphi_1 = \varphi_0 = 60°$; a) - without an interphase exchange; b) - taking into account an interphase exchange

It is clear that in the considered cases heat exchange in the core is determined by interaction of gas flows. The interphase exchange changes fields of temperatures only near to a fuel supply place, i.e. in order area in the size 0.2 R_0.

6. Conclusions

In all means of spraying, for the regimes considered it is possible to neglect the exchange of momentum between the gas and droplets and to judge the interaction of droplets with an air flow from temperature fields.

Injection of a fuel by a jet injector may cause a substantial change in the gas temperature. In the given case it occurs due to heat transfer from the gas to droplets and is spent on their heating and evaporation. In the case of pneumatic spraying of a fuel by a cold air jet the influence of interphase exchange is insignificant. Heat transfer is predominantly determined by the interaction of the main and spraying flows. During injection of a hot spraying air, when heat transfer inside the gas flow is less intense, the influence of the injection of a fuel on the formation of temperature fields again becomes appreciable. However, in this case the gas is cooled less than in jetty spraying. This effect is due to the fact that when droplets are injected into a stalling air flow, the rate of their evaporation is higher than during injection into a cocurrent flow.

In the case of the swirl the region of flow cooling significantly depends on the operating conditions. This effect is connected with the absence or presence of paraxial reverse zone.

The conclusions drawn confirm the necessity of taking into account the processes of interphase heat and mass exchange when investigating the mixture formation.

7. The further development of a calculation method

The further development of a computational technique should actuate the account of coagulation and breakage of droplets. The calculations resulted below illustrate the importance of turbulent coagulation of droplets of the spraying fuel behind injectors in combustion chambers.

The main assumptions of physical character imposed on system coagulation of particles, consist in the following. The number of particles is great enough, that it was possible to apply function of distribution of particles on weights and in co-ordinate space. Only binary collisions are considered, the collisions conserve the mass and volume, and the aerosol particles coagulate each time they collide. Within the Smoluchowsky's theoretical framework (see Friedlander at al., 2000), at any time, each aerosol particle could be formed by an integer number of base particles (or monomers), which would be the smallest, simple and stable particles in the aerosol, and the density of the number of particles with k monomers, n_k, as a function of time, would be the solution of the following balance equation:

$$\frac{dn_k}{dt} = \frac{1}{2}\sum_{i+j=k} K_{ij}n_i n_j - n_k \sum_{i=1}^{\infty} K_{ki}n_i \qquad (17)$$

Non-negative function K_{ij} is called as a coagulation kernel, it describes particular interaction between particles with volumes i and j. The first term at the right hand side of Eq. (17) is the production of the particles with k monomers due to collisions of particles with i and j monomers such that i + j = k, and the second term is the consumption of particles with k monomers due to collisions with other aerosol particles.

The majority of activities on coagulation research concern to atmospheric aerosols in which this process basically is called by Brown diffusion. Still in sprays behind injectors the main action calling increase of the sizes of drops, is turbulent coagulation. For such environments the coagulation kernel can be recorded in the form of (Kruis & Kusters, 1997)

$$K_{ij} = \sqrt{\frac{8\pi}{3}}\,(a_1 + a_2)^2\sqrt{W_s^2 + W_a^2} \qquad (18)$$

Here a_1 and a_2 - radiuses of particles i and j, W_s - relative particle velocity due to inertial turbulent effects and W_a - relative particle velocity due to shear turbulent effects.

The system of equations (17-18) was solved by the finite-difference method (Maiharju, 2005). As a result of the solution of the equations of turbulent coagulation it is investigated the influence of ambient medium properties on growth rate of droplets behind the front module. In particular the influence of speed of a dissipation of turbulent energy, the initial size of droplets and ambient pressure on distribution of droplets in the sizes on various distances behind an injector was investigated. The variation of the mean- median diameter of droplets on time (distance from an injector) for droplets of the initial size 5 and 10 microns and normal ambient pressure is shown in fig. 19. The researches carried out have shown that coagulation process can considerably change the sizes of droplets. The initial diameter of droplets essentially influences coagulation process. So, at increase in the initial size of drops with 5 μm to 10μm, the relative mean median diameter of droplets in 0.01 seconds is increased at 1.2 time (see fig. 19).

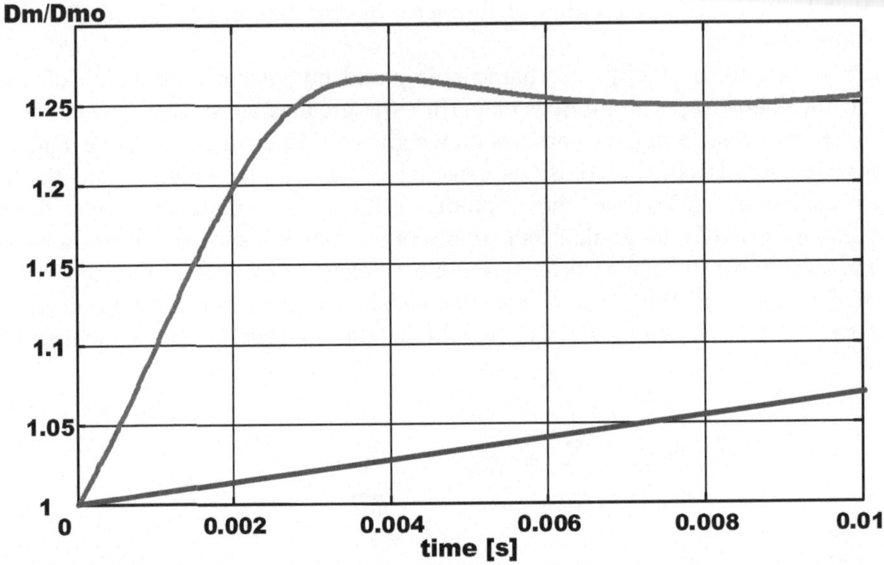

Fig. 19. The dependence of relative size of droplets in spray behind injector on coagulation time; blue line - D_{m0} = 5μm; read line - D_{m0} = 10 μm

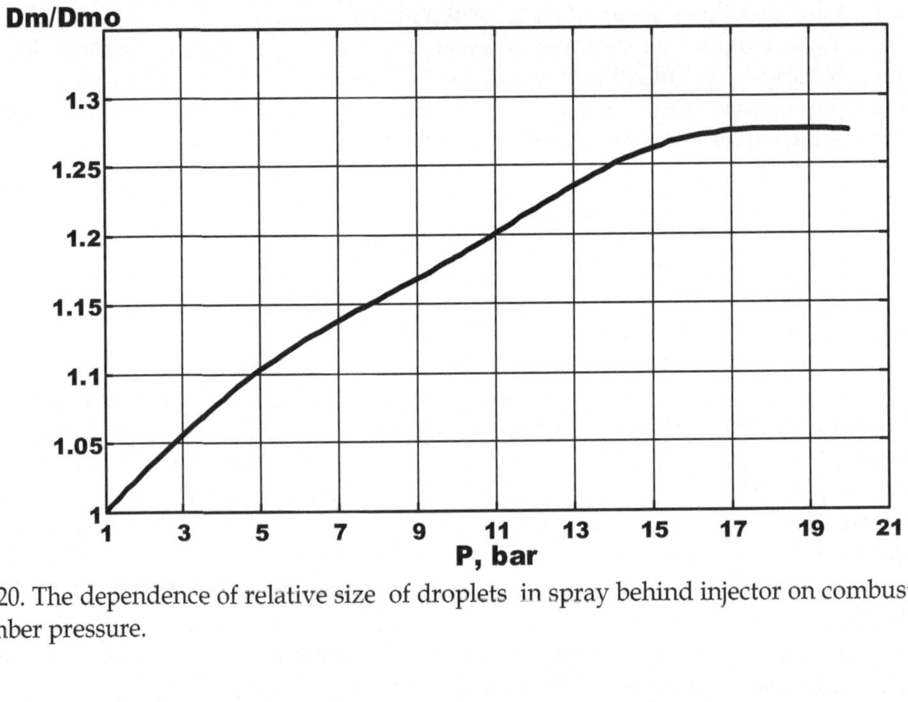

Fig. 20. The dependence of relative size of droplets in spray behind injector on combustion-chamber pressure.

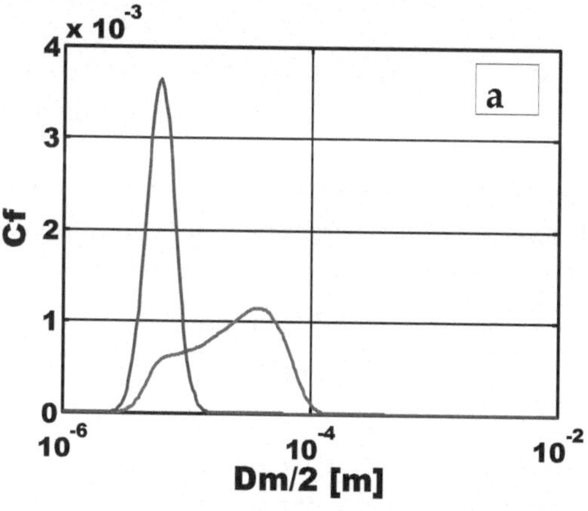

Fig. 21. The distribution of volumetric concentration on the sizes of droplets; blue lines - initial distribution; red lines - distribution in 0.01 seconds; a) - $\varepsilon = 1 m^2/s^3$; b) $\varepsilon = 100 m^2/s^3$

In fig. 20 data about influence of ambient pressure on coagulation of droplets of the kerosene spray are resulted. Calculations are executed at value of $\varepsilon = 10 m^2/s^3$ and initial $D_m = 5 \mu m$. It's evidently from the plot at pressure variation from 1 to 25 bar the mean size of droplets as a result of coagulation for 0.01 seconds is increased approximately at 30 %.
Rate of a dissipation of turbulent energy is the essential parameter determining a kernel of turbulent coagulation K (x, y). Estimations show, that behind front devices of combustion chambers the value of rate of a turbulent energy dissipation varies from 1 to 100 m^2/s^3. In drawings 21- a) and b) distributions of volumetric concentration C_f for two values of a rate of dissipation of turbulent energy are presented. The increase in dissipation leads to displacement of distribution of volumetric concentration in area of the big sizes.0 So the main fraction of drops of spraying liquid will fall to drops with sizes, 10 times magnitudes surpassing initial drops.
Thus, ambient pressure, rate of dissipation of turbulence energy and the initial size of the droplets leaving an injector make essential impact on coagulation of droplets.
It is necessary to note, that in disperse systems, except process of coagulation which conducts to integration of particles, there are cases when the integrated particle breaks up on small spontaneously or under the influence of external forces. Therefore coagulation process will be accompanied by atomization of drops as a result of aerodynamic effect of air. Thus as coagulation as breaking of droplets are desirable to take into account when calculating the mixture formation.

9. Notation

C_f, volumetric concentration of a liquid fuel, kg/m^3; c_f, coefficient of specific heat of liquid, J/(kg·K); c_{pg}, coefficient of specific heat of gas at constant pressure, J/(kg·K); C_R, coefficient of droplet resistance; C_v, concentration of fuel vapor per unit volume, kg/m^3; D_d, droplet diameter, m; D_m, droplet mean median diameter, m; H, channel height, m; h, specific total enthalpy, J/kg; k, energy of turbulence per unit mass, m^2/s^2; L, latent heat of evaporation, J/kg; m_d, mass of a droplet, kg; m_f, mass fraction of kerosene vapors; n_k, density of the number of particles with k monomers; $Pr = \mu_g c_{pg}/\lambda_g$, Prandtl number; $\vec{R}$, force of aerodynamic resistance; $Re = \rho_g D_d W/\mu_g$, Reynolds number of a droplet; S_Φ, internal source term in the equation of transfer of the variable Φ; T, temperature, K; t, time, s; $\vec{U}g$, vector of averaged gas velocity; U_{gi} (i = 1, 2, 3), components of the vector of averaged gas velocity, m/s; $\vec{V}_d$, vector of droplet velocity; $\vec{W} = \vec{V}_d - \vec{U}g$, vector of droplet velocity relative to gas; x, y, z, Cartesian coordinates; x, r, φ, cylindrical coordinates; α, summed coefficient of air excess; Γ_Φ, coefficient of diffusion transfer of variable Φ; Δt_d, time of droplet residence in the volume element, s; Δv, elementary volume, m^3; ε, rate of dissipation of turbulence energy, m^2/s^3; λ_g, thermal conductivity of gas, W/(m·K); μ_g, coefficient of dynamic viscosity of gas, kg/(m·s); ρ, density, kg/m^3; Φ, dependent variable; φ_1, φ_0, wane angles of swirlers in inner and outer channels, °. Subscripts and superscripts: 0, main flow; 1, spraying air; g, gas; f, liquid fuel; d, droplet; int, interphase; v, vapor-like fuel; i, individual droplet.

10. References

Chien K.J. (1982). Predictions of channel and boundary-layer flows with low-Reynolds-number turbulence model. *AIAA J.*, Vol. 20, 33–38.

Dityakin Yu. F., Klyachko L. A., Novikov B. V. and V. I. Yagodkin. (1977). *Spraying of Liquids* (in Russian), Mashinostroenie, Moscow.

Friedlander, S. K. (2000). *Smoke, Dust and Haze.* Oxford Univ. Press, Oxford.

Lefebvre A.H. (1985). *Gas Turbine Combustion*, Hemisphere Publishing corporation, Washington, New York, London.

Koosinlin M.L., Launder B.E., Sharma B.J. (1974). Prediction of momentum, heat and mass transfer in swirling turbulent boundary layers. *Trans. ASME, Ser. C.*, Vol. 96, No 2, 204.

Kruis F. E & Kusters K.A. (1997) The Collision Rate of Particles in Turbulent Flow, *Chem. Eng. Comm.* Vol. 158, 201-230.

Maiharju S.A (2005). *Aerosol dynamics in a turbulent jet, A Thesis the Degree Master of Science in the Graduate School of The Ohio State University.*

Maiorova A.I. & Tretyakov V.V. (2008). Characteristic features of the process of mixture formation upon fuel injection into a high-temperature air flow. *Journal of Engineering Physics and Thermophysics*, Vol. 81, No. 2, 264-273. ISSN: 1062-0125.

Patankar S. (1980). *Numerical Heat Transfer and Fluid Flow*, Hemisphere Publishing, New York.

8

Integrated numerical procedures for the design, analysis and optimization of diesel engines

1. Introduction

Both the design and analysis of a diesel engine requires the integration of accurate theoretical methods, resorting to 1D - 3D CFD modelling and vibro-acoustic engine analysis. In this chapter, the above numerical approaches will be deeply presented and integrated to perform a diesel engine design and/or analysis. As known in fact, the possibility to simulate the physical and chemical processes characterising the operation of internal combustion engines by using appropriate codes and high performance computers is continuously spreading. These simulations can predict, as an example, fuel consumption, toxic emissions and noise radiation. By varying the design and/or control parameters, different engine configurations or working conditions can be tested and their performances compared. Optimization techniques (Papalambros et al. 2000; Stephenson, 2008; Costa et al., 2009), properly matched with the various simulation procedures, are hence the most suitable tool to identify optimal solutions able to gain prescribed objectives on engine efficiency, power output, noise, gas emissions, etc.. The choice of the optimization goal, moreover, strictly depends on the application type and the definition of a compromise solution among the conflicting needs is in many cases required.

Concerning the design of a combustion engine, a complicated and multi-objective task is to be afforded, since it generally requires the fulfilment of various objectives and constraints, as high efficiency and power output, low noise and gas emissions, low cost, high reliability, etc. A tool for multi-objective optimization, therefore, can be considered as fundamental at the engine design stage, in order to gain insight into the complicated relationships between the physical entities involved in the design and design-dependent parameters. Ultimately, optimization can greatly reduce the time-to-market of new engine prototypes.

Optimization techniques can successfully be applied to analyze the operating conditions of existing engines, too. In this case, the optimization process can be focused on the selection of the control parameters in order to obtain an optimal engine behaviour. It is well known, in fact, that combustion development and emission production depend on a complex interaction among different parameters, namely injection modulation and phasing (Stotz et al. 2000), boost pressure, EGR fraction, swirl ratio, fuel properties, and so on. The optimal choice of a so large number of parameters depends on speed and load conditions, and it is related to the fulfilment of a number of contrasting objectives, like reduced NOx, Soot, HC, CO, fuel consumption and noise emissions.

In the present chapter the cited approach to the design and analysis of a diesel engine will be explained. The discussion will be organized in the following paragraphs, each regarding a different case study. In particular, the first paragraph is focused on the description of single methodologies and to their integration:

- A 1D simulation of the whole propulsion system is realized by means of a proprietary code. It allows to determine engine-turbocharger matching conditions and is able to compute pressure, temperature and gas composition at the intake valve closure. The latter data represent initial conditions for the successive 3D analysis.
- A 3D simulation of the engine cylinder is developed by exploiting geometrical information derived by the engine CADs. The in-cylinder pressure cycle during the closed valve period, is predicted, starting from the initial conditions provided by the 1D code.
- 1D or 3D computed pressure cycles are then utilized within vibro-acoustic analyses aiming to estimate the combustion radiated noise. Depending on the application, different approaches are followed:
    - FEM-BEM approach: FEM analysis is applied to determine the vibration of the engine skin surface; Direct Boundary Element Method (DBEM) solves the exterior acoustic radiation according to the ISO directives, to predict the radiated overall noise level. This method is utilised during the engine design phase.
    - Simplified approach: An analytical model based on the decomposition of in-cylinder pressure cycles is developed to estimate the radiated noise level. Some coefficients included in the above correlation are properly tuned to get a good agreement with the acoustic experimental data. This method is applied during the engine analysis phase.
- The numerical models are in different ways coupled to the optimization code, to identify the optimal design parameters or the injection strategies, to the aim of realizing the maximization of the engine performance, the reduction of the NOx and soot emissions, and the reduction of the radiated noise, at a constant load and rotational speed.

The second paragraph illustrates the design and optimization of a new two-stroke diesel engine suitable for aeronautical applications. The engine, equipped with a Common Rail fuel injection system, is conceived in a two-stroke uniflow configuration, aimed at achieving a weight to power ratio equal to one kg/kW. Both CFD 1D and 3D analyses are carried out to support the design phase and to address some particular aspects of the engine operation, like the scavenging process, the engine-turbocharger matching, the fuel injection and the combustion process. The exchange of information between the two codes allows to improve the accuracy of the results. Computed pressure cycles are also utilized to numerically predict the combustion noise, basing on the integration of FEM and BEM codes. The obtained results are suitable to be used as driving parameters for successive engine optimization. In order to improve the engine performance and vibro-acoustic behaviour, the

1D model, tuned with information derived from the 3D code, is linked to the optimization code. A constrained multi-objective optimization is performed to contemporary minimize the fuel consumption and the maximum in-cylinder temperature and pressure gradient directly related to the noise emission. In this way a better selection of a number of engine parameters is carried out (exhaust valve opening, closing and lift, intake ports heights, start of injection, etc).

The third paragraph, indeed, describes an environmental and energetic optimization of a naturally aspirated, light-duty direct injection (DI) diesel engine, equipped with a Common Rail injection system. An experimental campaign is initially carried out to gain information on performance and noise levels on the engine and to acquire the data required to validate the 1D, the 3D model and the combustion noise procedure. As in the previous case, a preliminary numerical simulation is carried out. Then, an optimization process is settled in order to identify the control parameters of a three pulses injection profile, for a constant overall mass of injected fuel. These parameters are assumed as independent variables of the multi-objective optimization tool and are selected with the aim of simultaneously minimizing fuel consumption, pollutant emissions and radiated noise.

2. Numerical Procedures

1D Simulation: the 1D simulation of the whole propulsion system is realized by means of the 1Dime software developed at the Mechanical Engineering Department of the University of Naples "Federico II" (Siano et al., 2008, Costa et al., 2009) and by exploiting geometrical information of the intake and exhaust system derived by the engine CAD. The whole engine is firstly schematized as a network of pipe and plenums, then, the 1D flow equations are solved in each pipe constituting the intake and the exhaust system. The gas inside the cylinder is indeed treated as a zero-dimensional thermodynamic system.
The code solves the 1D flow equations in the intake and exhaust pipes:

$$\mathbf{U}=\begin{Bmatrix}\rho\\ \rho u\\ \rho E\\ \rho x_r\\ \rho x_f\end{Bmatrix}\quad \mathbf{F}=\begin{Bmatrix}\rho u\\ \rho u^2+p\\ \rho u H\\ \rho u x_r\\ \rho u x_f\end{Bmatrix}\quad \mathbf{S}=-\begin{Bmatrix}\rho u\dfrac{1}{\Omega}\dfrac{d\Omega}{dx}\\ \rho u^2\left[\dfrac{1}{\Omega}\dfrac{d\Omega}{dx}+\left(2\dfrac{f}{D}+\dfrac{C_p}{2L}\right)\dfrac{u}{|u|}\right]\\ \rho u H\dfrac{1}{\Omega}\dfrac{d\Omega}{dx}-4\dfrac{q}{D}\\ \rho u x_r\dfrac{1}{\Omega}\dfrac{d\Omega}{dx}\\ \rho u x_f\dfrac{1}{\Omega}\dfrac{d\Omega}{dx}\end{Bmatrix} \tag{1}$$

where ρ, u, p, $E=c_vT+u^2/2$, $H=c_pT+u^2/2$ respectively represent the density, the velocity, the pressure, and the total energy and enthalpy per unit mass. The source term **S** takes into account the duct area variation along the flow direction, $d\Omega/dx$, the wall heat exchange, q, and the friction losses. The last two equations describe the scalar transport of chemical species, x_r and x_f being the residual gases and fuel mass fraction, respectively. These equations allow to compute the composition of the gases flowing in the intake and exhaust systems and therefore to estimate also the in-cylinder charge composition.

In the case of turbocharged engine, the performance maps of the turbocharger group are employed to compute the engine-turbocharger matching.

Concerning the modeling of the combustion process, a classical Wiebe equation is utilized to compute the heat release rate in the engine. Proper values of the combustion process duration during both premixed and diffusive phases are specified. This preliminary approach is substituted in the following by the more detailed 3D analysis later described.

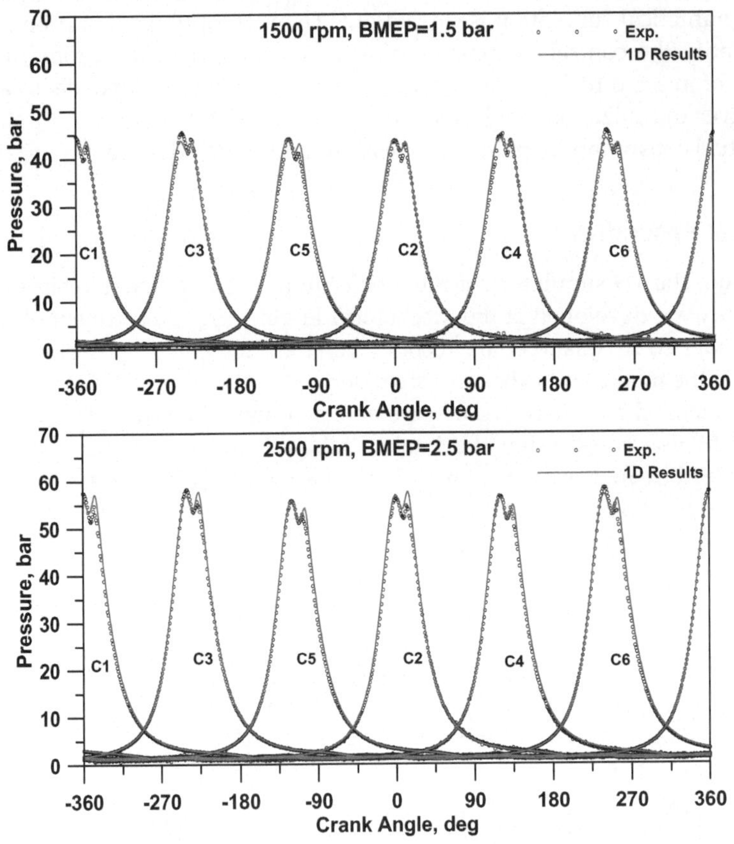

Fig. 1. 1D computed pressure cycles in different operating conditions

As an example, Figure 1 displays the comparison between computed and experimental pressure cycles in a turbocharged six cylinders engine, for two different operating conditions. The agreement along the compression stroke indicates that a good engine-turbocharger matching can be reached. Starting from the computed initial conditions at IVC, the 3D model is expected to further improve combustion phase analysis.

3D Simulation: The AVL FIRE™ 3D code is employed. It represents a multipurpose tool, specially conceived for engine applications. The first step of the analysis concerns the generation of the 3D domain representing the computational grid. This is effected by means of the semi-automatic procedure of the code, named Fame Advanced Hybrid, which allows to reach a good compromise between accuracy and reduced number of cells. The computational period is subdivided into intervals, each pertinent to a grid of a chosen size, that is deformed as the piston moves until predefined crank angles, starting from which new grids with different sizes are used. This occurs through a procedure of re-mapping of the computed variables, termed rezone, avoiding an excessive cells deformation. Within each cell, the Reynolds Averaged Navier-Stokes equations are numerically solved to compute the 3D flow field and the thermodynamic conditions inside the cylinder. The CFD analysis accounts for the fuel spray dynamics and for the subsequent chemical reactions, leading to the prediction of the rate of heat release (Colin and Benkenida, 2004) , pollutants formation and in-cylinder pressure cycle.
The fuel spray spatio-temporal dynamics is simulated according to a Discrete Droplets Model (DDM) (Liu and Reitz, 1993; O'Rourke. 1989; Dukowicz, 1980), where the Eulerian description of the gaseous phase is coupled with a Lagrangian approach to the study of the liquid flow. Modelling of spray accounts for primary and secondary atomization, evaporation, coalescence, turbulence effects and possible cavitation within the nozzle. Examples of spray, combustion and emission calculations will be presented in paragraphs 2 and 3, with reference to selected case studies.

Despite the modelling of the combustion and noxious emission, the 3D code can be also employed to improve the accuracy of the previously described 1D model through the theoretical evaluation of the discharge coefficients through valves or ports. The above values are usually derived from literature information in 1D modelling and, of course, the related accuracy is limited. Alternatively, the latter can be evaluated by numerically simulating the 3D air flow within the engine intake system. The computation is performed under the hypothesis of steady conditions, in such a way to reproduce a possible experiment realisable over a flow rate test bench. An example grid used for this kind of analysis is shown in Figure 2. It refer to a two-stroke engine more deeply described in paragraph 2 and clearly exhibits the geometrical characteristics of the air admission volume, with one inflow duct and the three cylinders of a bank placed with their axes in the direction orthogonal to that of the air inflow. Only the central cylinder is considered as opened. The volume corresponding to the intake ports of the central cylinder is meshed, those of the lateral cylinders are not included in the computational domain for the sake of simplicity. The total number of cells is 518346, 330746 of which are hexahedrical, thus assuring a certain grid regularity. The fourteen ports of the central cylinder form one block with an external cylindrical area, whose design follows the geometric characteristics of the cylinder jacket. The grid in this zone, quite well visible in Figure 2, is made particularly thick, since it comprehends 148415 cells. Computation is performed for fixed values of the entering air mass flow rate and always setting the static pressure and temperature at the outlet section equal to atmospheric conditions. The ports are opened to the 100%, 75%, 50% and 25% for a mass flow rate ranging from 0.03 to 0.15 kg/s.
Figure 3 shows a view of the velocity magnitude distribution obtained as a result of the calculation for intake ports completely opened. The velocity vector magnitude is considered

over a plane orthogonal to the cylinders axes, cutting the intake ports exit section in the middle. Air velocity distributes non-uniformly over the fourteen ports surfaces: the seven ports which are closer to the air inflow duct are better served and mainly contribute to fill the cylinder with the fresh charge. The velocity in correspondence of the other seven ports maintains lower. The shape of the ports, with their surfaces positioned in tangential direction with respect to the cylinder external surface, allows a swirl motion of the air entering the cylinder.

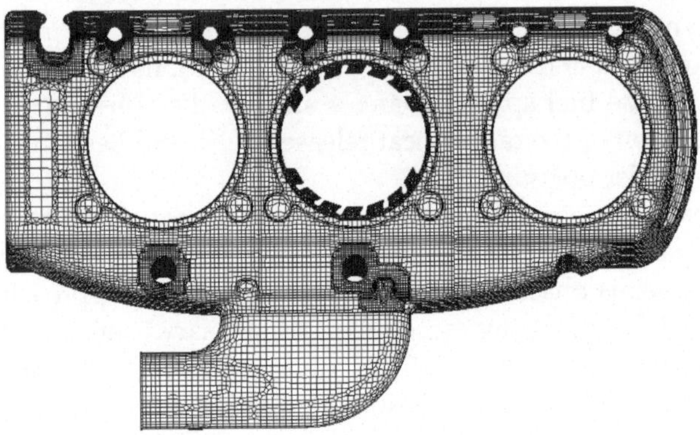

Fig. 2. Grid employed for the evaluation of the intake ports discharge coefficient.

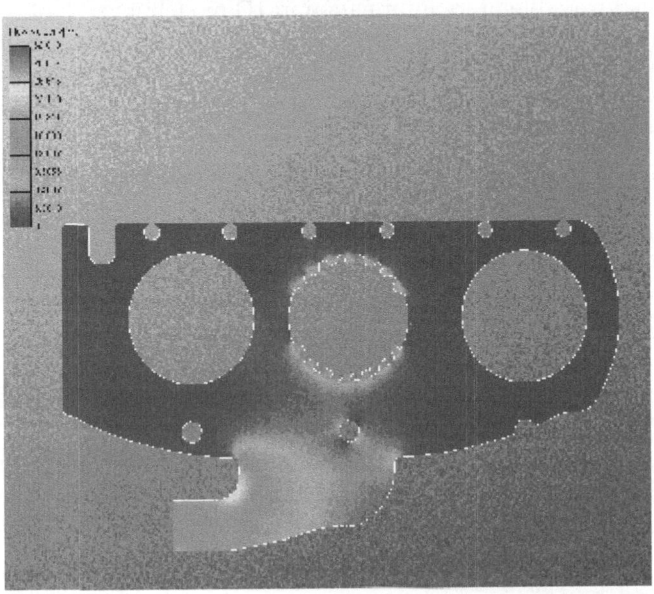

Fig. 3. Velocity field on a plane orthogonal to the cylinder axis

The results of the computations allow the determination of the discharge coefficient as the ratio between the effective mass flow rate, $\dot{m}_{eff}$, and the theoretical one, $\dot{m}_{th}$:

$$C_D = \frac{\dot{m}_{eff}}{\dot{m}_{th}} \tag{2}$$

The theoretical mass flow rate is evaluated as a function of inlet pressure and temperature and flow area, corresponding to the imposed effective mass flow rate.
Figure 4 summarises the results of the calculation for various surface percentages in the opening of the intake ports. The discharge coefficient maintains almost constant with the entering mass flow rate and ranges between about 0.6 and 0.98, due to the reduction of the exit section area used for the evaluation of the theoretical mass flow rate. These results will be directly employed in the 1D model, for accuracy improvement.

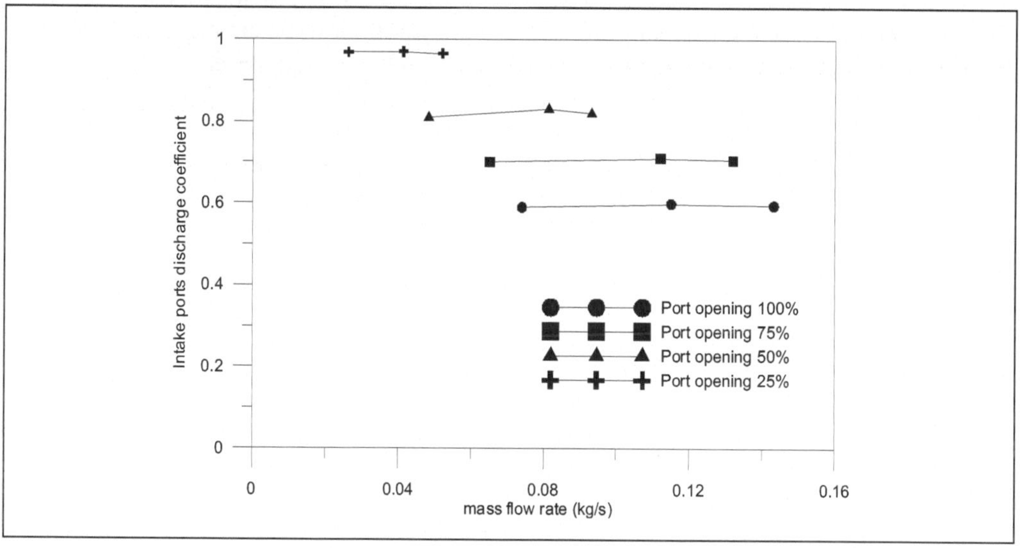

Fig. 4. Discharge coefficients evaluated for various opening of the intake ports.

Vibro-acoustic analyses: a FEM analysis is conducted in order to evaluate the engine surface vibrations induced by the combustion process evolution. The previously predicted pressure cycle is employed in this phase to compute the forces acting on the internal structure. The obtained vibrational output data represent the boundary conditions to be applied to the BEM code for the final evaluation of the radiated sound power. The selected codes to evaluate the sound radiation noise from the engine block surface are the MSC/Nastran™ and LMS/Sysnoise™. Both approaches require the development of a detailed 3D mesh and are very time-consuming. For this reason, this detailed methodology cannot be directly applied within the optimization loop. Additional insights of the above summarized approach will be given in paragraph 2.

Alternatively, a simplified and recently proposed methodology (Torregrosa et al., 2007; Payri et al., 2005) can be utilized for the prediction of the overall combustion noise, which includes in the correlation a strict dependency on the engine operating conditions and injection strategy. The main idea behind this technique is the decomposition of the total in-cylinder pressure signal according to three main contributions: compression-expansion, combustion and resonance pressures:

$$p_{tot} = p_{mot} + p_{comb} + p_{res} \tag{3}$$

The first contribution (also referred as pseudo-motored signal) is only related to volume variation, and is used as a reference signal. It is determined by a direct in-cylinder pressure acquisition during a fuel switch-off operation. The third term (resonance pressure) is indeed related to high-frequency pressure fluctuations, induced by the wave reflections in the combustion chamber, mainly occurring at autoignition time. It is computed through a high pass-band filter (above 4500 Hz) of the total pressure signal. By difference, the combustion pressure (second term in eq. 3) can be easily determined, too. The three terms in eq. 3 are reported for comparison in figure 5. Despite the presence of the previously discussed high-frequency amplitudes, the resonant pressure is significantly lower than other contributions. Nevertheless, it may still exert a non-negligible effect on the overall noise.

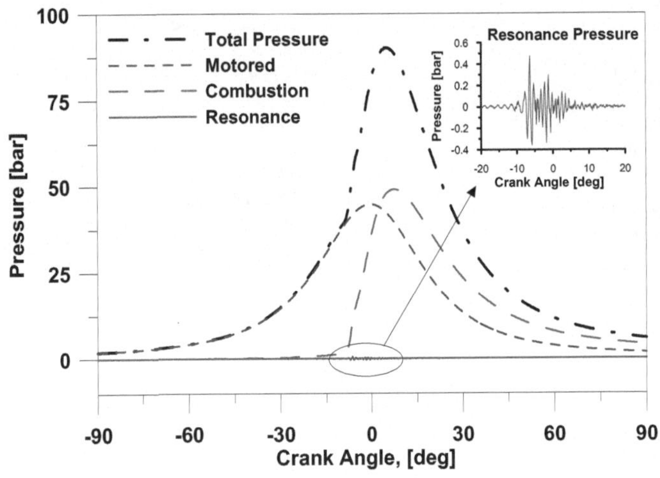

Fig. 5. Decomposition of the total pressure in motored, combustion and resonance contributions.

The three decomposed pressures are utilized to compute two characteristic indices I_1 and I_2 defined as:

$$I_1 = \frac{n}{n_{idle}} \left[\frac{\left(\frac{dp_{\max 1}}{dt}\right)_{comb} + \left(\frac{dp_{\max 2}}{dt}\right)_{comb}}{\left(\frac{dp_{\max}}{dt}\right)_{mot}} \right] \tag{4}$$

$$I_2 = \log_{10}\left[10^6\left(\frac{\int p_{res}^2 dt}{\int p_{mot}^2 dt}\right)\right] \quad (5)$$

The I_1 index is a function of the maximum pressure gradient of the combustion contribution, occurring after the pilot $(dp_{max1}/dt)_{comb}$ and the main injection $(dp_{max2}/dt)_{comb}$. The I_1 index is also non-dimensionalized over the maximum pressure gradient of the pseudo-motored pressure $(dp_{max}/dt)_{mot}$. In the case of a single-shot injection, a unique term is of course present in the eq. (4) numerator. The I_2 index takes into account the acoustic energies ($\int p^2 dt$) associated with resonance and motored pressure signals. An additional index In is finally defined accounting for mechanical noise contribution, related, as stated, to the sole engine speed:

$$I_n = \log_{10}\left(\frac{n}{n_{idle}}\right) \quad (6)$$

n being the engine speed and n_{idle} the idle rotational speed.
Basing on the above definitions, the Overall Noise (ON) can be finally computed as:

$$ON = C_0 + C_n \cdot I_n + C_1 \cdot I_1 + C_2 \cdot I_2 \quad (7)$$

C_i being proper tuning constants, depending on the engine architecture and size.
Following the relations (4-7), a Matlab routine was developed to properly process the in-cylinder pressure cycle and compute the various noise indices and the overall noise.
The method is validated on acoustic measurements taken on a commercial engine, as reported in figure 6. The agreement obtained is satisfactory at each engine speed. A maximum absolute error of about 1.3 dB is found at a medium engine regime.

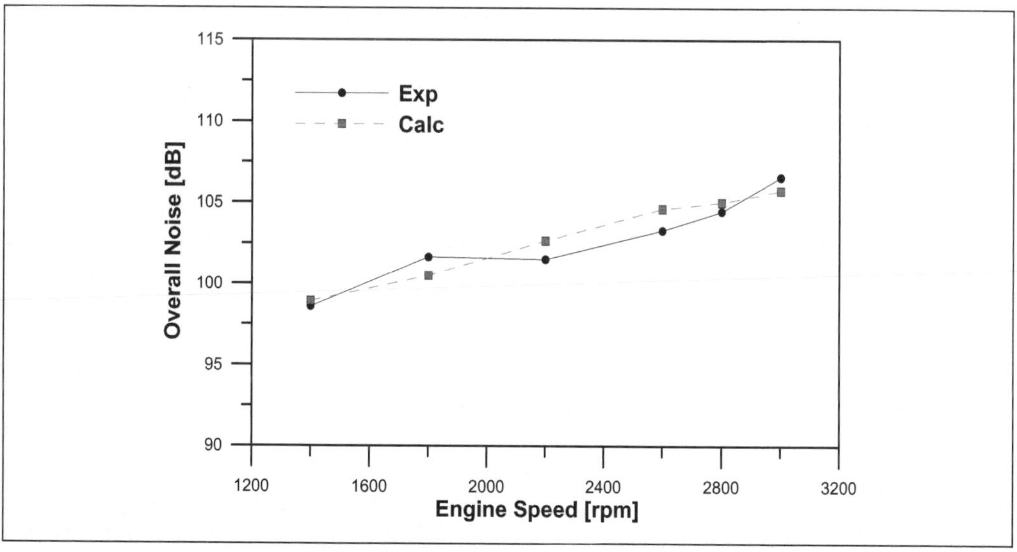

Fig. 6. Comparisons on the overall noise

The method can be applied to both experimental and numerical pressure cycles and strictly depends on the engine operating conditions and injection strategy. Once validated, this simplified approach is directly included in the optimization loop to predict the overall noise.

Optimization process: the briefly described 1D, 3D and acoustic tools are coupled together within an optimization loop searching for design or control parameters minimizing fuel consumption, gaseous emissions and radiated noise. The logical development of the optimization problem is developed within the ModeFRONTIER™ environment. For each set of the design or control parameters, the 1D, 3D and acoustic tools are automatically started. 1D results allow to run the 3D code from reliable conditions at the intake valve closure. Then, the 3D computed pressure cycle is automatically given in input to a Matlab™ routine computing the overall combustion noise. Simultaneously, the Indicated Mean effective pressure (IMEP), is returned back to the optimizer, together with the NO and soot levels at the end of the 3D run. A multi-objective optimization is so defined to contemporarily search the maximum IMEP, the minimum soot, the minimum NO and the minimum overall noise. To solve the above problem, genetic algorithms (Sasaki, 2005) are usually utilized, employing a range adaptation technique to overcome time-consuming evaluations. As usual in multi-objective optimization problems, a multiplicity of solutions is expected, belonging to the so-called Pareto frontiers. In order to select a single optimal solution among the Pareto-frontier ones, the "Multi Criteria Decision Making" tool (MCDM) provided in modeFRONTIER™ is employed. This allows the definition of preferences expressed by the user through direct specification of attributes of importance (weights) among the various objectives. Depending on these relations, the MCDM tool is able to classify all the solutions with a rank value. The highest rank solution is the one that better satisfies the preference set.

In the following paragraphs, two examples are presented where the described methodology is applied to perform the design of a Two-Stroke Engine for aeronautical application and to select an optimal fuel injection strategy for a light-duty automotive engine.

3. Optimal Design of a Two-Stroke Engine for aeronautical application

In this paragraph, some aspects concerning the development of a prototype of a diesel engine suitable for aeronautical applications are discussed (Siano et al., 2008). The engine aimed at achieving a weight to power ratio equal to one kg/kW (220 kg for 220 kW) is conceived in a two stroke Uniflow configuration and is constituted by six cylinders distributed on two parallel banks. Basing on a first choice of some geometrical and operational data, a preliminary fluid-dynamic and acoustic analysis is carried out at the sea level. This includes the engine-turbocharger matching, the estimation of the scavenging process efficiency, and the simulation of the spray and combustion process, arising from a Common Rail injection system. Both 1D and 3D CFD models are employed.

A CAD of the engine under investigation is shown in figure 7. Six cylinders are distributed on two parallel banks with separate air admission. The supercharging system consists of a dynamical turbocharger coupled to a mechanical one (of the roots type), serving the engine start-up, as well. An automotive derived roots compressor is chosen with a transmission-ratio equal to 5. As a first step, a preliminary 1D simulation of the entire propulsion system is realized by means of the previously described 1D software, and by exploiting geometrical information derived by the engine CAD. Figure 8 reports the engine 1D schematization including the three cylinders, the turbocharger group (C-T), the intercooler (IC) and the mechanical supercharger (C), coupled to the engine shaft. A waste-gate valve (BY) is also considered upstream of the turbine. Half engine is schematized, due to the symmetry property of the two engine banks. Each of the three cylinders is connected to the intake plenum through fourteen inlet ports and to the exhaust plenum through two exhaust valves. In the 1D computation, the 3D computed discharge coefficients are employed. Scavenging is indeed considered as in the middle between the two opposite limits so-called of perfect displacement and perfect mixing. In other words, a parameter, wmix, representing a relative weight factor between the occurrence of a perfect mixing and a perfect displacement process, is assumed equal to the value of 0.67. The above parameter results, once again, from accurate analyses carried out on the engine cylinder by means of the 3D code (figure 9).

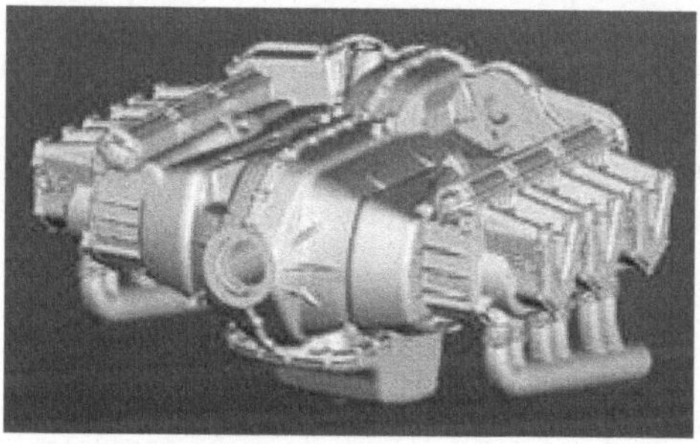

Fig. 7. 3D cad view of the 6 cylinder, two-stroke Diesel engine.

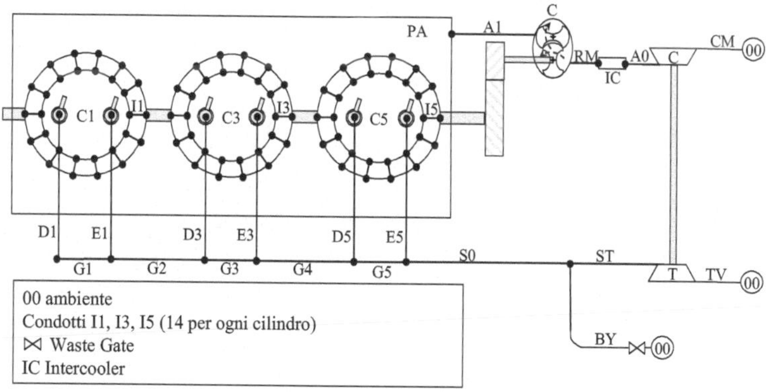

Fig. 8. 1D schematization of the AVIO3 engine

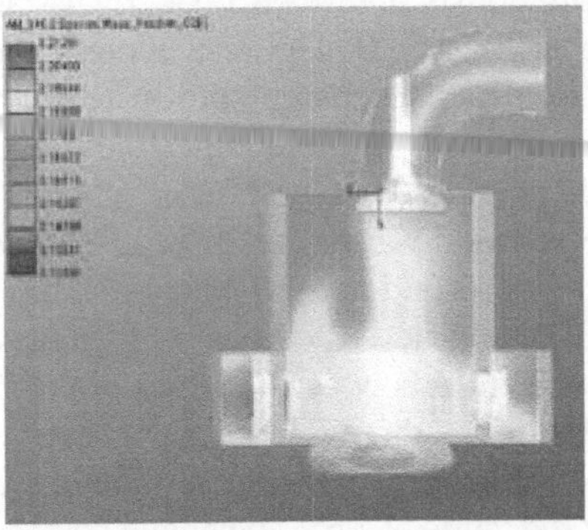

Fig. 9. 3D analysis for the calculation of the scavenging efficiency

The 3D analysis also provides the definition of a proper heat release law, to be included in the 1D model, assuming injection in one shot. Figure 10 shows the 3D computed pressure cycle in comparison with the results of the 1D model. An idea of how the injection strategy affects combustion is also given. It is evident that a too advanced injection makes for a too much high pressure peak, which may be dangerous in terms of mechanical stresses, whereas a late injection makes for a low cycle area, hence a low power output and high fuel consumption.

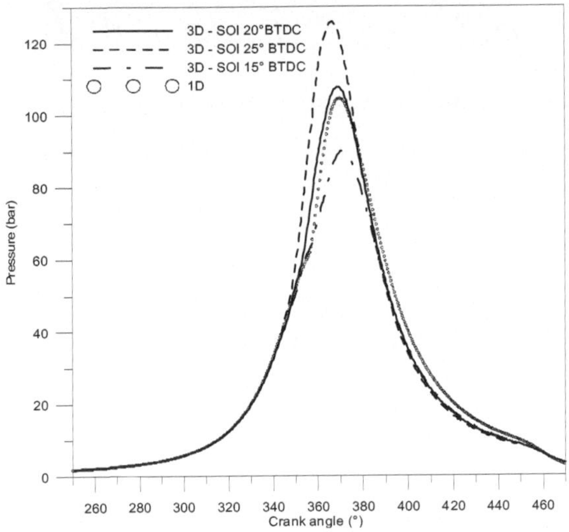

Fig. 10. Comparison of the in-cylinder pressure as obtained by the 1D and the 3D codes for SOI at 20° BTDC. The 3D simulations are also relevant to SOI at 15° and 25° BTDC.

In parallel to the 1D and 3D analyses, an acoustic study is also carried out to predict the combustion noise radiation following the FEM/BEM approach.

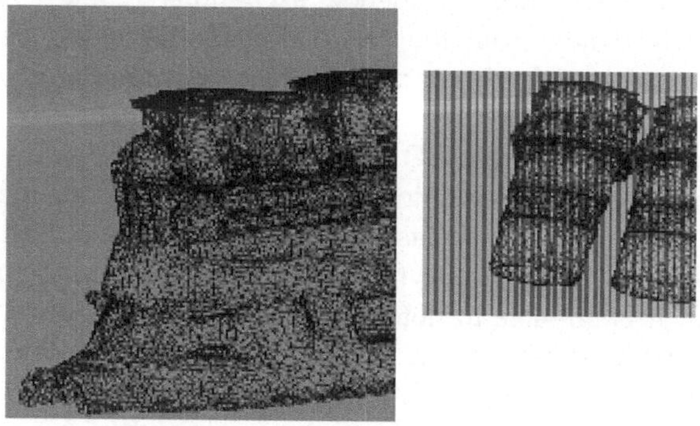

Fig. 11. Mesh models of the engine block and cylinder liners.

In particular, the FE model is developed subdividing the engine into single groups, each manually meshed and finally assembled. Two parts are mainly considered, as shown in Figure 11: the engine block and the cylinder liners. A non automatic meshing process is required to handle the great complexity of the cylinder geometry especially concerning the presence of the fourteen inlet ports. With the purpose of getting information about the skin surface vibrations, a frequency response analysis is conducted using as excitation the 1D computed pressure forces acting inside the cylinders during the combustion process at the 2400 rpm engine speed.

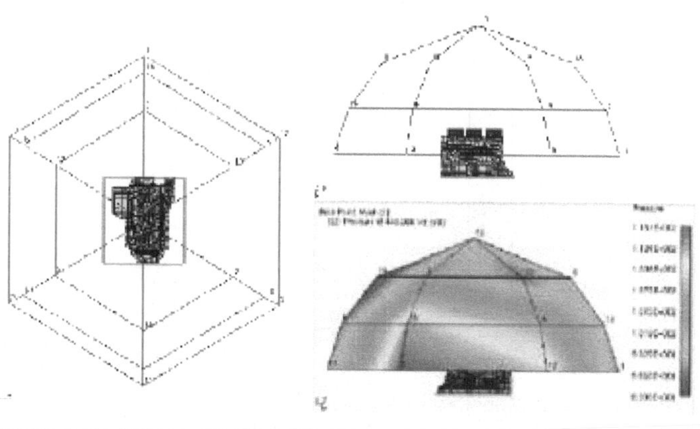

Fig. 12. Hemispherical surface with field points and Sound Power map according to the ISO 3746 directive.

Beside the calculation of the surface velocity, a boundary element mesh is realised with a reduced number of nodes and elements. The obtained vibrational output data represent the boundary conditions to be applied to the BEM for the final evaluation of the radiated sound power. The approach used is the ATV methodology (Acoustic Transfer Vectors). This technique, through the preliminary evaluation of the transfer functions of surface-receivers (microphones), allows to evaluate the answer to different boundary conditions, as the application of fine-loads or multi-frequency excitations (engine noise). The acoustic radiation can be so evaluated from the calculation of the sound pressure on a virtual measurement surface that completely contains the radiant surface.

For the measurement of the radiated power, an hemispherical surface is created around the engine model, according to normative ISO 3746.

Figure 12 shows the above surface, positioned at the distance of one meter from the engine, which includes the nineteen field points (virtual microphones) used to get information about the noise radiation. In the same figure the resulting sound pressure map is also plotted. In particular, it is possible to note that the major contribution to the overall noise comes from a lateral part, corresponding to the carter, which presents a smaller thickness with respect to the other engine parts. A non negligible contribution also comes from the engine top, excited by the subsequent combustion process events.

Figure 13 displays the frequency spectrum of the average sound power radiation on the surface. It is important to remark the presence of two tonal peaks at the frequencies of 440 Hz (118 dB) and 1060 Hz corresponding to a resonance phenomenon with the fundamental firing frequency at about 40 Hz (2400 rpm).

In conclusion, it can be stated that a great noise radiation is revealed in correspondence of resonance conditions.

This kind of integration of different numerical procedures allows to predict, with a good accuracy, the engine radiated noise and can be used in a pre-design phase in order to characterize the acoustic behaviour of the engine structure.

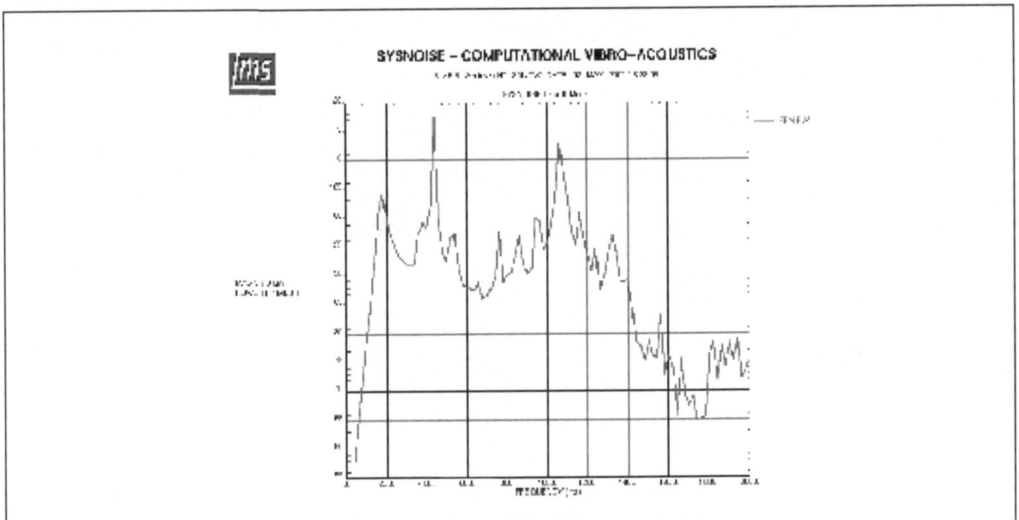

Fig. 13. Average Sound Power radiation on the hemispherical surface.

The iterative exchange of information between the 1D and 3D codes allows to define the main performance outputs of the engine under development. Although the numerical analysis confirms the possibility to reach the prescribed power output with the imposed limitation on the maximum pressure (126 bar), it also puts into evidence the occurrence of a high value of the Brake Specific Fuel Consumption (BSFC = 258 g/kWh). The acoustic analysis also estimates the presence of a high combustion noise level, with a sound power peak of about 118 dB, strictly related, as known, to the maximum in-cylinder pressure gradient reached during the combustion process.

In order to improve the overall performance characteristics of the engine, an optimization procedure is carried out to the aim of finding a better selection of some geometrical and operating parameters. In particular, a different phasing of both exhaust valves and intake ports is considered, together with a different phasing of the injection law. The above parameters actually affect also the supercharging level, and, for this reason, the 1D code must be mandatory utilized in the optimization procedure. The 1D analysis, however, includes the details of the previous 3D study in terms of both scavenging efficiency, discharge coefficients and heat release rate.
Figure 14 displays the logic chart of the optimization procedure, developed in the ModeFrontier graphical environment. The independent variables considered are:

- EVO: Exhaust Valve Opening, deg
- EVD: Exhaust Valve Duration, deg
- EVL: Exhaust Valve Lift, mm
- IPO: Intake Port Opening, deg
- IPL: Intake Port width, mm
- THJ: Start of injection, deg

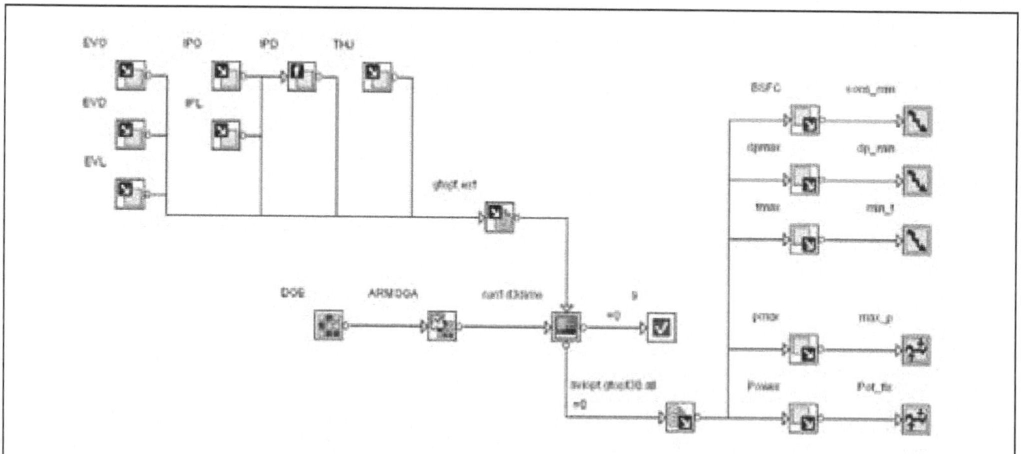

Fig. 14. Logic chart of the optimization procedure.

At each iteration, the values of the above variables are automatically written in the input file of the 1D code. ModeFrontier then runs the 1D code and extracts the required output results. After that, the independent variables are iteratively changed within prescribed intervals to the aim of finding the minimum fuel consumption. Additional objectives are also specified concerning the minimization of the pressure gradient and the minimization of the maximum average temperature inside the cylinder. In this way both noise and NOx emissions are expected to be reduced. Of course, each set of the independent variables must also guarantee the possibility to reach the prescribed power output (110 kW per bank) with a maximum pressure limited to 126 bar. These two additional requirements are fulfilled through the definition of two constraint variables in the logic scheme of figure 14.

Summarizing, a multi-objective constrained optimization problem is set-up, as follows:

Objective 1: min (BSFC)
Objective 2: min ($dp/dtheta_{max}$)
Objective 3: min (T_{max})
Constrain 1: $p_{max} < 126$ bar
Constrain 2: Power > 108 kW

To solve the above problem, the ARMOGA algorithm is utilized. The latter belongs to the category of genetic algorithms and employs a range adaptation technique to carry out time-consuming evaluations.
The specification of 3 objectives determines the existence of a two-dimensional Pareto frontier (Pareto surface) including all the solutions of the optimization problem.
Different sections of the Pareto surface are represented in figure 15 that highlights the presence of a clear trade-off between the three specified objectives. Due to the strong correlation between the maximum pressure and maximum temperature, a similar trade-off behaviour is found between the fuel consumption and the maximum pressure.
All the displayed points, however, respect the specified Constrain 1. The initial design point obtained in the previously discussed preliminary simulation, is located far away from the Pareto frontiers, as highlighted in the Figure 15. A relevant improvement of all the three objectives, hence, is surely realized.

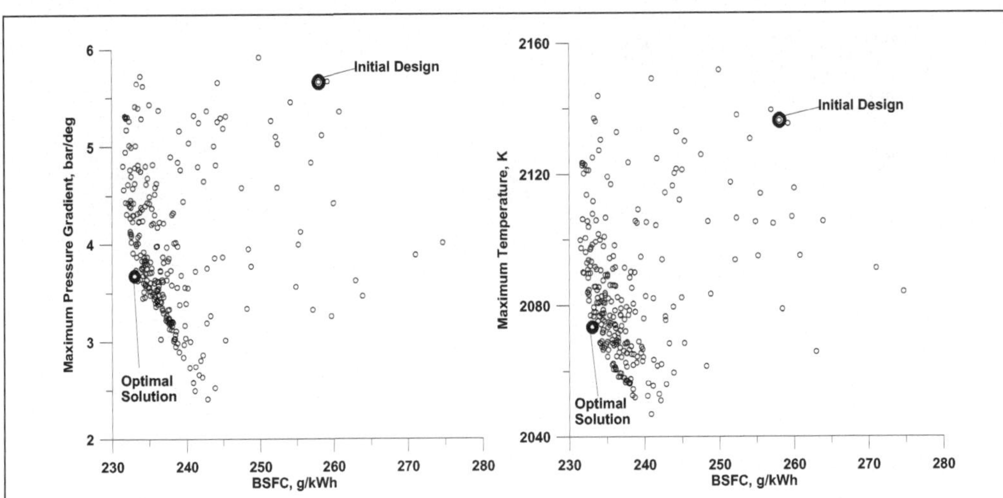

Fig. 15. Optimization results. Trade-off analysis.

In order to select a single solution among the ones located on the Pareto frontiers, the "Multi Criteria Decision Making" tool (MCDM) provided in modeFRONTIER™ is employed. It allows the definition of preferences expressed by the user through direct specification of attributes of importance (weights). BSFC and pressure gradient were considered as the most relevant parameters. Depending on the above relations, the MCDM tool is able to classify all the solution with a rank value. The solution which obtains the highest rank, therefore, can be identified. Basing on the described methodology, the solution with the highest rank value is the one characterized by the identification number (ID) 238. The latter is also depicted along the Pareto frontiers in Figure 15.

Design ID	0	238	238-0
Input Variables	**Value**	**Value**	**Delta**
EVO, deg ATDC	80.00	94.17	14.17
EVD, deg	135.00	155.96	20.96
EVL, mm	12.00	14.66	2.66
IPO, deg ATDC	111.50	119.74	8.24
THJ, deg ATDC	347.49	351.69	4.2
IPL, mm	9.520	12.126	2.606
Transfer Variables	**Value**	**Value**	**Delta**
IPD, deg	137.00	120.53	-16.47
EVC, deg ATDC	215.00	250.13	35.13
IPC, deg ATDC	248.50	240.27	-8.23
IPH, mm	26.98	20.99	-5.99
Objectives	**Value**	**Value**	**% Var**
Min(BSFC), g/kWh	257.98	233.01	-9.68 %
Min(dP/dth), bar/deg	5.665	3.676	-35.11 %
Min(Tmax), K	2136.3	2073.5	-2.94 %
Constraints	**Value**	**Value**	**Delta**
Pmax < 126 bar	125.83	97.43	-28.4
Power Output > 108 kW	110.03	110.00	----

Table 1. Comparison between initial solution (ID=0) and "global optimum" (ID=238)

The position of the optimal solution also puts into evidence that the MCDM procedure effectively realizes a compromise between the conflicting needs, quantified by the attributes of importance described. In addition, this procedure defines a standardized method for the selection of the "global" optimum.

Table 1 reports a comparison between the initial and optimal solutions in terms of both independent (or input) variables, objectives parameters and constraints. Some other "transfer" variables, directly derived from the input data, are also listed.

The table puts into evidence that a BSFC improvement higher than 9% can be reached, together with a relevant reduction of both pressure gradient, maximum temperature and maximum pressure. This means that both a lower noise and NOx emission are expected, together with well lower thermal and mechanical stresses on the engine.

The above results are obtained thanks to a delayed opening of the exhaust valve and to an increased duration of exhaust phase. Contemporarily, a lower height and a greater width of the 14 intake ports are also selected by the optimization procedure.

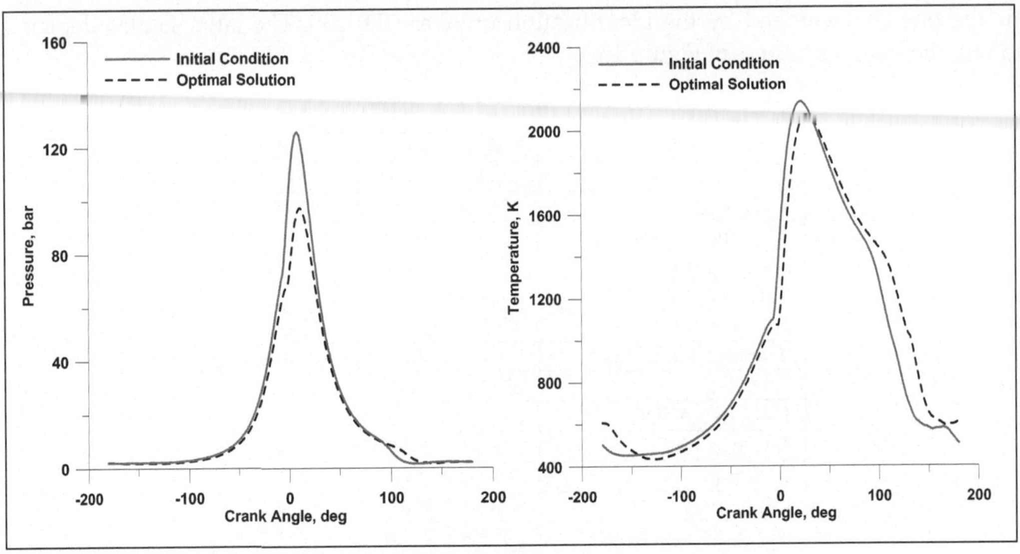

Fig. 16. Initial and optimal pressure and temperature cycles.

The delayed opening of the exhaust valve also produces an increased expansion work, as clearly observable in the in-cylinder pressure cycle plotted in Figure 16. The same figure highlights that a very lower pressure peak is obtained as a consequence of a lower supercharging level and a delayed injection start (see THJ variable in Table 1). Similar considerations can be draw looking at the average in-cylinder temperature profile.
Despite the lower boost pressure, the net shaft power remains the same, as requested by the Constrain 2, mainly due to a lower mechanical energy absorbed by the roots compressor.

It is worth putting into evidence that each modification to the engine geometry also determines a change in the operating conditions in terms of the super-charging level. This, together with a different power absorption of the roots, requires a control of the waste-gate opening in order to reach the prescribed power output at the engine shaft. In this sense, the optimization design regards the whole propulsion system, since it keeps into account the complex interaction between the various engine components.

4. Optimal selection of fuel injection strategies for a light-duty automotive engine

In this paragraph, a 3D modeling and an optimization procedure is applied to a naturally aspirated light-duty diesel engine (505 cm3 displacement). The engine is equipped with a mechanical Fuel Injection System (FIS) and is originally designed for non-road applications. Starting from the above base engine, a new prototype, equipped with a Common Rail (CR) FIS, is developed for being installed on small city-cars. The behavior of the CR injection system is firstly experimentally analyzed, in order to define the spray structure and injection rate realized under different operating conditions. As an example, in figure 17, the injection rates related to three different load conditions are compared. They are measured by an AVL Injection Gauge Rate System working on the Bosch tube principle. In addition, experimental data on the spray tip penetration are available from the analysis of the liquid fuel spray images, carried out by image processing procedures (Alfuso et al., 1999; di Stasio et al., 1999). These data are employed to validate the spray model in the 3D CFD analysis (Allocca et al. 2004).

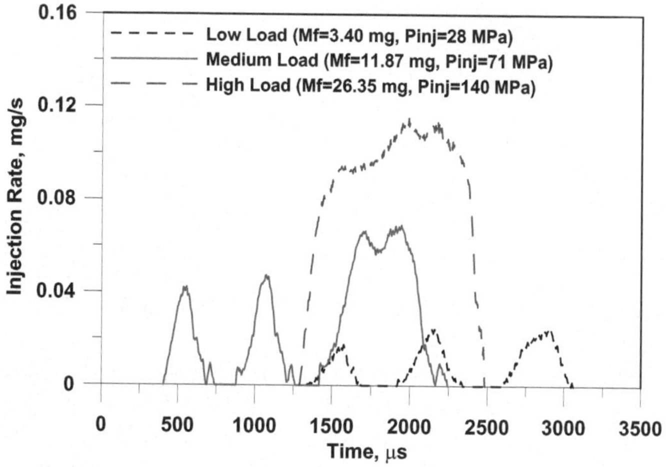

Fig. 17. Experimental injection rate of the CR-FIS.

Figure 18 summarizes the results of the preliminary numerical tuning of the spray break-up model, by comparing the experimentally measured penetration length and the numerical results. The Huh-Gosman and the Wave model are both tested and tuned by a change in the constants determining the aerodynamic break-up time, C2 and C1. Even with a value of 40 for the C2 constant, the Huh-Gosman model underestimates the spray penetration length, whereas quite reliable results are achieved by activating the Wave model with C1=60.

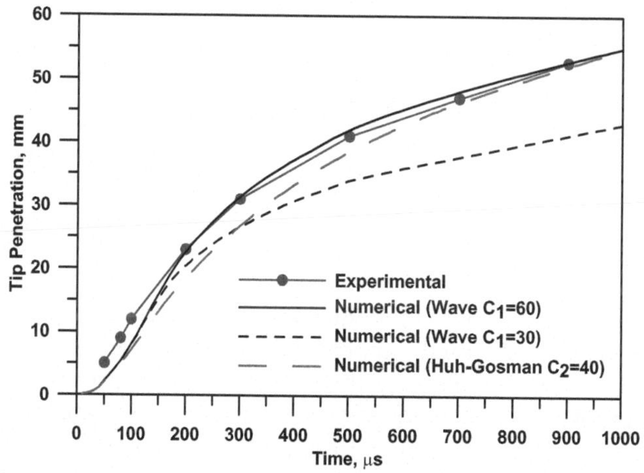

Fig. 18. Numerical and experimental spray penetration length.

The tuned spray model is part of a more complete 3D CFD analysis. Figure 19 shows a top view of the unstructured grids employed in the calculations.

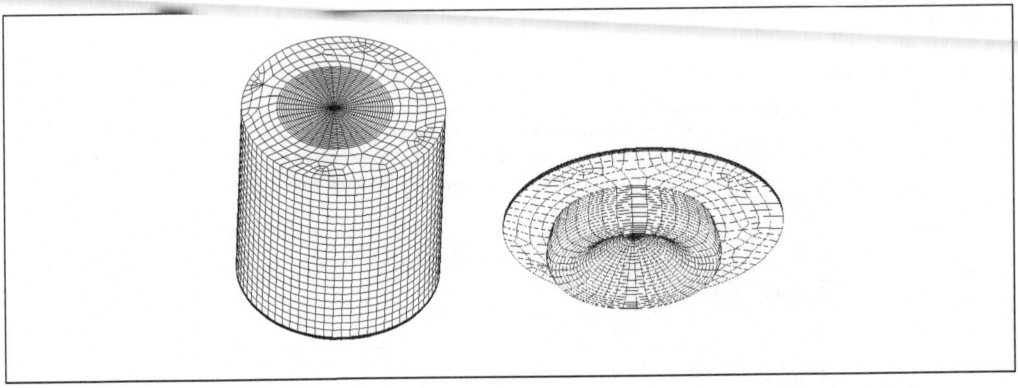

Fig. 19. A top view of the grid at the BDC, and a bottom view of the grid at the TDC

During the 3D analysis, a three pulses injection strategy is specified as shown in Figure 20, compared to the actual experimental profile. Five degrees of freedom - namely the start of pilot injection (soip), the dwell time between the first and second pulse (dwell_1), the dwell time between the second and third pulse (dwell_2), and the percentages of fuel mass injected during the first two pulses - completely define the overall injection profile.

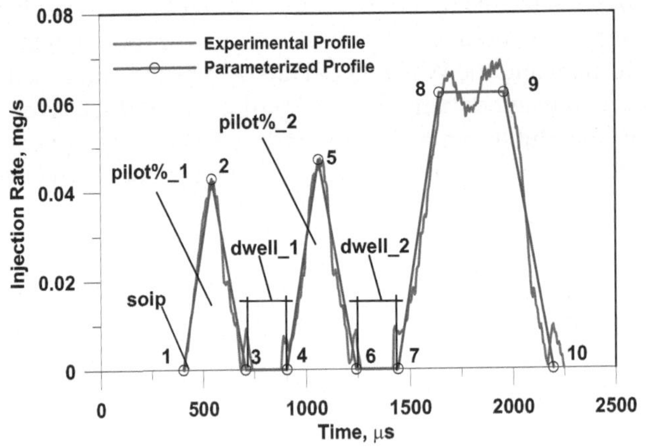

Fig. 20. Parametric Injection strategy at medium load

In this way, by varying the above 5 parameters, different combustion developments and noxious emissions arises. Each predicted pressure cycle is also processed to estimate the combustion-radiated noise, with the simplified approach previously described.

The optimization problem is settled in order to identify the 5 control parameters with the aim of simultaneously minimizing fuel consumption, pollutant emissions and radiated noise. The logical development of the optimization problem within the ModeFRONTIER™ environment is explained in figure 21.

Figure 22 displays the scatter charts of the 440 points computed along the optimization process, highlighting the complex interactions among the various objectives. A clear trend exists between the IMEP and the Overall Noise. A greater dispersion of the results is found looking at the trade-off between NO and soot mass fractions.

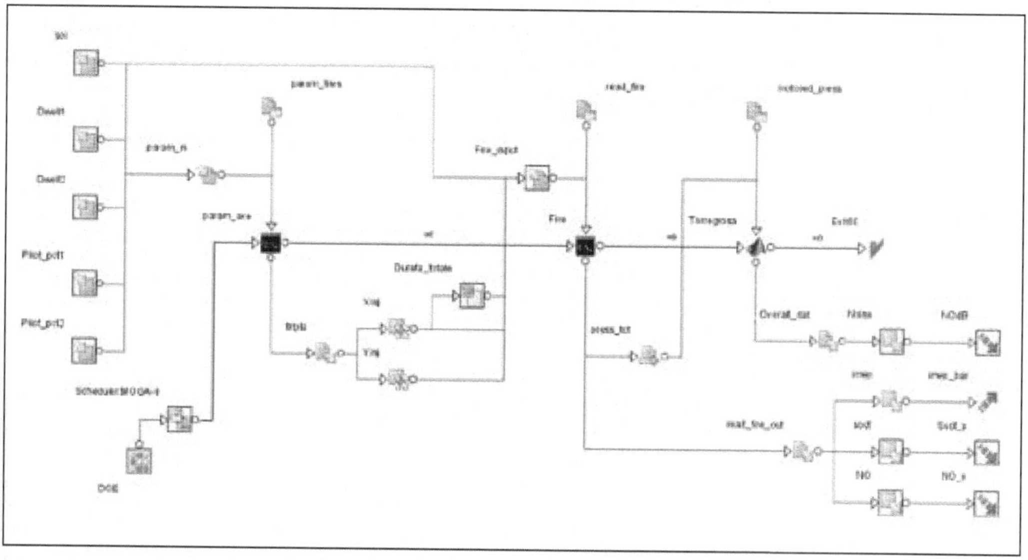

Fig. 21. Logic scheme of the optimization process within ModeFRONTIER

The "Multi Criteria Decision Making" tool (MCDM) provided in modeFRONTIER™ is finally employed to select single solutions among the ones reported in figure 22.

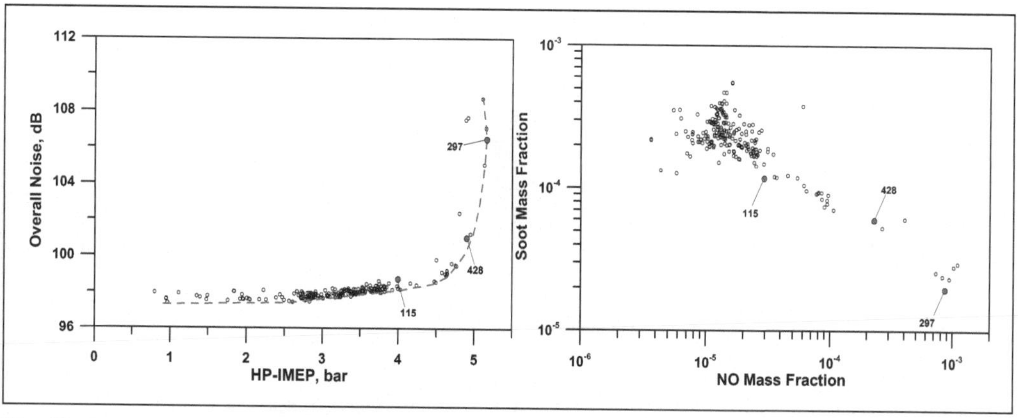

Fig. 22. Scatter charts of the optimization process

Three different solutions are identified, the first one selecting the IMEP and soot as the most important parameters (solutions #297). In the second and third one, the importance of NO emission and Overall Noise are more and more increased (solutions #428 and #115, respectively).

Figure 23 compares the related optimal injection strategies, while figure 24 finally shows the pressure cycles, the heat release rates, and the NO and soot production. High IMEP and low soot are obtained with a very advanced start of both pilot and main injections (solutions #297). This strategy determines the highest pressure peak and IMEP, while at the same time producing the highest heat release peak and pressure gradient, responsible of increased NO amounts and Noise level. Although introducing some IMEP and soot penalization, a radically lower Noise level and a better NO emission is found with a delayed soip and smaller dwell times (solution #428).

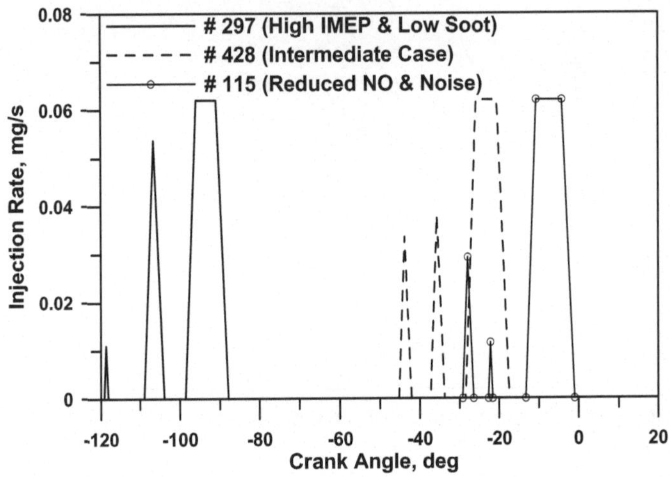

Fig. 23. Optimal injection strategies

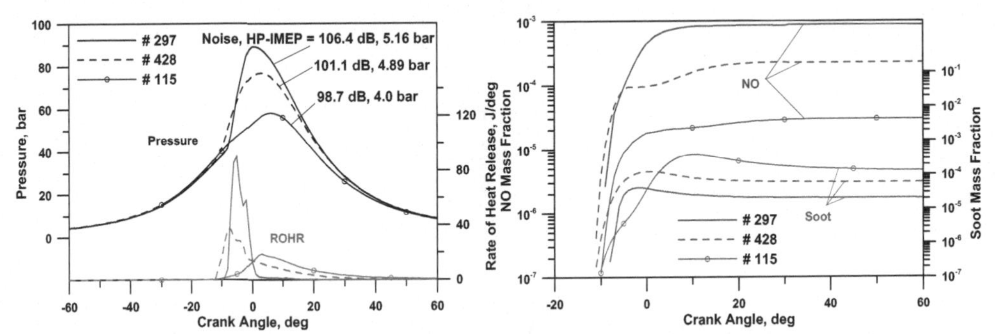

Fig. 24. Comparisons of the pressure cycles, heat release, NO and soot emission

In this case a very smoother heat release is found (fig. 24). The main path for a substantial noise reduction is indeed the specification of a much delayed pilot and a still more delayed main injection (solution #115). Both these effects, in fact, contribute in reducing the heat release rate and the pressure gradient during the combustion process, even if a non-negligible mechanical output loss has to be expected. This last strategy probably represents the best compromise solution in terms of pollutant species production and radiated noise.

However, being the present results obtained at a constant overall injected fuel mass, the lower power output obtained in this case directly impacts on the specific fuel consumption and on the CO2 emission. This demonstrates the difficulties in identifying an optimal injection strategy, able to complying with so many conflicting needs, even when a highly modulated injection process is considered.

As a conclusive remark, the optimization procedure is able to capture the expected effects of the injection parameters on the overall performance and radiated noise and provides a method for the choice the best compromise solution. The methodology can be easily extended to multiple operating conditions and can include additional variables (injection pressure, swirl ratio, boost pressure, EGR rate) and objectives (CO and HC production). In this way a variable combustion mode, like standard premixed-diffusive, HCCI (Bression et al, 2008; Zavala et al. 2001), PLTC (Shimazaki et al. 2003; Kalghatgi, 20009), etc. can be realized, depending on operating conditions and selected objectives.

5. Conclusion

The present chapter presented two different examples for the design and analysis of modern CR equipped Diesel engines. A tool for multi-objective optimization was found very useful at the engine design stage to reduce the time-to-market of new engine prototypes, thanks to the possibility to effectively evaluate the influence of geometrical and operating engine parameters. In addition, a similar technique was applied to find the optimal selection of the control parameters (namely the injection strategy) in order to obtain an better engine behaviour. To reach the above objectives, different simulation techniques were employed, resorting to 1D, 3D and acoustic analyses. More accurate or simplified approaches were presented and differently used along the different phases of the engine development process. In each case, a continuous exchange of information among the various methods allowed to improve the overall simulation accuracy and results reliability. The whole procedure hence represents a very useful tool to reduce the huge experimental activity usually required to design and develop a modern CR diesel engine.

6. References

Alfuso, S., Allocca, L., Corcione, F.E., di Stasio S. (1999) "Image Diagnostics of Common Rail Diesel Sprays Evolving in Nytrogen Ambient at Different Densities", ICE'99 Internal Combustion Engines: Experiments and Modeling.

Alfuso, S., Allocca, L., Auriemma, M., Caputo, G., Corcione, F.E., Montanaro, A., Valentino, G. (2005) "Analysis of a High Pressure Diesel Spray at High Pressure and Temperature Environment Conditions", SAE Paper 2005-01-1239

Allocca, L., Corcione, F.E., Costa, M. (2004) "Numerical and Experimental Analysis of Multiple Injection Diesel Sprays", SAE Paper 2004-01-1879

Bression, G., Soleri, D., Dehoux, S., Azoulay, D., Hamouda, H., Doradoux, L., Guerrassi, N. Lawrence, N.J. (2008) "A Study of Methods to Lower HC and CO Emissions in Diesel HCCI", SAE Paper 2008-01-0034

Colin, O., and Benkenida, A. (2004) "The 3-Zones Extended Coherent Flame Model (ECFM3Z) for Computing Premixed/Diffusion Combustion", Oil & Gas Science and Technology, IFP, Vol. 59, N. 6, pp. 593-609

Costa, M., Siano, D., Valentino, G., Corcione, F.E., Bozza, F. (2009) "Prediction and Optimization of the Performances, Noxious Emissions and Radiated Noise of a Light Duty Common-Rail Diesel Engine", proceedings of 9th International Conference on Engines and Vehicles (ICE2009)

di Stasio, S., Alfuso, S., Allocca, L., Corcione, F.E. (1999) "Experimental Study on the Atomization Mechanism for Fuel Sprays Evolving in Atmospheres of Different Nature and Density" - ImechE Seminar Publication 1, 17, pp.241-255

Dukowicz, J.K. (1980) "A Particle-Fluid Numerical Model for Liquid Sprays", J. Comp. Physics, 35, 229-253

Kalghatgi, G. (2009) "Is Gasoline the Best Fuel for Advanced Diesel Engines? - Fuel Effects in "Premixed-Enough" compression ignition Engines", Towards Clean Diesel Engines, TCDE2009

Liu, A.B. and Reitz, R.D. (1993) "Modeling the Effects of Drop Drag and Break-up on Fuel Sprays", SAE 930072

O'Rourke, P.J. (1989) "Statistical Properties and Numerical Implementation of a Model for Droplet Dispersion in Turbulent Gas", J. Comput. Physics 83

Papalambros, P.V., and Wilde, D.J.(2000) "Principles of Optimal Design Modeling and Computation", Cambribde University Press, Cambridge

Payri, F., Broatch, A., Tormos, B., Marant, V., (2005) "New methodology for in-cylinder pressure analysis in direct injection diesel engines—application to combustion noise", Meas. Sci. Technol. 16 540–547 doi:10.1088/0957-0233/16/2/029

Sasaki, D. (2005) "ARMOGA, An efficient Multi-Objective Genetic Algorithm", Technical Report

Shimazaki, N., Tsurushima, T., Nishimura, T. (2003) "Dual Mode Combustion Concept with Premixed Diesel Combustion by Direct Injection Near Top Dead Center", SAE 2003-01-0742

Siano, D., Bozza, F., Costa, M. (2008) "Optimal Design of a Two-Stroke Diesel Engine for Aeronautical Applications Concerning both Thermofluidynamic and Acoustic Issues", IMECE2008-68713, Proceedings of 2008 ASME International Mechanical Engineering Congress and Exposition, Boston, Massachusetts, USA.

Stephenson, P.W. (2008) "Multi-Objective Optimization of a Charge Air Cooler using modeFRONTIER and Computational Fluid Dynamics", SAE Paper 2008-01-0886

Stotz, M., Schommers, J., Duvinage, F., Petrs, A., Ellwanger, S., Koynagi, K., Gildein, H. (2000) "Potential of Common-Rail Injection System for Passenger Car Di Diesel Engines", SAE Paper 2000-01-0944

Torregrosa, A.J., Broatch, A., Martın J., Monelletta, L. (2007) "Combustion noise level assessment in direct injection Diesel engines by means of in-cylinder pressure components", Meas. Sci. Technol., 18 2131-2142, doi:10.1088/0957-0233/18/7/045,

Zavala, P.A.G., Pinto, M.G, Pavanello, R., Vaqueiro J. (2001) "Comprehensive Combustion Noise Optimization", Sae Paper 2001-01-1509

Hydrogen fuelled scramjet combustor - the impact of fuel injection

1. Introduction

The scramjet engine is one of the most promising propulsive systems for future hypersonic vehicles. Over the last fifty years the scramjet engine technology has been intensively investigated and several such engines have been flight-tested in recent years (Neal, Michael, & Allan, 2005; Paul, Vincent, Luat, & Jeryl, 2004). Research on supersonic combustion technologies is of great significance for the design of the engine and many researchers pay significant attention to the hypersonic airbreathing propulsion. The mixing and diffusive combustion of fuel and air in conventional scramjet engines take place simultaneously in the combustor (Huang, Qin, Luo, & Wang, 2010). Since the incoming supersonic flow can stay in the combustor only for a very short period of time, i.e. of the order of milliseconds (Aso, Inoue, Yamaguchi, & Tani, 2009; Huang *et al.*, 2010; Hyungseok, Hui, Jaewoo, & Yunghwan, 2009), and the whole process of combustion has to be completed within this short duration, this is a significant restriction to the design of the scramjet engine. In order to solve this problem, hydrogen, one of the most promising fuels for the airbreathing engine with ~10 times faster reaction than hydrocarbons, is widely used in the scramjet combustor.

In recent years, a cavity flameholder, which is an integrated fuel injection/flame-holding approach, has been proposed as a new concept for flame holding and stabilization in supersonic combustors (Alejandro, Joseph, & Viswanath, 2010; Chadwick *et al.*, 2005; Chadwick, Sulabh, & James, 2007; Daniel & James, 2009; Gu, Chen, & Chang, 2009; Jeong, O'Byrne, Jeung, & Houwong, 2008; Kyung, Seung, & Cho, 2004; Sun, Geng, Liang, & Wang, 2009; Vikramaditya & Kurian, 2009). The presence of a cavity on an aerodynamic surface could have a significant impact on the flow surrounding it. The flow field inside a cavity flameholder is characterized by the recirculation flow that increases the residence time of the fluid entering the cavity, and the cavity flame provides a source of heat and radicals to ignite and stabilize the combustion in the core flow.

However, so far, the flow field in the scramjet combustor with multiple cavity flameholders has been rarely discussed, and this is an important issue as it can provide some useful guidance for the further design of the scramjet combustor. Multi-cavity flameholder can produce larger drag forces on the scramjet combustor, as well as improve the combustion efficiency of the combustor. A balance between these two aspects will be very important in the future design of the propulsion system in hypersonic vehicles. At the same time, the combustor configuration, i.e. the divergence angle of each stage, makes a large difference to the performance of the combustor. Researchers have shown that (Huang, Li, Wu, & Wang, 2009) the effect of the divergence angles of the posterior stages on the performance of the scramjet combustor is the most important, and the effect of the divergence angle on the first stage is the least important. When the location of the fuel injection moves forward, the effect of the divergence angle of the former stages becomes more important.

In this chapter, the two-dimensional coupled implicit Reynolds Averaged Navier-Stokes (RANS) equations, the standard *k-ε* turbulence model (Huang & Wang, 2009; Launder & Spalding, 1974) and the finite-rate/eddy-dissipation reaction model (Nardo, Calchetti, Mongiello, Giammartini, & Rufoloni, 2009) have been employed to investigate the effect of the location of the fuel injection on the combustion flow field of a typical hydrogen-fueled scramjet combustor with multi-cavities.

2. Physical model and numerical method

The engine investigated adopts the single-expanded combustor and fractional combustion mode, and it consists of an isolator and three staged combustors, see Fig. 1. There are four cavity flame holders located on the upper and lower walls of the first and the second staged combustors, respectively. Hydrogen is injected from the slot, located at 5mm from the leading edge of the four cavity flame holders on both the upper and lower walls of the first and the second staged combustor. The width of the slot is 1mm.

Assuming that the height of the isolator H_i is 1 unit, the distance between the upstream forward face of the cavity flameholder in the upper wall and that in the lower wall of each staged combustor is 0.183 along the x axis. The dimensions of the components of the scramjet combustor are shown in Table.1, where L_i, L_{c1}, L_{c2} and L_{c3} are the lengths of the isolator, the first staged combustor, the second staged combustor and the third staged combustor, respectively. The divergence angles of the first staged combustor, β_1, the second staged combustor, β_2 and the third staged combustor, β_3 are 2.0 degree, 3.5 degree and 4.0 degree, respectively.

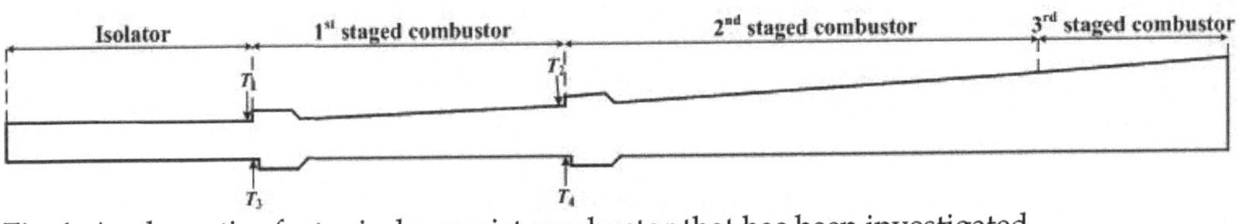

Fig. 1. A schematic of a typical scramjet combustor that has been investigated.

H_i	L_i	L_{c1}	L_{c2}	L_{c3}	$\beta_1/(°)$	$\beta_2/(°)$	$\beta_3/(°)$
1.0	7.0	8.8	12.8	5.8	2.0	3.5	4.0

Table 1. Geometrical dimensions of the scramjet combustor.

The primary geometry parameters of the cavity flameholder: the length of the cavity flameholder L=1.376, the height of the leading edge D_u=0.275, the ratio of length-to-height

L/D_u=5.0, the swept angle θ=45° and the height of the trailing edge D_d=0.275. A schematic diagram of a typical cavity flameholder that has been investigated is shown in Fig. 2.

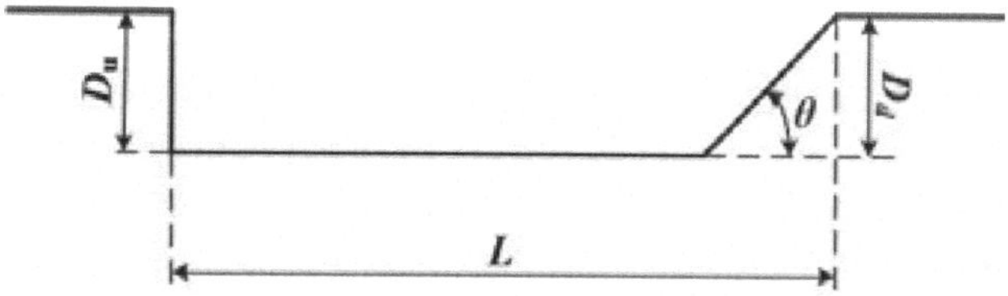

Fig. 2. A schematic of a typical cavity flameholder that has been investigated.

Table.2 shows the boundary conditions employed in the computational fluid dynamics (CFD) models. The ratio of the oxygen gas mol fraction to the nitrogen gas mol fraction at the entrance of the combustor is 23:77, with the Mach number being 3.2, the total pressure 2.9MPa and the total temperature 1505.0K. The hydrogen is injected into the core flow with sonic velocity, as shown in Table.2. The static pressure and temperature of the injection are 1060KPa and 250K, respectively.

	Ma	P_e/KPa	T_e/K	Y_{N2}	Y_{O2}	Y_{H2}
The entrance of the combustor	3.2	58.66	493.77	0.77	0.23	0
The exit of the injection	1.0	1060	250	0.0	0.0	1.0

Table 2. Boundary conditions for the numerical model.

In the CFD model, the standard k-ε turbulence model is selected. This is because of its robustness and its ability to fit the initial iteration, design lectotype and parametric investigation. Further, because of the intense turbulent combustion effects, the finite-rate/eddy-dissipation reaction model is adopted. The finite-rate/eddy dissipation model is based on the hypothesis of infinitely fast reactions and the reaction rate is controlled by the turbulent mixing. Both the Arrhenius rate and the mixing rate are calculated and the smaller of the two rates is used for the turbulent combustion (FLUENT, 2006). While a no-slip condition is applied along the wall surface, at the outflow all the physical variables are extrapolated from the internal cells due to the flow being supersonic.

3. Model validation

In order to validate the present numerical method for computing these complex fluid flows in the scramjet combustor with multi-cavities, three computational cases are investigated, namely, the problems of an injection flow, a cavity flow and a fuel-rich combustion flow. The grids for the geometries are structured and generated by the commercial software Gambit, and the grids are distributed more densely near the walls and in the vicinity of the shock wave generation in order to resolve the boundary layers.

3.1 Injection flow

In this first case, the physical model that was experimentally investigated by Weidner et al.(Weidner & Drummond, 1981) is employed since the model has a good two-dimensional structure and it can be used to validate the correctness of the injection phenomenon in the scramjet combustor.

The experimental test investigates the phenomenon of the traverse injection of helium into parallel air flow, namely θ=90°, and the setup of the experiment is schematically shown in Fig. 3. The air stream is introduced from the left hand side of a rectangular channel which is 25.4cm long and 7.62cm high. The static pressure of the air stream is P=0.0663MPa, the static temperature is T=108.0K and the March number is M=2.9. The helium is injected at sonic condition from a 0.0559cm slot into an air stream from the bottom surface of the rectangular channel at a location which is 17.8cm downstream from the entrance of the channel. The flow conditions for the helium at the slot exit are P=1.24MPa, T=217.0K and M=1.0.

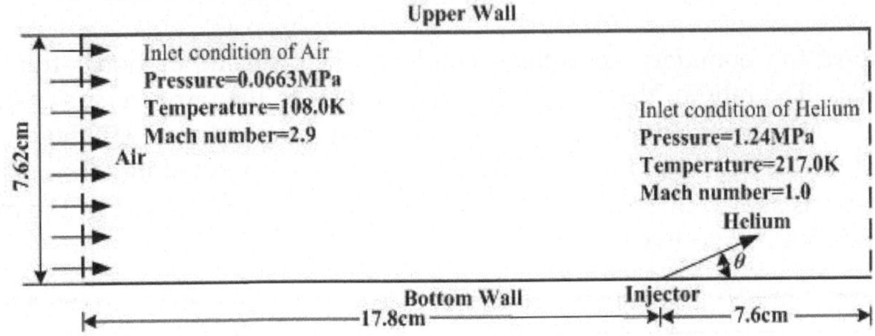

Fig. 3. Schematic of the physical model investigated for injection flow.

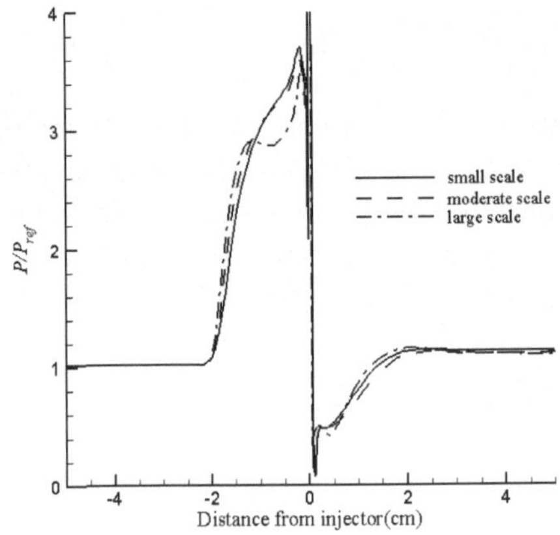

Fig. 4. Static pressure distribution along the bottom wall of the channel for the different grid systems.

In order to investigate grid independency of the numerical simulations, three sets of mesh with different numbers of cells have been employed, namely approximately 19,200, 38,080 and 76,230 cells, respectively. Fig. 4 shows the static pressure distribution along the bottom wall of the channel for the three different grids. It is observed that the shock wave can be captured accurately for all three different grid scales, and the pressure distributions along the bottom wall of the channel in the downstream region of the injection slot are almost the same for the three grids employed. With different grid scales, the location of the disappearance of the reattachment region and the location of the generated shock wave can be predicted reasonably accurately when compared with the experimental data, see Fig. 5. This means that the difference in the three grid systems employed in the simulations makes only a small difference to the numerical predictions for the interaction between the air stream and the injection.

Fig. 5 shows a comparison between the experimental data and the computational predictions for the pressure along the bottom wall. The reference pressure P_{ref} is 0.0663MPa. It is observed that the computational results obtained in this investigation show good qualitative agreement with the experimental data for both the upstream and downstream regions of the injection.

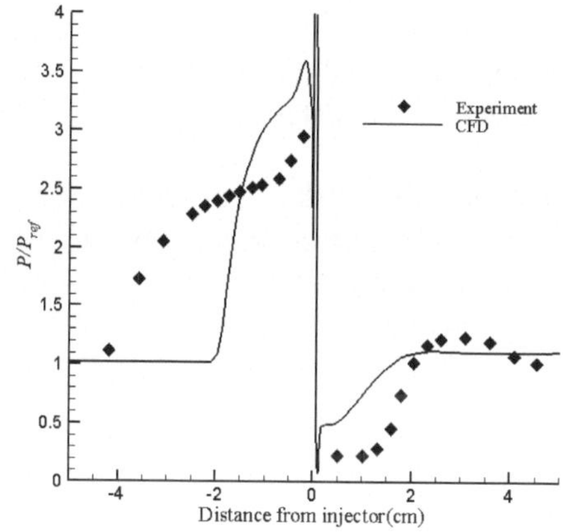

Fig. 5. Comparison between the experimental data of Weidner et al. (Weidner & Drummond, 1981) and the predicted computational pressures along the bottom wall.

Fig. 6 shows a comparison between the experimental data and the predicted computational pressures at a distance of 3.81cm downstream of the injection slot when the reference pressure is 0.21MPa and the reference height is 7.62mm. It is observed that there is a rapid pressure drop at a distance of about 1.524cm (i.e. y/h=0.2) from the bottom wall, and this is the location where the separated region disappears downstream of the injection slot. This rapid pressure drop is followed by a pressure rise in the central region of the channel, and this is the intersection point between the shock wave and the transverse line at this location. At the same time, we observe that there are also some discrepancies between the experimental data and the calculated results because of the complex flow field in the vicinity of the injection exit and the inaccuracy of the k-ε turbulent model to simulate the separation region generated just upstream and downstream of the injector.

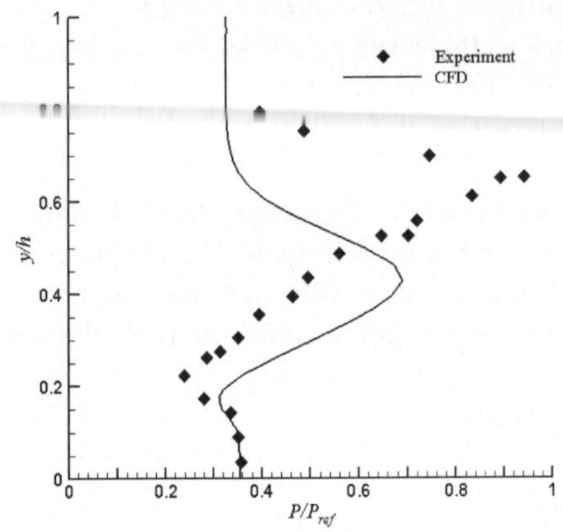

Fig. 6. Comparison between the experimental data of Weidner et al. (Weidner & Drummond, 1981) and the computational pressures at a distance of 3.81cm downstream of the injector.

The helium mass fraction distribution at a distance of 3.81cm downstream of the injector, as obtained from the computational model, agrees reasonably well with the experimental data, see Fig. 7, although there is a slight underprediction by the numerical simulation. It should be noted that the height is nondimensionalized by the height of the channel, namely

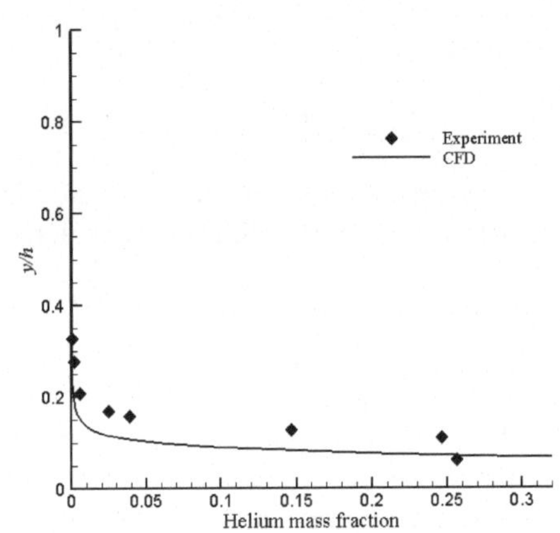

Fig. 7. Comparison between the experimental data of Weidner et al. (Weidner & Drummond, 1981) and the computed value for the helium mass fraction at a distance of 3.81cm downstream of the injector.

h=7.62cm.

From the results presented in Figs. 5, 6 and 7, it is found that the mathematical and computational model can reasonably accurately simulate the interaction between the air stream and the injection. In particular, the model can capture the shock wave and predict the parametric distribution. Therefore we conclude that the mathematical and computational model can be used with confidence to investigate the flow field of the scramjet combustor.

3.2 Cavity flow

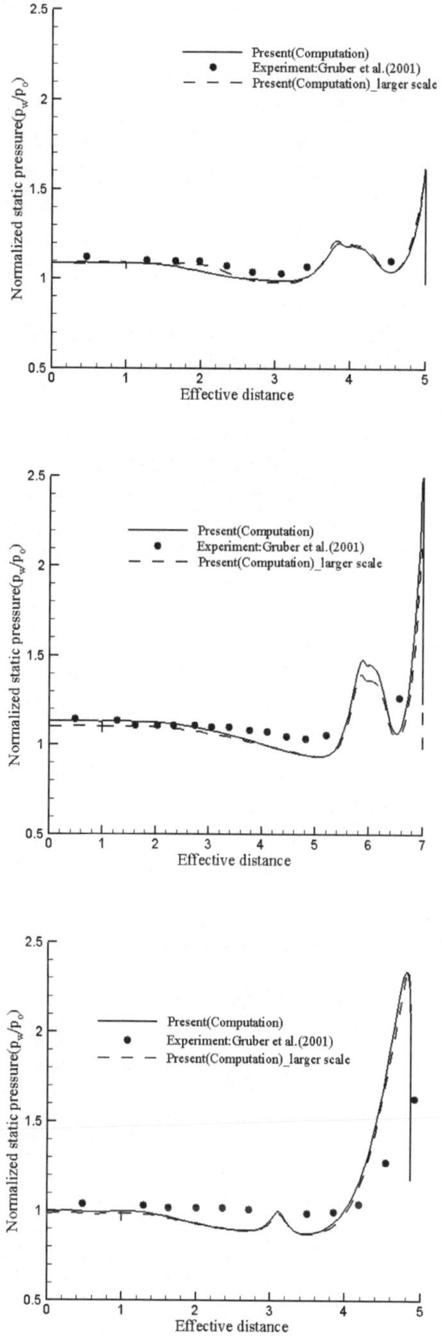

Fig. 8. Wall static pressure distributions for: (a) L/D=3 and no swept angle; (b) L/D=5 and no swept angle; and (c) L/D=3 with the swept angle 30°.

The second model considered follows the experimental work of Gruber et al. (Gruber, Baurle, Mathur, & Hsu, 2001) who studied several cavity configurations for an unheated flow at Mach 3. Cavities with a depth of 8.9mm were used in the experimental work and for the conditions of L/D=3, L/D=5 without a swept angle, and L/D=3 with the swept angle (θ) of 30°, see Fig. 2. In addition, the stagnation temperature (T_0) and stagnation pressure (P_0) of the free stream are 300K and 690kPa, respectively. This physical model is used to validate the correctness of the predicting flow past the cavity flameholder in the scramjet combustor. Fig. 8 shows the wall pressure distributions for L/D=3, L/D=5 without a swept angle, and L/D=3 with the swept angle 30°. Two sets of mesh, with different number of cells, have been employed in order to investigate the grid independency of the numerical simulations, namely approximately 36,400 and 147,200 cells have been employed.

In Fig. 8, the effective distance comprises of the cavity upstream leading edge from the separation corner, the cavity floor and the cavity trailing edge (Kyung *et al.*, 2004). A good agreement is observed between the computed and experimental results, and the difference in the two numbers of grids employed in the simulations produces prediction that makes almost no difference for the unheated cavity flow. We observe that the numerical method employed in this investigation can be used with confidence to simulate the flow field of the scramjet combustor with multi-cavities, and investigate the effect of the fuel injection location on the performance of the scramjet combustor.

3.3 Fuel-rich combustion flow field

The third model considered follows the experimental configuration and flow conditions for the case investigated by Wang Chun et al. (Wang, Situ, Ma, & Yang, 2000), and this model is used to validate the correctness of the combustion model employed in this investigation.

The geometry consists of a straight channel with a length of 370mm followed by a divergent channel with a divergent angle of 3.6°. There is a clapboard between the entrance of the air and the entrance of hot gas, see Fig. 9, and the length of the clapboard is 6mm. All the dimensions used in the CFD model are exactly the same as in the experimental configuration. The air and hot gas flow conditions are presented in Table.3.

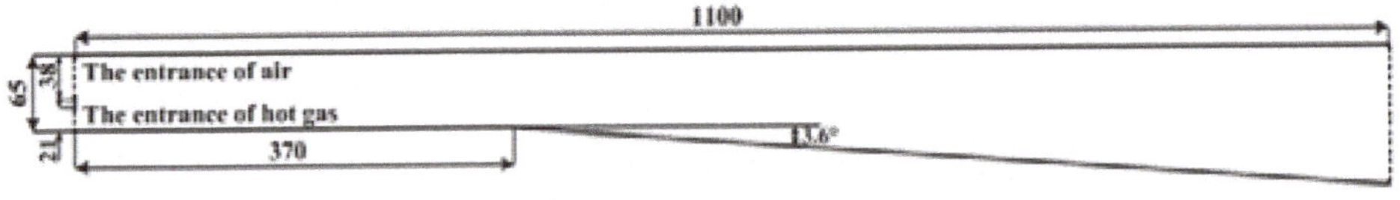

Fig. 9. The geometry of the combustor investigated (Unit: mm)(Wang et al., 2000).

Flow	P_s/MPa	T_s/K	*Ma*	Mass fraction				
				C_2H_4	O_2	CO_2	H_2O	N_2
Air	0.0977	491.9	2.09	-	0.2330	-	0.0520	0.7150
Hot gas	0.1731	1771.9	1.25	0.1059	0.0103	0.1205	0.1566	0.6067

Table 3. Parameters at the entrance of the supersonic combustor(Wang *et al.*, 2000).

Computational simulations have been performed with a coarse and a fine computational mesh consisting of 8,700 (CFD1) and 16,900 cells (CFD2), respectively. Fig. 10 shows the comparisons of the wall pressure distributions obtained from the present CFD calculations and the experimental data of Wang Chun et al. (Wang et al., 2000). The solid line represents the numerical results from the coarse mesh, CFD1, and the dashed line is for CFD2. It can be observed that the static pressure distributions on the top and bottom walls obtained by the CFD results show good qualitative agreement with the experimental results. The CFD model captures the shock wave reasonably well in terms of both the location and strength of the wave system. The pressure disturbance on the top and bottom walls is due to the compression and expansion of the flow that occurs alternately in the mixing and expansion sections of the combustor caused by the shock wave system. At the entrance to the mixing section of the combustor, due to the differences in the flow parameters in the two supersonic flows of air and hot streams, and the effect of the clapboard, the expansion wave appears during flow expansions. When the two flows intersect, the flow direction changes, and the two flows become compressed (Situ, Wang, Niu, Wang, & Lu, 1999). It is concluded that the CFD approach used in this investigation can reasonably accurately simulate these physical phenomena in the scramjet combustor.

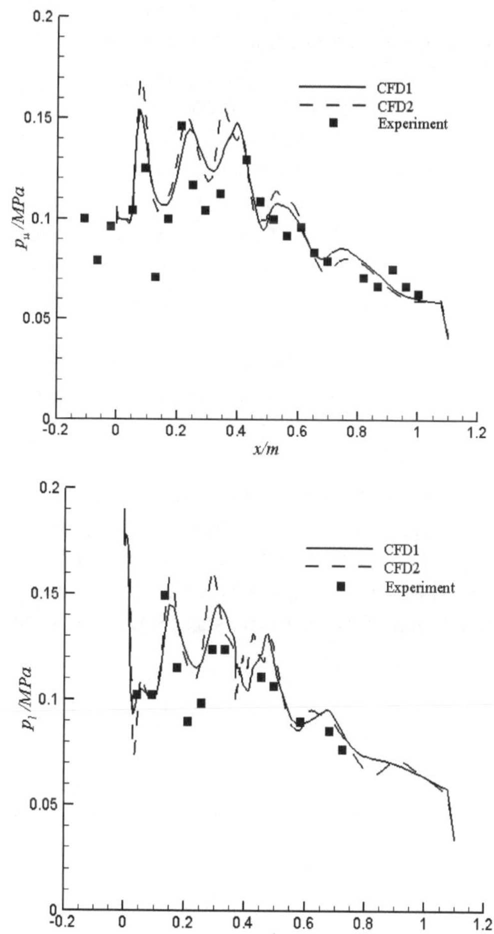

Fig. 10. Wall pressure comparisons of the CFD calculations and the experimental results of Wang Chun et al. (Wang et al., 2000): (a) top wall; and (b) bottom wall.

4. Results and discussion

In order to discuss the influence of the fuel injection location on the flow field of the scramjet combustor with multiple cavity flameholders, three sets of the fuel injection location are employed in this investigation, namely, T_2, T_4 and both T_2 & T_4, in Fig. 1. The other fuel injection locations are not considered here, i.e. T_1 or T_3, because placing the fuel injection location closer to the entrance of the combustor and more concentrated in a certain distance can be of much assistance in the optimization of the performance of the combustor, but the fuel injection location being excessively close to the entrance of the combustor can cause the interaction between the isolator and the combustor to occur more easily and push the shock wave forward, and this will cause the inlet unstart (Wu, Li, Ding, Liu, & Wang, 2007).

Figs. 11-13 show the parametric contours of the cases with the hydrogen injected from T_2, T_4 and both T_2 & T_4, respectively. When the hydrogen is injected from both T_2 and T_4, the shock wave in the combustor is pushed forwards into the isolator by the intense combustion and a high static pressure region formed between the first upper cavity flameholder and the second upper cavity flameholder, see Fig. 13 (a). Then if the fuel injection location moves forward, i.e. T_1 or T_3, the shock wave is pushed out of the isolator into the inlet and this causes the inlet unstart.

There exits a complex shock wave system in the combustor. When the hydrogen is injected from T_2, the shock waves generated from the leading edges of the first upper and lower cavity flameholders interact and form a high pressure region, see Fig. 11 (a). At the same time, we observe that the high pressure region exists mainly in the vicinity of the injection due to the fuel combustion. There is a low Mach number region generated on the upper wall of the combustor due to the fuel injection, see Fig. 11 (b). Meanwhile, due to the interaction between the shock wave and the boundary layer, there exists a separation region on the lower wall of the combustor, see Fig. 14 (a). The fuel injection makes the vortices in the cavity flameholder become larger and it deflects into the core flow. The shear layer formed on the leading edge of the second upper cavity flameholder impinges on its trailing edge, and there are almost no vortices in the first upper and lower cavity flameholders. The region in the cavity flameholders acts as a pool to provide the energy to ignite the fuel and prolong the residence time of the flow in the combustor. The Mach number in the cavity flameholders is much lower than that in any other place of the combustor, except in the separation regions, see Fig. 11 (b), and the static temperature in the cavity flameholders is slightly higher than that in the core flow, see Fig. 11 (c). If we change the geometry of the cavity flameholder, it can act as an ignitor in the scramjet combustor, but we should

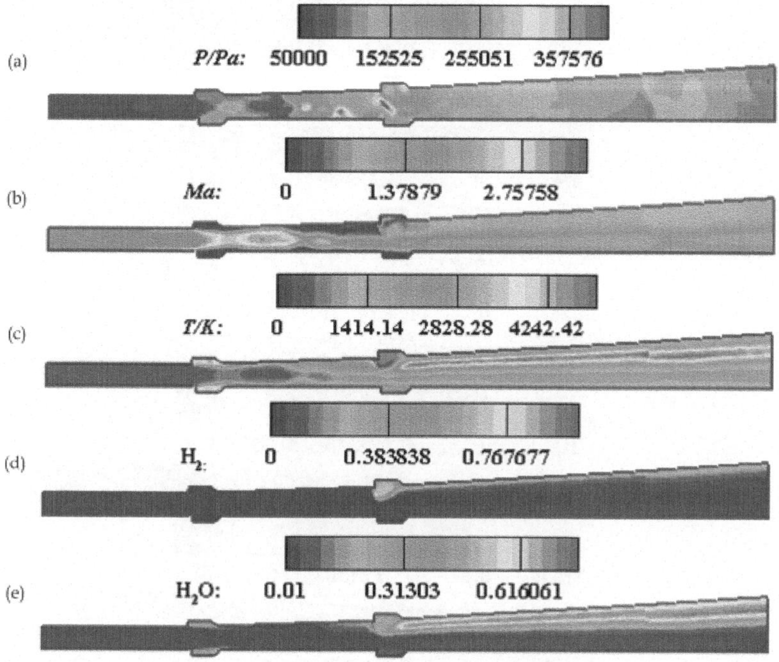

Fig. 11. Parametric contours of the case with hydrogen injected from T_2: (a) static pressure; (b) Mach number; (c) static temperature; (d) H_2 mass fraction; and (e) H_2O mass fraction.

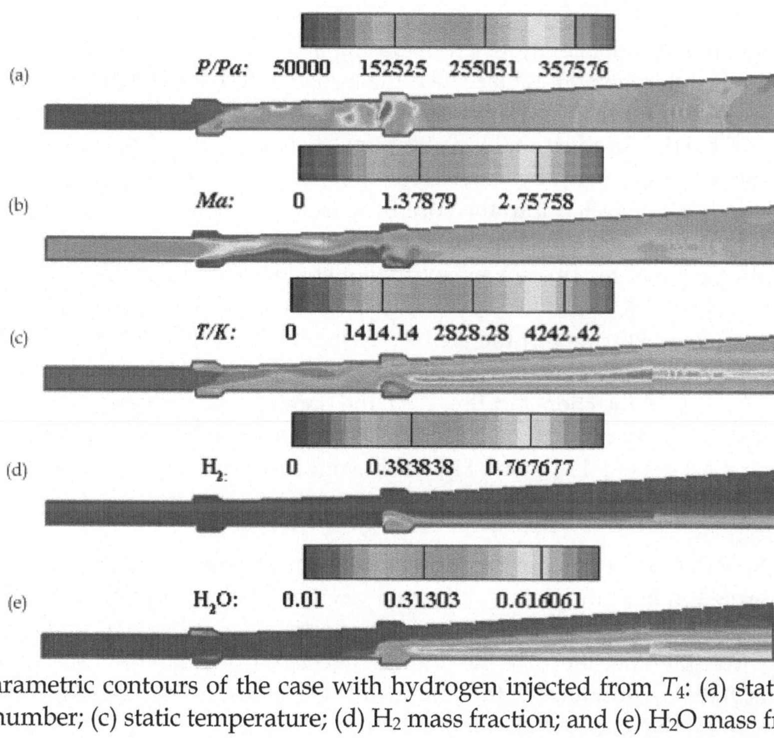

Fig. 12. Parametric contours of the case with hydrogen injected from T_4: (a) static pressure; (b) Mach number; (c) static temperature; (d) H_2 mass fraction; and (e) H_2O mass fraction.

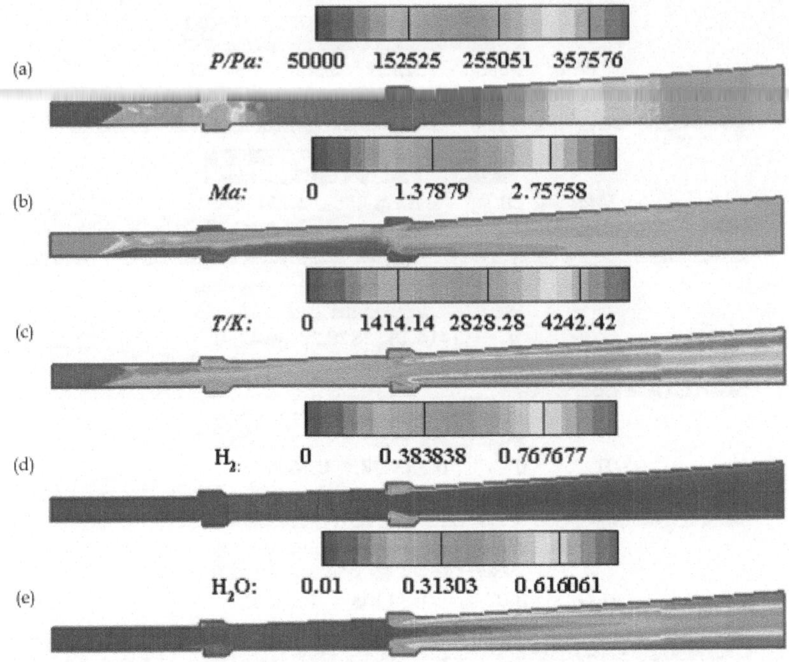

Fig. 13. Parametric contours of the case with hydrogen injected from both T_2 and T_4: (a) static pressure; (b) Mach number; (c) static temperature; (d) H_2 mass fraction; and (e) H_2O mass fraction.

consider the material of the cavity when operating at such high temperatures. Further, the combustion of the hydrogen takes place near the upper wall of the combustor, see Fig. 11 (d), and the combustion product, namely, H_2O mainly distributes along the upper wall. There is also a small combustion production in the first upper and lower cavity flameholders, see Fig. 11 (e), and it is brought forward by the recirculation zone.

When the hydrogen is injected into the core flow from T_4, the shock wave generated from the leading edge of the first upper cavity flameholder is much weaker than that generated from the leading edge of the first lower cavity flameholder, and this makes the shock wave, after the interaction, deflect into the upper wall of the combustor. Further, we can observe a high pressure region generated in the vicinity of the upper wall, see Fig. 12 (a), and this is different from the case with the hydrogen injected from T_2. The reason may lie in the differences in the fuel injection locations. At the same time, we observe two low Mach number regions on the lower wall of the scramjet combustor and this has been caused by the recirculation zones, see Fig. 12 (b) and Fig. 14 (b), and because of the interaction of the shock wave and the boundary layer, there also exists a separation area in the vicinity of the upper wall of the combustor.

Because of the variation in the fuel injection location and the effect of the shock wave, small eddies are formed in both the upper and lower cavities of the first flameholders, and it lies on the rear edge of the cavity, see Fig. 14 (b). The vortices can act as a recirculation zone for the mixture. At this condition, if the fuel is injected from the first staged combustor simultaneously, the performance of the combustor will be improved since the residence time is longer than in the case when the hydrogen is injected from T_2. Meanwhile, the distributions of the fuel and the combustion production are opposite to the case when the hydrogen is injected from T_2, and they mainly distribute along the lower wall of the scramjet combustor because of the fuel injection location, see Fig. 12(d) and (e). Due to the fuel injection being before the cavity flameholder, the eddy generated in the second lower cavity flameholder become larger than before, see Fig. 14 (b), namely the case without fuel injection before the cavity flameholder. The eddy is deflected into the core flow, and the shear layer generated at the leading edge of the second lower cavity flameholder impinges on its trailing edge.

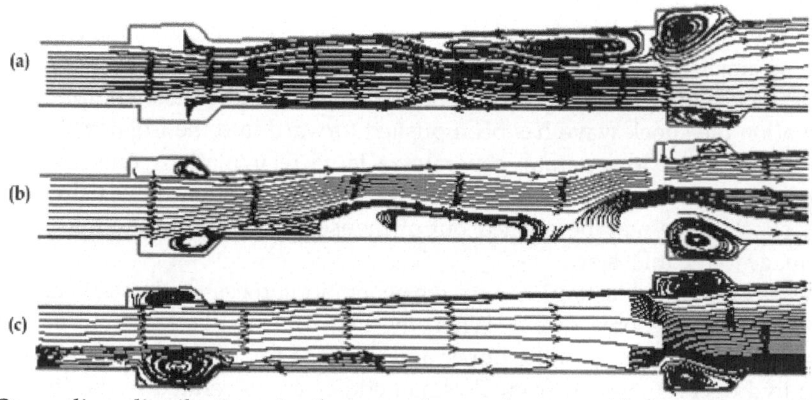

Fig. 14. Streamline distributions in the scramjet combustor with hydrogen injected from different locations: (a) T_2; (b) T_4; and (c) T_2 and T_4.

When the hydrogen is injected from both T_2 and T_4, the flow field is the most complex in the combustor, see Fig. 13. At this condition, the shock wave is pushed out of the combustor because of the intense combustion, and a larger low Mach number region is generated on the lower wall of the combustor because of the stronger interaction between the shock wave and the boundary-layer, see Fig. 13 (b), and it spreads forward to the lower wall of the isolator. A higher static pressure is obtained in the region between the first and the second cavity flameholder, see Fig. 13 (a), and this is the main cause for the spreading forward of the shock wave. Due to the hydrogen injected from both T_2 and T_4, the fuel and the combustion product distribute both on the upper and lower walls of the combustor, see Fig. 13 (d) and (e), and the combustion occurs mainly in the vicinity of the walls. This illustrates that the injection pressure is not high enough to make the fuel penetrate deeper. The recirculation zone generated at this condition is much larger than that formed in the other two cases, and thus the flow can stay in the combustor much longer, see Fig. 14(c). While travelling over the cavity, the injected hydrogen interacts with the strong trailing edge shock wave, which plays an important role in the combustion. The trailing edge shock wave can improve the static pressure and the static temperature of the flow in the vicinity of the trailing edge of the cavity flameholder, and this can also benefit the combustion.

5. Conclusion

In this chapter, the two-dimensional coupled implicit RANS equations, the standard k-ε turbulence model and the finite-rate/eddy-dissipation reaction model are introduced to simulate the combustion flow field of the scramjet combustor with multiple cavity flameholders. The effect of the fuel injection location on the flow field of the combustor has been investigated. We observe the following:

- The numerical methods employed in this chapter can be used to accurately simulate the combustion flow field of the scramjet combustor, and predict the development status of the shock wave.
- The fuel injection location makes a large difference to the combustion flow field of the scramjet combustor with multiple cavity flameholders. The flow field for the case with hydrogen injected from both T_2 and T_4 is the most complex, and in this situation the shock wave has been pushed forward into the isolator. This causes the boundary layer to separate, generates a large recirculation zone and reduces the entrance region of the inflow. If the fuel injection location moves slightly forward, the shock wave may be pushed out of the isolator, and into the inlet. This will do damage to the inlet start.
- The fuel injection location changes the generation process of the vortices in the cavity flameholders to some extent. When the hydrogen is injected from T_2, there is no vortex formation in both the upper and lower cavity of the first flameholder. When the hydrogen is injected from T_4, small eddies are generated in the first upper and lower cavity flameholders. Further, if the hydrogen is injected from both T_2 and T_4, the eddies in the first upper and lower cavity flameholders become larger, and this is due to the spread of the shock wave pushed by the higher static pressure because of the more intense combustion.
- The fuel injection varies the dimension of the eddy generated in the nearby cavity flameholder. Due to the fuel injection, the eddy generated in the nearby cavity flameholder becomes larger, over the cavity and deflects into the core flow. This makes a larger recirculation zone than the case without fuel injection.
- The cavity is a good choice to stabilize the flame in the hypersonic flow, and it generates a recirculation zone in the scramjet combustor. Further, if its geometry can be designed properly, it can act as an ignitor for the fuel combustion, but the material of the cavity flameholder should be considered for operating at those high temperatures.

7. References

Alejandro, M. B., Joseph, Z., & Viswanath, R. K. (2010). Flame stabilization in small cavities. *AIAA journal, 48*(1), 224-235.

Aso, S., Inoue, K., Yamaguchi, K., & Tani, Y. (2009). A study on supersonic mixing by circular nozzle with various injection angles for air breathing engine. *Acta Astronautica, 65*, 687-695.

Chadwick, C. R., James, F. D., Kuang-Yu, H., Jeffrey, M. D., Mark, R. G., & Campbell, D. C. (2005). Stability limits of cavity-stabilized flames in supersonic flow. *Proceedings of the Combustion Institute, 30*, 2825-2833.

Chadwick, C. R., Sulabh, K. D., & James, F. D. (2007). Visualization of flameholding mechanisms in a supersonic combustor using PLIF. *Proceedings of the Combustion Institute, 31*, 2505-2512.

Daniel, J. M., & James, F. D. (2009). Combustion characteristics of a dual-mode scramjet combustor with cavity flameholder. *Proceedings of the Combustion Institute, 32*, 2397-2404.

FLUENT, I. (2006). *FLUENT 6.3 User's Guide*. Lebanon, NH: Fluent Inc.

Gruber, M. R., Baurle, R. A., Mathur, T., & Hsu, K. Y. (2001). Fundamental studies of cavity-based flameholder concepts for supersonic combustors. *Journal of Propulsion and Power, 17*(1), 146-153.

Gu, H.-b., Chen, L.-h., & Chang, X.-y. (2009). Experimental investigation on the cavity-based scramjet model. *Chinese Science Bulletin, 54*(16), 2794-2799.

Huang, W., Li, X.-s., Wu, X.-y., & Wang, Z.-g. (2009). Configuration effect analysis of scramjet combustor based on the integral balanceable method. *Journal of Astronautics, 30*(1), 282-286.

Huang, W., Qin, H., Luo, S.-b., & Wang, Z.-g. (2010). Research status of key techniques for shock-induced combustion ramjet (shcramjet) engine. *SCIENCE CHINA Technological Sciences, 53*(1), 220-226.

Huang, W., & Wang, Z.-g. (2009). Numerical study of attack angle characteristics for integrated hypersonic vehicle. *Applied Mathematics and Mechanics(English Edition), 30*(6), 779-786.

Hyungseok, S., Hui, J., Jaewoo, L., & Yunghwan, B. (2009). A study of the mixing characteristics for cavity sizes in scramjet engine combustor. *Journal of the Korean Society, 55*(5), 2180-2186.

Jeong, E. J., O'Byrne, S., Jeung, I. S., & Houwong, A. F. P. (2008). Investigation of supersonic combustion with angled injection in a cavity-based combustor. *Journal of Propulsion and Power, 24*(6), 1258-1268.

Kyung, M. K., Seung, W. B., & Cho, Y. H. (2004). Numerical study on supersonic combustion with cavity-based fuel injection. *International Journal of Heat and Mass Transfer, 47*, 271-286.

Launder, B. E., & Spalding, D. B. (1974). The numerical computation of turbulent flows. *Computer Methods in Applied Mechanics and Engineering, 3*(2), 269-289.

Nardo, A. D., Calchetti, G., Mongiello, C., Giammartini, S., & Rufoloni, M. (2009). *CFD modeling of an experimental scaled model of a trapped vortex combustor*. Paper presented at the ECM 2009 Fourth European combustion meeting, Vienna, Austria.

Neal, E. H., Michael, K. S., & Allan, P. (2005). *Flight data analysis of HyShot 2*. Paper presented at the 13th AIAA/CIRA International Space Planes and Hypersonic Systems and Technologies Conference, USA.

Paul, L. M., Vincent, L. R., Luat, T. N., & Jeryl, R. H. (2004). NASA hypersonic flight demonstrators-overview, status, and future plans. *Acta Astronautica, 55*, 619-630.

Situ, M., Wang, Z.-c., Niu, Y.-t., Wang, C., & Lu, H.-p. (1999). Investigation of supersonic combustion of hydrocarbon fuel-riched hot gas. *Journal of Propulsion Technology, 20*(6), 75-79.

Sun, M.-b., Geng, H., Liang, J.-h., & Wang, Z.-g. (2009). Mixing characteristics in a supersonic combustor with gaseous fuel injection upstream of a cavity flameholder. *Flow Turbulence Combust, 82*, 271-286.

Vikramaditya, N. S., & Kurian, J. (2009). Pressure oscillations from cavities with ramp. *AIAA journal, 47*(12), 2974-2984.

Wang, C., Situ, M., Ma, J.-h., & Yang, M.-l. (2000). Numerical simulation on supersonic combustion of fuel-rich hot gas. *Journal of Propulsion Technology, 21*(2), 60-63.

Weidner, E. H., & Drummond, J. P. A. (1981). *Parametric study of staged fuel injector configurations for scramjet applications*. Paper presented at the 17th AIAA/SAE/ASME Joint Propulsion Conference, United States.

Wu, X.-y., Li, X.-s., Ding, M., Liu, W.-d., & Wang, Z.-g. (2007). Experimental study on effects of fuel injection on scramjet combustor performance. *Chinese Journal of Aeronautics, 20*(6), 488-494.

Plasma flame sustained by microwave and burning hydrocarbon fuel: Its applications

1. Introduction

Thermal plasma torches have been developed for various industrial applications. Industries require them to be high power, contaminant-free, low-maintenance, low-cost, and large-volume. Principally, the plasma torch is a device to produce an arc plasma column between two electrodes. There are several kinds of plasma torches, including dc arc torch, induction torch, and high-frequency capacitive torch. The dc arc torch is operated by the dc electric field between two electrodes at a severe environment of high arc current in the range from several tens to thousands of amperes. Therefore, their electrodes are replaced often due to their limited lifetime, in particular an oxidative environment. Almost all radio frequency torches are inductively coupled discharges. Their typical thermal efficiencies (% of power effectively dissipated in the plasma forming gas) are in the range of 40-50% (Fauchais & Vardelle, 1997). These conventional torches also have a small volume of plasma, high operational cost and require many expensive additional systems for operation. Although the conventional plasma torches are used in many industrial applications, the wide acceptance of these processes is limited by economic, competitive, reliability, and other concerns. From the reason above-mentioned, they may not be useful in environmental applications.

In order to overcome problems related with the conventional plasma torch, an electrodeless microwave plasma torch at atmospheric pressure was developed (Hong et al., 2003; Kim et al., 2003). Microwave plasmas operated at the atmospheric pressure, especially waveguide-based plasma, have been subject of increased attention during the last decade (Margot, 2001; Moisan et al., 2001). Such an interest comes from their potential and actual use in various applications, including excitation sources for elemental analysis, lighting, and purification or remediation of gas effluents detrimental to the environment (Hartz et al., 1998; Woskov & Haddi, 1999). The microwave plasma torch can easily be made by modifying typical household microwave ovens as inexpensive method (Kim et al., 2003). Therefore, the microwave plasma torch is simple, compact and economical. Furthermore, in previous works, the microwave plasma torch has been investigated in various applications such as the abatement of CF_4, NF_3 and SF_6, the elimination of chemical and biological warfare agents, and synthesis of carbon nanotube, titanium dioxide, titanium nitride, and zinc oxide (Hong et al., 2004; Kim et al., 2007). Although the microwave plasma torch in air discharge provides high plasma density and high gas temperature in inexpensive ways (Green, 2001), the plasma volume and temperature of the microwave plasma torch decrease drastically outside the discharge tube, thereby limiting its capability of bulk treatment of waste. For example, the gas flow rates in the treatment experiments of CF_4 and phosgene were only tens of liters per minute (lpm), although high destruction and removal efficiency more than 96% had been accomplished (Hong et al., 2003; Hong et al., 2005). In this context, plasma flames made of a microwave plasma and a fuel burning flame have been developed for producing an enlarged high-temperature plasma flame by injecting a hydrocarbon fuel into the microwave plasma torch. The complete combustion of a hydrocarbon fuel in the microwave plasma torch causes the increase of plasma flame volume, high temperature zone, and residence time of target materials in high temperature zone.

There are many experimental investigations related to the fuel combustion using arc plasma torch and microwave. Since the experimental research in the 1970s (Weinberg et al., 1978), studies on plasma torch igniters have been extensively conducted. The advantages of a plasma torch in combustion are its role as a source of a radical pool and high temperature. Recently, Takita et al. showed that a considerable increase in burning velocity by addition of radicals occurred only in the case when the mixture temperature was high and the mixture included a large number of radicals (Takita, et al., 2001). Also, Masuya *et al.* investigated the ignition of H_2, H_2/N_2, H_2/air, and O_2/N_2 in high-temperature airflow by an arc torch (Masuya et al., 2002). The effects of microwave radiation on combustion were experimentally investigated by many researchers. For instance, Ogawa et al. investigated the influence of microwaves on CH_4/air laminar flames in a cavity resonator excited at 2.45 GHz by measuring the burned-gas temperature, brightness, and electron temperature (Ogawa et al., 1998). Finally, it is concluded that the combustion enhancement by microwave is due to the microwave heating of the bulk gases in the flame zone, and thus yielding a greater flame temperature. Therefore, the microwave plasma burning system can provide a near perfect combustion of a hydrocarbon fuel gas with air or oxygen gas due to the high-temperature plasma and a large quantity of radicals in the microwave plasma. For instance, in destructing fluorinated compound gases (FCs) using a microwave plasma torch (Hong et al., 2003), the impact process of electrons dissociates or ionizes FC molecules, and other plasma constituents convert into benign or more treatable products by injecting into the center part of the microwave plasma torch flame sustained from any gas mixture. However, the microwave torch abated the contaminants only in 10-20 lpm nitrogen gas contaminated by CF_4, SF_6, and NF_3 gases. The reason for the abatement limitation of the microwave plasma torch is due to the small volume of the torch plasma and the short residence time of contaminants in the reactor. Actually, the volume of the plasma flame sustained by the microwave plasma and burning hydrocarbon fuel is much larger than that of the microwave plasma torch. Therefore, the large-volume, high temperature plasma flame is expected to overcome the abatement limitation at high flow rate.

This chapter contributes the combustion enhancement of a hydrocarbon fuel augmented by the microwave plasma torch. Also, this chapter describes the configuration of plasma flame generator made of a microwave plasma and a fuel-burning flame, investigates physical properties containing its temperature and optical emission, and lastly shows the experimental results in abating fluorinated compound gases, decontaminating chemical and biological warfare agents, and eliminating odorous chemical agents.

2. Microwave plasma torch

Figure 1 displays a schematic configuration of the atmospheric pressure microwave plasma system. The design and operation of the atmospheric microwave plasma torch are briefly summarized here for completeness, although they have been reported in detail in previous articles (Hong et al., 2003; Kim et al., 2003). It is comprised of the 2.45 GHz microwave generator, WR-340 waveguide components, including an isolator, a directional coupler, a 3-stub tuner, and a microwave plasma torch as a field applicator. The WR-340 waveguide (86 mm × 43 mm) used in the microwave plasma torch was tapered to a reduced cross-section of 86 mm × 20 mm. In the unloaded waveguide without plasma, the reduction of the waveguide height provides an increase in the electric field strength even with the same microwave power. Due to the Poynting theorem, we get about a $\sqrt{2}$ times higher electric field strength in the plasma torch region. The discharge tube was inserted vertically, perpendicular to the wide wall of the waveguide. The discharge tube was located at a 1/4 wavelength away from the shorted end of the waveguide. The highest-intensity electric field occurs at this location confirmed by a high frequency structure simulator code (Kim et al., 2003). The microwave radiation generated from magnetron passes through the 3-stub tuner, is guided through the tapered waveguide, and enters the discharge tube made of fused silica. The center axis of the discharge tube with an outer diameter of 30 mm is located one-quarter wavelength from the shorted end of the waveguide. The tube penetrates through the wide waveguide walls, as shown in Fig. 1. The igniters not shown in Fig. 1 with their terminal electrodes inside the discharge tube initiate the plasma. The plasma generated inside the discharge tube is stabilized by a swirl gas, which enters the discharge tube sideways, creating a vortex flow (Gutsol et al., 1998) in the tube. The impedance of the plasma and the field applicator to the characteristic impedance of the WR-340 waveguide were matched by tuning the 3-stub tuner. The reflected power adjusted by the 3-stub tuner is almost zero. Even with all the tuning stubs completely withdrawn, reflected power is typically less than 10% of the forward power (Hong et al., 2003). The forward and reflected powers are monitored by the directional coupler.

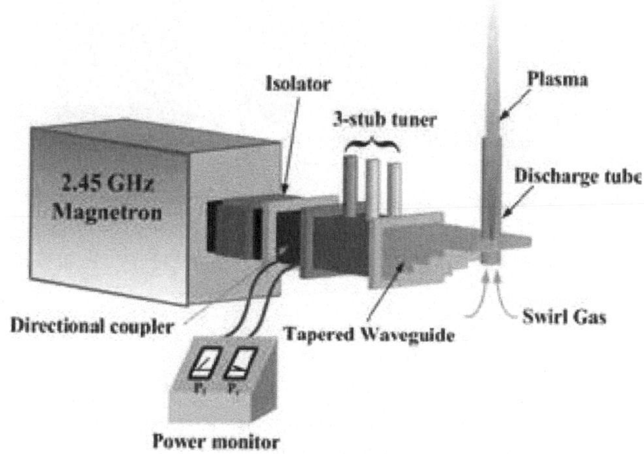

Fig. 1. Schematic presentation of the 2.45 GHz microwave system components and the microwave plasma torch.

A number of experimental results in atmospheric microwave plasmas have been reported (Green et al., 2001; Moon et al., 2002). For example, Green et al. measured the torch flame temperature inside a discharge tube by making use of the Fe I emission lines in the 370-377 nm range (Green et al., 2001). The microwave plasma torch is similar or almost same with ours. The temperature profiles are almost flat out to the largest measurable plasma radius of 10 mm with a maximum value of 6550 ± 350 K on axis at an air of 28 lpm and 1.4 kW power. The flame temperature at the 10 mm radius is still 80% of its value on axis. Generally, the plasma column length in the microwave plasma torch depends on the amount of swirl gas. In previous literature (Kim et al., 2003), the plasma column length was about 20-30 cm for 1 kW microwave power, for a discharge tube with 27 mm inner diameter and for 20 lpm of air swirl gas. The plasma column length was reduced to 10 cm when the swirl gas increased from 20 to 80 lpm. The microwave plasma torch can be operated in various gases. In this context, Figs. 2(a)-(f) reveal the microwave discharge plasmas in 10 lpm argon, 1 lpm argon, 10 lpm helium, 10 lpm nitrogen, 10 lpm air, and mixture of 5 lpm nitrogen and 10 lpm helium, respectively.

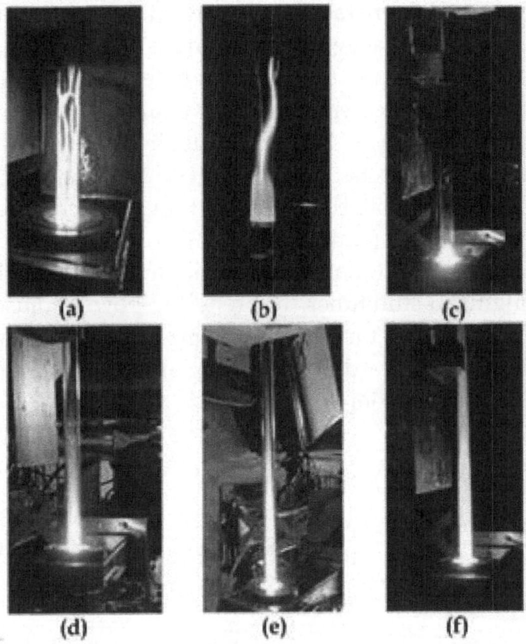

Fig. 2. Various atmospheric pressure microwave plasmas at (a) 10 lpm argon, (b) 1 lpm argon, (c) 10 lpm helium, (d) 10 lpm nitrogen, (e) 10 lpm air, and (f) mixture of 5 lpm nitrogen and 10 lpm helium. Then, the applied microwave power is approximately 1 kW.

3. Plasma flame generator

3.1 Arrangement of plasma flame generator

Figure 3(a) shows the schematic view for the plasma flame generator made of the microwave plasma and a fuel-burning flame. The main parts of experimental configuration for the plasma flame generator, as shown schematically in Fig. 3(a), consist of the microwave plasma torch, a fuel-injector, and a plasma flame exit. Air, oxygen or a mixture of air and oxygen can be used as a swirl gas. Therefore, the swirl gas provides atomic oxygen and molecular singlet oxygen of high-density (Lai et al., 2005) for near perfect combustion of hydrocarbon fuels, which is sprayed from the fuel injector in Fig. 3(a). The fuel injector, which is a typical fuel nozzle used in home-boilers, is equipped with the stainless steel tube and provides fuel for plasma. The injector is installed contiguously to the upper side of the waveguide, as shown in Fig. 3(a). The inner diameter of stainless steel tube has the same inner size as the discharge tube and is installed on the tapered waveguide to sustain a steady vortex flow of the swirl gas. The hydrocarbon fuel injected into plasma mixes with the swirl gas (air or oxygen) and extends the plasma flame to the open air, evaporating instantaneously and breaking down the molecular structure by energetic electrons and high temperature. The temperatures at different positions of the plasma flame were measured by a thermocouple of *R* type. The marks L_0 ~ L_{15} in Fig. 3(a) indicate the temperature-measurement points. For example, the mark L_6 represents a measurement point 6 cm away from the plasma flame exit. The mark $L_{-1.5}$ is the position corresponding to a measurement point 1.5 cm away the direction of the wide-wall waveguide. Fig. 3(b) is the cross-sectional view of swirl generator marked with dotted line in Fig. 3(a). The swirl gases are injected towards tangential direction through four tangential holes. The holes for the swirl gases are inclined towards axial direction by 30°.

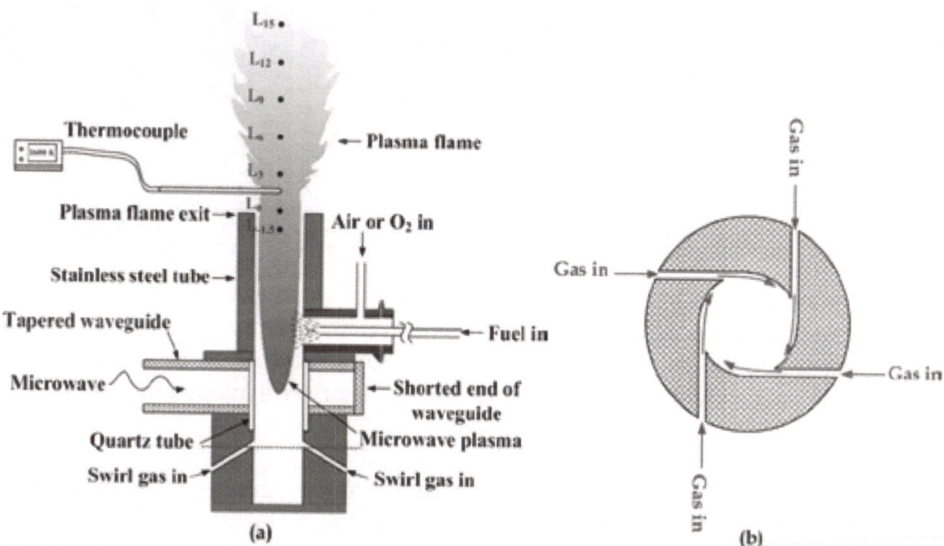

Fig. 3. (a) Schematic view shows a plasma flame generator with the microwave plasma torch. The fuel injector was installed adjacent to the upper side of the waveguide. (b) Cross-sectional view displays a swirl gas generator with four tangential holes for the dotted line in (a)

3.2 Temperature profile of plasma flames

3.2.1 Plasma flame from kerosene

Generally, flames already contain a weakly ionized plasma with typical density greater than 10^{10} ions/cm^3. For example, the oxidation of methanol by atomic oxygen is 10 million times faster than that by oxygen molecules at the gas temperature of 1300 K (Uhm, 1999). If so, as mentioned earlier, because the microwave plasma torch has high plasma density of $10^{13}/cm^3$ in air discharge and high temperature of about 6500 K at the center axis (Green et al., 2001), we expect that the microwave plasma torch can accomplish near perfect combustion of fuel. In this regard, Figs. 4(a) and (b) show the microwave plasma-burner flames before and after the fuel injection at the applied microwave power of 1.5 kW, respectively. In Fig. 4(a) and (b), a mixture of 50 lpm air and 10 lpm oxygen as a swirl gas was injected into the microwave plasma torch, while 50 lpm air as a swirl gas and 10 lpm oxygen with a fuel through the fuel injector were injected in Fig. 4(c). Figure 4(a) is a picture of the plasma torch flame without fuel injection. The flame was not expanded to the exit of the stainless steel tube of 10 cm in length. However, as shown in Fig. 4(b) and (c), the burner flame shot out through the exit of the stainless steel tube when 0.025 lpm kerosene was injected as a fuel into the microwave plasma torch. The burner flame diameter and length from the flame exit were about 8 cm and 40 cm, respectively. As a matter of fact, the fuel injector in this work was installed just above the waveguide and was 2 cm away from the waveguide excitation region, as shown in Fig. 3(a). When 10 lpm oxygen gas was added to the microwave plasma torch, not shown in Fig. 4, it was observed that the plasma flame color changed from a yellowish white to a bluish white, implying the phenomenon of near perfect combustion.

Fig. 4. Microwave plasma-burner flames before (a) and after (b and c) a fuel injection at the applied microwave power of 1.5 kW. A mixture of 50 lpm air and 10 lpm oxygen as a swirl gas was injected into the microwave plasma torch in (a) and (b), while 50 lpm air as a swirl gas and 10 lpm oxygen with a fuel through the fuel injector were injected in (c) (Hong et al., 2006).

The temperature of the microwave plasma torch flame at the center of the flame exit (mark L_0 in Fig. 3) measured by a thermocouple was only 550 K, when 60 lpm air as a swirl gas was injected. However, the temperature of the burner flame with the addition of 0.025 lpm kerosene drastically increased to about 1380 K. Moreover, the temperature of the burner flame with the addition of 0.025 lpm kerosene and 10 lpm oxygen gas drastically increased to about 1700 K. Temperature distributions along the radial and axial directions at different kerosene flow rates were measured with the addition of oxygen. Figure 5(a) shows the radial temperature profiles at marks L_0, L_3, and L_6 when a mixture of 40 lpm air and 20 lpm oxygen as a swirl gas and 0.025 lpm kerosene were injected. The length of the flame was about 30 cm. Each line marked by the rectangles, circles, and triangles indicates the radial temperature distribution at 0, 3, and 6 cm from the stainless steel tube. In comparison with the temperature distributions at L_3 and L_6, the distribution at L_0 decreases more drastically owing to the small inner diameter of the exit. Generally speaking, the volume of the burner flame decreases as oxygen flow rate increases, while the flame temperature at the center axis increases. As shown in Fig. 5(a), temperatures at L_0, L_3, and L_6 distribute from 1750 to 1850 K, revealing approximately uniform temperature-distribution in the axial direction. Figure 5(b) presents the axial temperature profiles at different kerosene flow rates, when a mixture of 50 lpm air and 10 lpm oxygen as a swirl gas was injected. Each line marked by the rectangles, circles, and triangles indicates the axial temperature profiles at kerosene flow rates of 0.031, 0.025, 0.019 lpm, starting from the stainless steel tube. The fuel injectors of 0.031 and 0.025 lpm spray the fuel in a shape of a conic shell with 80° spraying angle. The injector of 0.019 lpm sprays the fuel in the shape of a solid cone with 60° spraying angle. As shown in Fig. 5(b), it is observed that the trends of the temperature profiles at axial positions 0 and 3 cm depend on the type and angle of fuel injection. The difference of the mixture of swirl gas and fuel in the stainless steel tube causes different trend in temperature profiles. Generally speaking, flame temperature and volume increase as the kerosene flow rate increases.

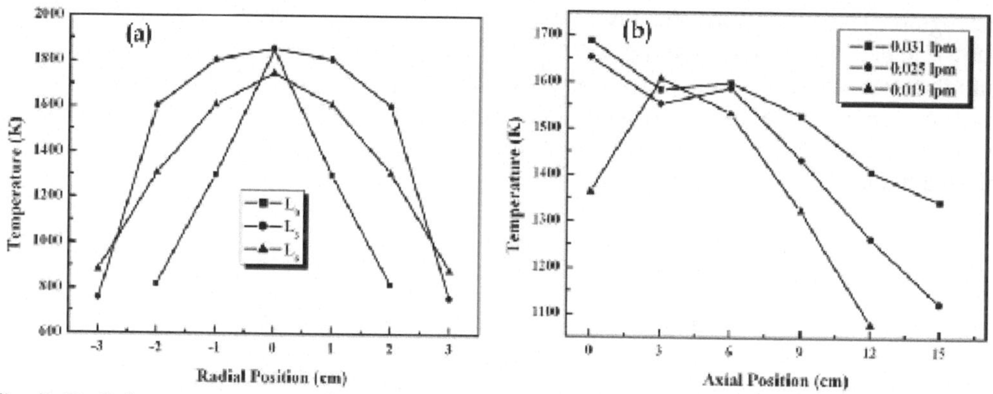

Fig. 5. Radial temperature profile (a) when a mixture of 40 lpm air and 20 lpm oxygen as a swirl gas and 0.025 lpm kerosene were injected into the microwave plasma torch and axial temperature profile (b) at kerosene flow rates of 0.031, 0.025, and 0.019 lpm when a mixture of 40 lpm air and 20 lpm oxygen as a swirl gas was injected (Hong et al., 2006).

3.2.2 Plasma flame from diesel

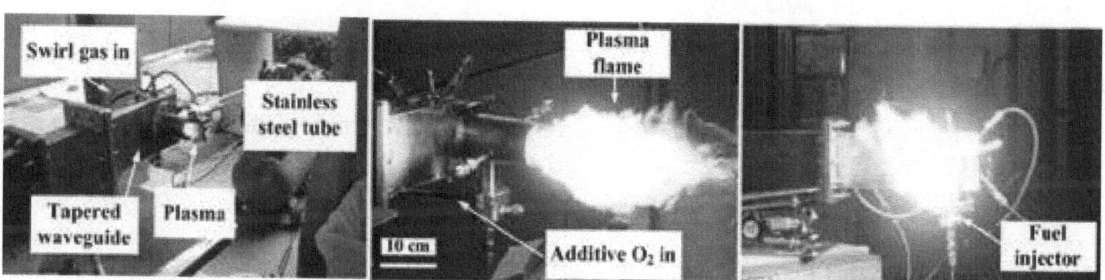

(a) (b) (c)

Fig. 6. Diesel microwave plasma-burner flames (a) before and (b and c) after a fuel injection at the applied microwave power of 1.2 kW. 50 lpm air as a swirl gas and 10 lpm oxygen with 0.019 lpm diesel through the fuel injector were injected, where (c) is a front view of picture (b) (Hong & Uhm, 2006).

In Fig. 6, the plasma flames before and after the injection of diesel fuel were compared. Similar to the kerosene microwave plasma burner reported in the previous work (Hong, et al., 2006), the diesel microwave plasma flame also shows that the volume is more than 50 times that of the torch plasma, burning diesel fuel instantaneously. Figure 6(a) is a picture of the microwave plasma torch flame operated at 1.2 kW microwave power, 50 lpm air as a swirl gas, and 10 lpm oxygen through the fuel injector without diesel. Figure 6(b) is a picture of the plasma flame generated by 50 lpm air as a swirl gas and 10 lpm oxygen with 0.019 lpm diesel injection through the fuel injector. Figure 6(c) is a front view of Fig. 3(b). While the torch flame in Fig. 6(a) is weakly expanded to the exit of the stainless steel tube of about 10 cm in length, the burner flames shot out through the exit of the stainless steel tube when 0.019 lpm diesel was injected as a fuel into the microwave plasma torch, as shown in Figs. 6(b) and (c). The plasma flame diameter and length from the flame exit were about 8 cm and 30 cm, respectively. The liquid diesel can evaporate instantaneously, breaking down the molecular structure by the microwave plasma column with high gas temperature, and burn immediately with oxygen. The fuel injector in this work was installed just above the waveguide and was approximately 2 cm away from the waveguide excitation region, as shown in Fig. 3(a). When 20 lpm oxygen gas with 0.019 lpm diesel was additionally injected into the microwave plasma torch, not shown in Fig. 6, it was observed that the plasma flame color changed from a yellowish white to a bluish white, indicating a high temperature at oxygen injection. In this context, we measured the gas temperature of the diesel microwave plasma flame in terms of oxygen content.

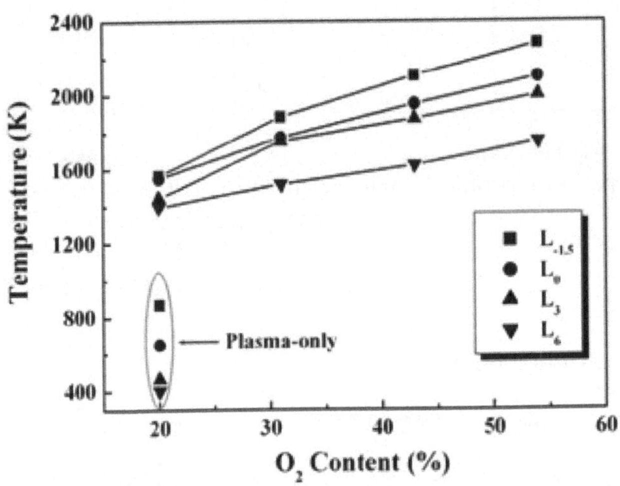

Fig. 7. Axial temperature profiles of the plasma flame in terms of O_2 content (%) at 0.019 lpm diesel. Total flow rate of mixture of air and O_2 was 70 lpm and the applied microwave power was 1.2 kW. The mixture was injected as a swirl gas (Hong & Uhm, 2006).

Figure 7 exhibits the gas temperature profiles of the diesel microwave plasma flames measured axially by a thermocouple in terms of oxygen content (%) at $L_{-0.5}$, L_0, L_3, and L_6 in Fig. 3(a). The total flow rate of the mixture composed of air and oxygen was fixed at 70 lpm and the mixture was injected into the microwave plasma torch as a swirl gas. For example, O_2 content of 54% represents the mixture composed of 40 lpm air and 30 lpm O_2. In the case of 20% O_2, while the gas temperatures of the microwave plasma torch flame at four measurement positions distribute in the range of 400-850 K, the gas temperatures of the plasma flame with 0.019 lpm diesel are in the range of 1400-1600 K. Moreover, when O_2 content at $L_{-1.5}$ position increases from 20% to 54%, the gas temperature significantly increases from 1580 K to 2210 K. In general, the diesel plasma flame shows similar properties with the kerosene plasma flame such as Fig. 4.

3.2.3 Plasma flame from methane

Figure 8(a) shows the microwave plasma torch flame at applied power of 1.2 kW when 60 lpm air as a swirl gas was injected. As shown in Fig. 8(b), however, once CH_4 (10 lpm in this test) as a hydrocarbon fuel is injected into the microwave plasma torch through the fuel injector in Fig. 3(a), volume of the CH_4 plasma burner flame significantly increases, emitting milky white lights. Fig. 8(c) is the CH_4 flame picture without the microwave plasma. Unlike Fig. 8(b), Fig. 8(c) shows the flame with light blue color. The flame temperature of Fig. 8(b) and (c) at the point of L_0 in Fig. 3(a) is ~1590 K and 1250 K, and the visual flame length of Fig. 2(b) and (c) is 18 cm and 11 cm, respectively. From another experiment result of CH_4 microwave plasma burner not shown in this chapter, it is identified that CH_4 injection rate up to 30 lpm (near 0.36 g/s) is reasonable at 1.2 kW microwave plasma torch and stoichiometric fuel/air mixture. For practical application such as power plant, a microwave plasma torch with 915 MHz or 896 MHz microwave system or multiple microwave plasma torches may be suitable for obtaining high power.

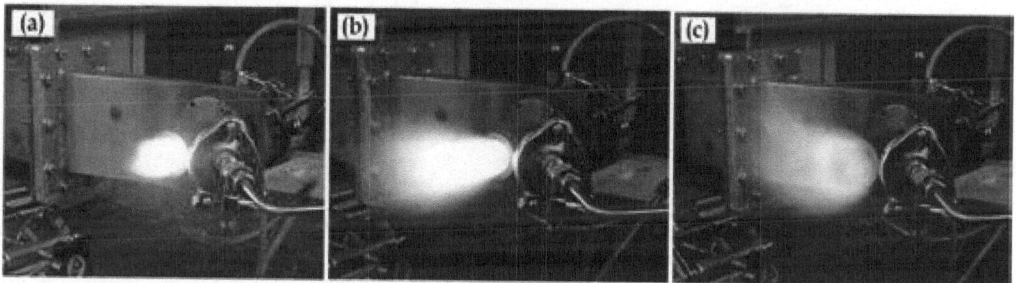

Fig. 8. Plasma-burner with (a) microwave plasma only, (b) microwave plasma + CH_4 injected at 10 lpm and (c) CH_4 flame only (Bang et al., 2006).

In this regard, we measured temperatures of the CH_4 microwave plasma burner flame at different positions presented in Fig. 3(a). The rectangular marks in Fig. 9(a) represent the axial temperature profile of the plasma burner flame at marks L_0-L_{15} when 60 lpm air as a swirl gas, and a mixture of 10 lpm CH_4 and 40 lpm air through the fuel injector were injected the microwave plasma torch. Therefore, the value of mixture ratio of air/CH_4 is 10 : 1. The temperature of the microwave plasma torch flame at the center of the flame exit (mark L_0 in Fig. 1) was only 600 K without CH_4. As shown in Fig. 9 (a), the temperature of the plasma burner flame increased to about 1890 K when 10 lpm CH_4 was injected. And then the visual length of the burner flame was about 24 cm. It is well-known that the adiabatic flame temperature of a CH_4/air flame is about 2222 K. 1890 K in this test is flame temperature measured at the point away 5 cm from the fuel injector in Fig. 1. Therefore, the burner flame temperature near a region of fuel injection may be as high as that of CH_4/air flame in adiabatic condition. The circular marks in Fig. 9(b) indicate the axial temperature profile of the burner flame when 40 lpm air as a swirl gas, and a mixture of 10 lpm CH_4 and 60 lpm air was injected through the fuel injector. With 10 lpm CH_4, temperature of the burner flame increased from 600 K to 1680 K. And then the visual length of the burner flame in Fig. 9(b) was approximately 30 cm. Compared Fig. 9(a) with (b), the temperature profile in Fig. 9(a) falls rapidly at axial position of 6 cm, whereas the temperatures in Fig. 9(b) reduce gently along with axial direction. In this context, Fig. 9 implies that the temperature and length of the burner flame can be controlled by injection way or mixing rate of air and fuel. In general, it is recognized that the use of a thermocouple for measurement of flame temperatures may encounter some problems. Also, flames already contain a weakly ionized plasma with typical density greater than 10^{10} ions/cm^3 (Uhm, 1999). However, the thermocouple used in this test is perfectly covered with alumina (Al_2O_3). So plasma impacts in temperature measurements may be neglected.

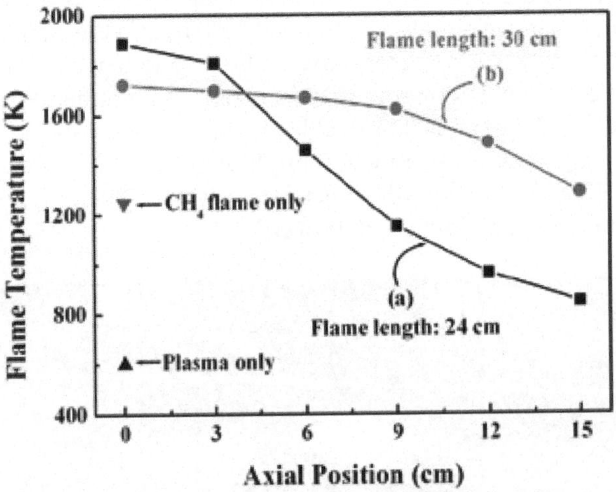

Fig. 9. Axial temperature profiles of the CH_4 augmented microwave plasma burner measured at positions L_0-L_{15} as denoted in Fig. 3(a). (a) 60 lpm swirl air + mixture of 10 lpm CH_4 and 40 lpm air. (b) 40 lpm swirl air + mixture of 10 lpm CH_4 and 60 lpm air (Bang, et al., 2006)

Figure 10 shows the radial temperature profile of the CH_4 augmented microwave plasma burner at marks L_0 in Fig. 1. The rectangular marks in Fig. 10(a) indicate the radial temperature profile of the burner flame when 60 lpm air as a swirl gas, and a mixture of 10 lpm CH_4 and 40 lpm air through the fuel injector were injected the microwave plasma torch. As shown in Fig. 10(a), the temperature of the burner flame decreased to about 1180 K, rapidly. The circular marks in Fig. 10 (b) indicate the radial temperature profile of the burner flame when 40 lpm air as a swirl gas, and a mixture of 10 lpm CH_4 and 60 lpm air through the fuel injector were injected the torch. The temperature of the burner flame decreased to about 1370 K, slowly. Figs. 9 and 10 showed the axial and radial temperature profiles in CH_4 augmented microwave plasma burner flame, respectively. The performance of CH_4 microwave plasma burner significantly depends on the physical and chemical properties of microwave plasma torch. The theoretical description of the microwave plasma torch is beyond the scope of the present study. However, one can refer the previous articles (Kim et al., 2003; Margot, 2001; Moon et al., 2002) describing the atmospheric pressure microwave plasma torch.

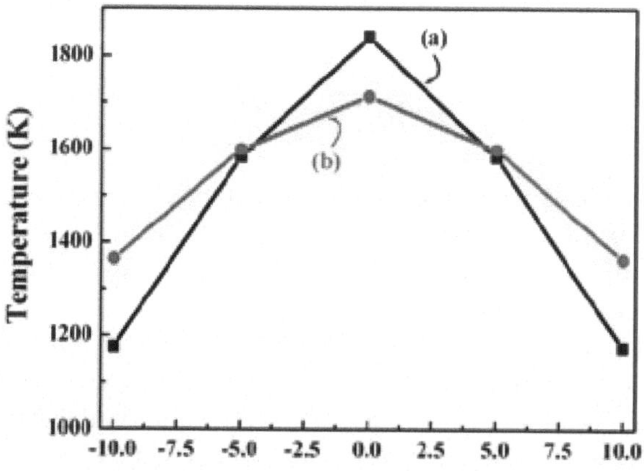

Fig. 10. Radial temperature profiles of the CH_4 augmented microwave plasma burner measured at position L_0 in Fig. 3(a). (a) 60 lpm swirl air + mixture of 10 lpm CH_4 and 40 lpm air. (b) 40 lpm swirl air + mixture of 10 lpm CH_4 and 60 lpm air (Bang et al., 2006).

The temperature profiles in Figs. 9 and 10 were changed with addition of the same CH_4 quantity at different swirl air flow rates and air flow rates through the injector. When a swirl air flow rate is more than that through the injector, the vortex flows inside the stainless steel tube in Fig. 3(a) can survive against air flow through the injector, increasing the combustion time of CH_4, confining the CH_4 flames axially, and thus increasing the temperature at L_0 point. On the other hand, when a swirl air flow rate is less than that through the injector, an air flow through the injector can suppress vortex flow by swirl air injection and increase axial flow velocity and flame length. Therefore, even though the same CH_4 flow rate is injected, each temperature profile in Figs. 9 and 10 can be changed due to different gas injection methods.

3.3 Simple description of atomic oxygen density in plasma flames

Principally, a discharge plasma and a high temperature environment generate many chemically active radicals. For example, oxygen atoms can be generated by the plasma and thermal dissociations of oxygen molecules, i.e., $O_2 \rightarrow O + O$. Plasma dissociation includes dissociative recombination of molecular oxygen ions, electron impact dissociation of oxygen molecules, and dissociative attachment of oxygen negative ions (Uhm, 1999). Thermal dissociation of oxygen molecules has reaction constant (Hong & Uhm, 2006) $k = 2.7 \times 10^{11} (T_R/T_P)^2 \exp(-59429/T_P)$ s^{-1}, where T_R and T_P represent room and plasma flame temperature, respectively, in units of Kelvin. The oxygen atoms recombine with recombination coefficient $\alpha = 2.3 \times 10^{-14} (T_R/T_P)^2$ cm^{3}s^{-1}, forming oxygen molecules (Hong & Uhm, 2006). The oxygen atom may also form ozone with oxygen molecule but ozone dissociates rapidly due to high plasma temperature. Therefore, ozone from the microwave plasma torch is not produced. The rate equation of oxygen atom density n_O is given by

$$\frac{dn_O}{dt} = 2kn_{O2} - \alpha n_O^2, \tag{1}$$

with the solution

$$n_O(t) = \sqrt{\frac{2kn_{O2}}{\alpha}} \tanh(\frac{t}{\eta}), \tag{2}$$

where $\eta = 1/\sqrt{2k\alpha n_{O2}}$, n_{O2} is the molecular oxygen density and the factor 2 in front of k represents 2 atoms from one molecular dissociation. For instance, the plasma flame temperatures from the fuel injection point to L_9 point in Fig. 3(a) range in T_P = 4000-1500 K (Hong, et al., 2006). The oxygen atom formation by the plasma may be significant, but it is difficult to find the plasma effects in the plasma flame. Neglecting the plasma effects and assuming T_P = 2000 K as an average value, we find η = 1.5 s and $n_O(t = \infty) = 1.3 \times 10^{15}/cm^3$ for $n_{O2} = 6 \times 10^{17}/cm^3$. Assuming the residence time t = 0.06 s in the stainless steel tube in Fig. 3(a) as a typical value, the oxygen atom density is calculated to be $n_O = 5.7 \times 10^{13}/cm^3$ from Eq. (2), which effectively combusts hydrocarbon fuels. The oxygen atom density increases drastically with the high plasma-flame temperature originated from the plasma torch.

3.4 Influence of microwave plasma in plasma flame

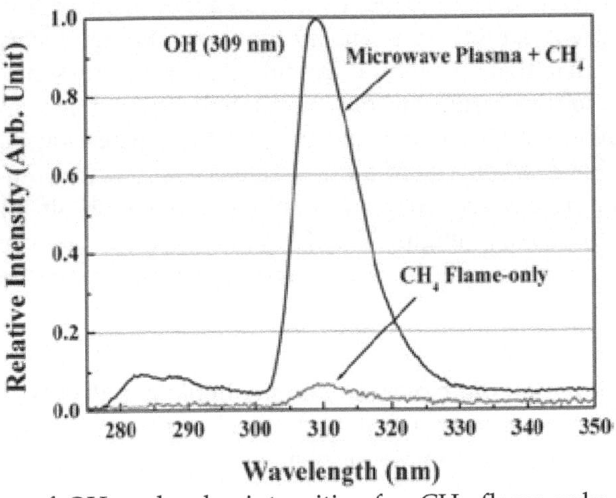

Fig. 11. Comparison of OH molecular intensities for CH_4 flame-only (gray) and plasma flame with CH_4 (black) (Hong & Uhm, 2006).

As above-mentioned, we compared CH_4 flame-only and the plasma flame with CH_4 by providing visual changes of flame color, flame lengths, and flame temperature by a thermocouple. These may be under the influence of the microwave plasma on the combustion flame. Here, we present the influence on the microwave plasma by observing hydroxyl (OH) molecules in an emission spectrum as a supporting data. OH molecular spectrum is necessarily observed in many kinds of flames and hot gases containing oxygen and hydrogen. In this sense, we compared the emission intensities of OH radical for CH_4 flame-only and CH_4 flame combined with the microwave plasma at L_0 point in Fig. 3(a). Experimental parameters correspond to the curve in Fig. 10(a). Figure 11 compares the relative emission intensity of OH for CH_4 flame-only (gray line) and CH_4 flame combined with the microwave plasma (black line). Two OH emission intensities were normalized by the intensity of CH_4 flame combined with the microwave plasma. In general, perfect combustion of hydrocarbon fuels produces gaseous water and carbon dioxide as final resultant products. OH species are essential intermediates during the process of water production. OH emission intensity strongly depends on the density of atomic oxygen. According to the simple description for atomic oxygen mentioned earlier, the atomic oxygen density was estimated to be $n_O = 5.7 \times 10^{13}/cm^3$ from Eq. (2), which effectively combusts hydrocarbon fuels. Ultimately, oxygen atoms produced by the microwave plasma in air are very helpful for CH_4 combustion, thereby exhibiting strong OH intensity as shown in Fig. 11. On the contrary, OH intensity of CH_4 flame-only is very weak in the comparison with the plasma flame of CH_4 combustion. This difference of OH intensity reflects one of rotational temperature (T_{rot}) of OH molecules to be equal to the gas temperature of the flames (de Izarra, 2000).

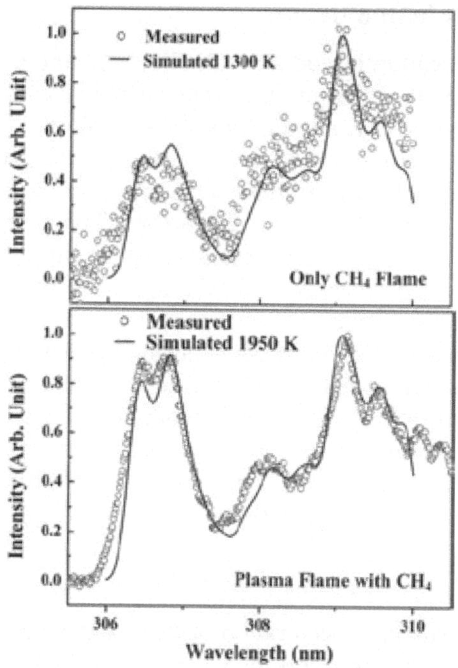

Fig. 12. Comparison of the measured data with the simulated OH spectrum for CH_4 flame-only and plasma flame with CH_4 yielding the rotational temperatures (T_{rot}) of 1300 K and 1950 K at L_0 point in Fig. 3(a) (Hong & Uhm, 2006).

In this context, Fig. 12 shows the unresolved OH molecular lines ($A\ ^2\Sigma^+,\ _{v=0} \rightarrow X\ ^2\Pi,\ _{v'=0}$) observed in the wavelength of 306-310 nm with a spectral resolution of 0.35 nm. T_{rot} was determined in this study by comparing the simulated OH spectrum with the measured spectrum obtained at a relatively low spectral resolution. The method for obtaining simulated OH spectrum at a given temperature was provided in previous articles (de Izarra, 2000). The gas temperatures for CH_4 flame-only and the plasma flame with CH_4 in Fig. 12

are determined to be approximately 1300 K and 1950 K, respectively, showing the influence of the microwave plasma on CH_4 combustion.

A large, high-temperature plasma flame may be suitable for a bulk material treatment, in particular environmental application. As mentioned earlier, the microwave plasma torch in air discharge has small volume and its temperature decreases drastically in the axial direction. For example, CF_4, NF_3, and SF_6 abatement experiments conducted in our research group showed destruction efficiency more than 90% only in contaminant flow of 20 lpm (Hong et al, 2003). In order to overcome treatment limitation of the microwave plasma torch, a tool for an enlarged high-temperature plasma flames was designed. As shown in Fig. 9, the temperature difference between the CH_4 flame-only and the CH_4 microwave plasma burner flame is approximately 640 K. Therefore, the plasma constituents, such as atomic oxygen and molecular singlet oxygen, produced in the microwave plasma torch can be very helpful for hydrocarbon fuel combustion and may be useful in the thermal treatment processes.

4. Applications using plasma flames

4.1 Mass purification of contaminated air with chemical and biological warfare agents

The elimination experiment of any chemical warfare agent is almost impossible in an ordinary laboratory due to safety issues. Considering thus, the experimentalists customarily carry out a simulated experiment by making use of toluene gas. For this same reason, the biological warfare agents are not used in an ordinary laboratory (Hong, et al., 2004). The airborne biological warfare agents like microbes or bacteria are attached to organic or inorganic aerosols and are spread when aerosol particles float around. Consequently, the elimination of soot from a diesel engine as the simulated carrier aerosol of biological agents was carried out.

The reaction chamber for mass treatment of contaminated air was designed specially, providing the necessary residence time for the best decontamination effects. The detail explanation of the reaction chamber has been reported in previous article (Uhm et al., 2006). The toxic warfare agents contaminating air enters the inner compartment through slits from the outer compartment and are eliminated mainly by oxidation process exposed to the high-temperature plasma flame with abundant oxygen atoms in the inner compartment. The destruction model of the chemical and biological warfare agents can be expressed as (Hong et al., 2004)

$$\frac{X}{X_0} = \exp(-\frac{E}{\beta}), \tag{3}$$

where X represents the leftover concentration of the warfare agents after the plasma flame treatment and X_0 is the initial concentration before the treatment, E denotes the energy density (in units of joules per liter) deposited on the contaminated air by the plasma flame during the treatment and β represents the energy density required for bringing down the concentration to 1/e of its initial concentration; i.e. the energy density needed for 63 % decomposition. Designating R as the flow rate of the contaminated air, we note RE = constant for specified physical parameters of the decontamination system. In other words, the energy density E deposited by the plasma flame during the treatment is inversely proportional to the airflow rate R. Assuming that X_1 and X_2 correspond to the leftover concentrations for the flow rates R_1 and R_2, respectively, we find the relationship

$$\frac{R_1}{R_2} = \frac{\ln(X_0 / X_2)}{\ln(X_0 / X_1)}, \tag{4}$$

which relates the leftover concentration X to the airflow rate R. We can find the leftover concentration X_2 in terms of R_2 if we know the concentration X_1 in terms of R_1.

As an example, we used toluene (C_7H_8) as a simulated chemical warfare agent, and kerosene and methane were used as the hydrocarbon fuels. A liquid fuel is better than a gaseous fuel when pertaining to the instance of compactness and mobility. A suction fan supplied the contaminated air with evaporated toluene to the reaction chamber. The airflow rate was R = 5,000 lpm. 40 lpm of the compressed air was supplied to the swirl gas. The injection rates of the kerosene in this experiment were 1.15 kg/hr (≈ 0.3 gal/hr), 1.46 kg/hr and 1.87 kg/hr. The 0.3 gal/hr nozzle is the smallest fuel nozzle ever found. The methane flow rates were 5 lpm, 10 lpm, 15 lpm, 20 lpm and 30 lpm. The energy contained in kerosene and in methane are 10^7cal/kg and 9.52×10^6 cal/m^3, respectively. The fuel injection rate can be translated into watts. For example, a 1.15 kg/hr injection rate of kerosene is 13.3 kW and a 20 lpm injection rate of methane is also 13.3 kW. The energy density E in Eq. (3) can be calculated by making use of the fuel power and airflow rate. The microwave power was 1.4 kW and the initial toluene concentration was X_0 = 170 particulates per million (ppm). The kerosene injection rates 1.15 kg/hr, 1.46 kg/hr and 1.87 kg/hr with the microwave power of 1.4 kW correspond to the energy density E = 176.4 J/L, 219.4 J/L, and 276.3 J/L, respectively, for R = 5,000 lpm. The size of the reaction chamber used in the experiment was 22 cm diameter and 30 cm long. The compactness and lightweight of the decontamination system are the key issues for a quick and easy application in life-threatening situations. Therefore, the reaction chamber must be as small as possible for a specified airflow rate. The reaction chamber of 22 cm diameter and 30 cm length is good for the airflow rate of 5,000 lpm. The leftover concentration X of the toluene had been measured by making use of detector tubes from the GASTECH Company. The gas chromatography (GC) or the Fourier transform infrared (FTIR) can be used for more accurate data. In spite of this, those diagnostic tools may give completely wrong measurement values, because toluene is in liquid form at the room temperature of one atmospheric pressure. A sample leading to the diagnostic tools can easily be spoiled by toluene condensation. The measurement by detector tubes can be done at the flame exit of the reaction chamber without any delay or any interference. Therefore, the detector tube may reliably measure the leftover toluene, although the data may have a large error bar. Figure 13 shows the leftover toluene-concentration rate in terms of energy density for kerosene (closed square dots) and methane (open square dots) fuel injections. Each data point in Fig. 13 represents the average of 8 repeated measurements. The rectangular dots at E = 16.7 J/L represent toluene decomposition only by the microwave torch plasma with 1.4 kW. The typical error in the measurement as shown in the open square dot at E = 16.7 J/L of Fig. 13 is about 5 % associated with the detector tube. The error bars of most other data are smaller than the dot size in Fig. 13. The toluene curve in Fig. 13 for kerosene was obtained from Eq. (3) with the β-value that was least-squared fitted to the data points (closed square dots). The β-value of the toluene decomposition by the plasma flame is β = 84.76 J/L for kerosene, which is much less than β = 393 J/L by the pulse corona (Penetrante et al., 1997) and β = 173 J/L by the microwave plasma torch (Hong et al., 2004). The toluene curve in Fig. 13 for methane was obtained from Eq. (3) with the β-value that was least-squared fitted to the data points (open square dots). The β-value of the toluene decomposition by the plasma flame is β = 62.74 J/L for methane. Clearly, the toluene decomposition by the high-temperature plasma flame is far more efficient than that by the pulse corona or by the microwave torch. Furthermore, the present decomposition system is very compact and light to be handy for various applications. The temperature of the reaction chamber wall and the exit gas is not hot due to a large amount of airflow. In fact, the outer wall of the reaction chamber only feels warm.

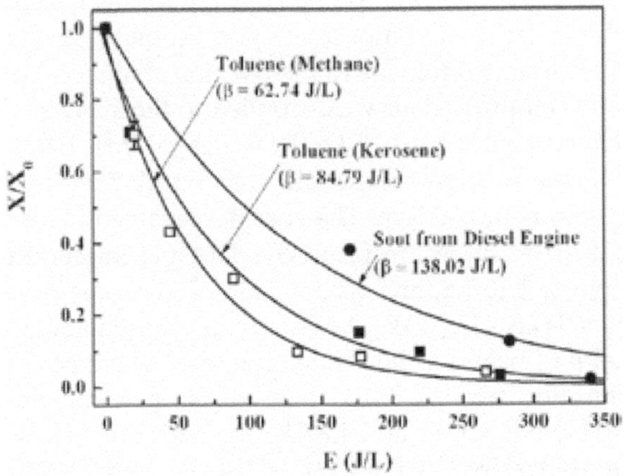

Fig. 13. Leftover toluene and soot concentrations in terms of the energy density E. The closed and open square dots represent measurement data for kerosene and methane injections, respectively, and the closed circular dots are the soot concentration data for methane injection (Uhm et al., 2006).

An elimination experiment of the airborne biological warfare agents is very difficult because of the complexity of detecting the agents before and after the plasma flame treatment. Spores of the biological warfare agents are usually attached to aerosol particles. The elimination of aerosol particles may indirectly show elimination of the airborne biological warfare agents. Elimination of soot from the diesel engine, which can be seen as airborne aerosol particles, was observed in the experiment. The burning kerosene may generate its own soot, which may interfere with the observation of the diesel engine soot, hence the gaseous fuel of methane was used in the experiment. The methane injection rate was 15 lpm, 25 lpm and 30 lpm in the soot elimination. The discharge gas from a 10,000cc bus diesel engine at 800 rpm was used as the contaminated air with soot. The airflow rate at the engine exit was 8,000 lpm, which is estimated to be 3,500 lpm at the end of the tail pipe due to the cooling of the ambient air. The energy density therefore was calculated by the methane injection into the airflow of 3,500 lpm. White filters captured soot from the discharge gas. A smoke meter from BOSCH, which determines opacity, measured the captured soot-amount in the filter. The remaining soot (closed circular dots) in relative to the untreated case is plotted in Fig. 13 in terms of the energy density for methane injected into plasma. The soot was almost completely eliminated at E = 340 J/L corresponding to the 30 lpm methane injection. The β-value of the soot elimination was determined by the least-squared-fitted to the experimental data (closed circular dots) in Fig. 13 and is given by β = 138.02 J/L. The plasma flame is an effective mean to eliminate the soot from the diesel engine. This means that the plasma flame may effectively eliminate airborne aerosol particles. Most of the aerosols are made of hydrocarbon materials, which can easily be oxidized at a high-temperature plasma flame with the temperature higher than 1000 degrees Celsius. The biological agents consisting of bacteria and virus may not survive as they go through the high-temperature plasma flame. Therefore, the plasma flame may effectively eliminate the airborne biological warfare agents. A different experimental observation confirmed that the plasma flame of the kerosene or diesel injected into the torch plasma does not produce its own soot. In this context, the plasma flame can also be useful for the elimination of soot from diesel engines in trucks, in buses, in trains and in ships.

It is noted from Eq. (4) that the airflow rate can increase by restricting the decomposition rate. For example, the leftover concentration of toluene at the kerosene fuel rate of 1.87 kg/hr corresponding to E = 276.3 J/L in Fig. 2 is X_1/X_0 = 0.02 for R_1 = 5,000 lpm. Substituting these numbers into Eq. (4), we find that R_2 = 19,560 lpm for X_2/X_0 = 1/e. About 20,000 lpm of the contaminated air with toluene can be treated if the treatment is at a 63 percent elimination requirement.

As mentioned earlier, the compactness and lightweight of the decontamination system are critical issues for rapid mobility and quick installation in life threatening situations. The reaction chamber size used in the examples presented earlier is 22 cm diameter and 30 cm long, which limits the airflow rate. The linear dimension of the waveguide and discharge tube in the plasma torch system is proportional to the wavelength of microwaves. Therefore, the torch plasma volume is inversely proportional to the square of the microwave frequency. For example, the torch plasma volume increases 7 times by changing the microwave frequency from 2.45 GHz to 915 MHz with an additional power. The larger volume of the plasma flame in an increased reaction chamber with low-frequency microwaves and additional fuel means the more treatment of the airflow rate. The treatment volume can easily be enhanced by increasing the size of the plasma flame in an enlarged reaction chamber. Therefore, there will be no scientific problem to extend the treatment volume to 100,000 lpm, although the system size may increase accordingly.

4.2 Elimination of air contaminated with odorous chemical agents

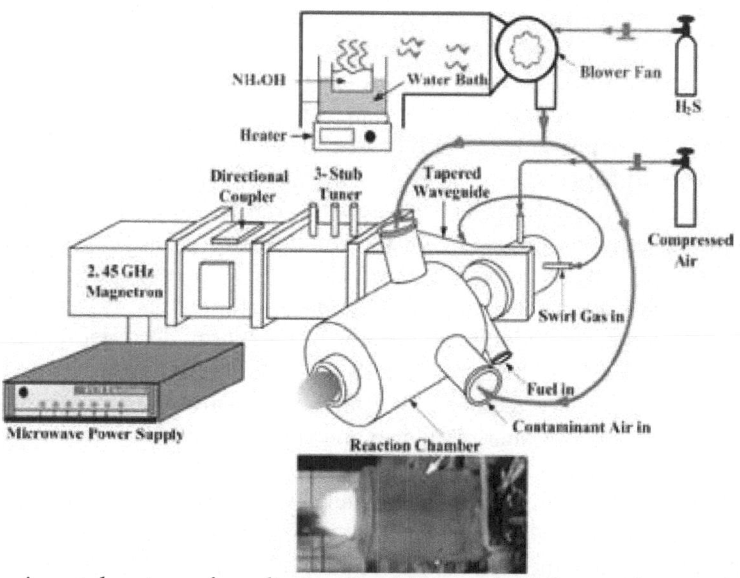

Fig. 14. Experimental set-up for eliminating NH_3 and H_2S as odor-causing chemical materials by making use of a microwave plasma burner. The inset is the picture of the kerosene plasma flame (Hong et al., 2007).

The inset in Fig. 14 shows the picture of the plasma flame produced from the microwave plasma burner at 1.4 kJ/s plasma power with no reflected power and 1.15 kg/hr kerosene. In Fig. 14, the blower fan connected to the reaction chamber by four stainless steel bellows sucks up air contaminated odorous gases, and transfers the contaminants into the reaction chamber. It can suck and blow airflow more than 5 000 litters per minute (lpm) at least. The reaction chamber consists of inner and outer compartment, providing a space between them. The contaminated air was injected into the reaction chamber via four injection ports in tangential direction installed on the outer compartment, thereby rotating in the space. In turn, the rotating airflows enter the inner compartment with tangential slits, which are also in tangential directions along the inner surface of the inner compartment wall, mixing with the plasma flame made of atmospheric microwave plasma and a fuel-burning flame. The dimensions of the reaction chamber used in the experiment were 22 cm diameter and 30 cm long. The plasma flame and the contaminated air in the inner compartment rotate in the same direction, providing the necessary residence time for the best elimination effects. These sequential processes then eliminate the odorous chemical agents in the passing air.
Aqua ammonia (NH_4OH) was used to obtain NH_3 gas in the simulated experiment for eliminating NH_3 and was maintained at 60 °C by a vaporization device in Fig. 14. On the other hand, in case of the simulated experiment of eliminating H_2S, gas-phase H_2S was directly injected into the blower fan and was mixed with air. The blower fan suck up air contaminated with NH_3 and H_2S gas, and transferred the contaminants into the reaction chamber. And then the total air-flow rate was approximately 5 000 lpm. 40 lpm of the compressed air as a swirl gas was injected into the microwave plasma torch. The injection rates of the kerosene were 1.15 kg/hr, 1.46 kg/hr and 1.87 kg/hr. The 1.15 kg/hr nozzle is the smallest fuel nozzle ever found. The methane flow rates were 5 lpm, 10 lpm, 15 lpm, 20 lpm and 30 lpm. The energy contained in kerosene and in methane are 10^7cal/kg and 9.52×10^6 cal/m^3, respectively. The fuel-flow rates injected can be translated into joules per second. The power of 1.15 kg/hr kerosene energy corresponds to 13.3 kJ/s and that of 20 lpm methane energy is also 13.3 kJ/s. The detailed simulated experiments for eliminating NH_3 and H_2S were carried out in terms of the input energy density of the microwave plasma burner. For instance, the kerosene injection rates of 1.15 kg/hr, 1.46 kg/hr, and 1.87 kg/hr with the 1.4 kJ/s plasma power correspond to the input energy densities 176.4 J/L, 219.4 J/L, and 276.3 J/L, respectively, for the total air-flow rate of 5 000 lpm.
In this work, the experimental results were presented by making use of a simple first order decay model for eliminating target chemicals. The destruction model (Hong et al., 2004) of the odorous chemicals can be expressed as $X/X_0 = \exp(-E/\beta)$, where X represents the leftover concentration of the odorous chemicals after the plasma flame treatment and X_0 is the initial concentration before the treatment, E denotes the input energy density (in units of joules per liter) deposited on the contaminated air by the plasma flame during the treatment and β represents the energy density required for bringing down the concentration to 1/e of its initial concentration; i.e. the energy density needed for 63 % destruction. The leftover concentrations of NH_3 and H_2S were measured by employing detector tubes from the GASTECH Company in Japan. The measurement by detector tubes was done at the flame exit of the reaction chamber without any delay or any interference. Therefore, the detector tube may reliably measure the leftover NH_3 and H_2S, although the data may have a large error bar. The data points in Fig. 15 indicate the average leftover NH_3 (open circle dots) and H_2S (open square dots) concentrations obtained from the repeated measurements in terms of

the input energy densities by means of the methane plasma burner. The closed square dots are the leftover H_2S concentrations by means of the kerosene plasma burner. The initial concentrations of NH_3 and H_2S was X_0 = 159 ppm and 120 ppm, respectively. The curves in Fig. 15 represent the least squared fits to the experimental data points for the microwave plasma burner. Eventually, the β-values of the NH_3 and H_2S elimination by the methane plasma burner are 39.69 J/L and 56.45 J/L, respectively. On the other hand, the β-value of H_2S elimination by the kerosene plasma burner is 46.52 J/L. The β-values of NH_3 and H_2S elimination by methane plasma burner are considerably less than 62.74 J/L for toluene and 138.02 J/L soot elimination, which were reported in the previous document (Uhm et al., 2006). In the recent article for decomposition of H_2S and NH_3 using a plate-to-wire pulse corona reactor (Huang et al., 2001), the β-values of H_2S (X_0 = 148 ppm) and NH_3 (X_0 = 58 ppm) decomposition were 65 J/L and 60 J/L, respectively. Gliding arc discharges (Dalaine et al., 1998; Czernichowski A. 1994) have been used as other example of H_2S depollution. Czernichowski (Czernichowski, 1994) reported that 7 Nm^3/h of air contaminated with 0.7% H_2S was completely purified at the energy consumption of 0.14 kWh per Nm^3 without any preheating. The energy in bringing down the concentration of its initial concentration to zero was estimated to be 540 J/L. In Fig. 15, the energy is approximately 300 J/L. Even though the initial concentrations are different for H_2S elimination, this work reveals that the kerosene microwave plasma burner may be more effective than the pulse corona reactor (Shi et al., 2005) and the gliding arc discharge (Czernichowski, 1994) in a standpoint of energy consumption. From the simple description for atomic oxygen produced in the microwave plasma burner (Hong & Uhm, 2006), the atomic oxygen density n_o was calculated to be n_o = $5.7 \times 10^{13}/cm^3$, which effectively combusts hydrocarbon fuels. It is also emphasized that a large volume of air can be treated by a compact apparatus in this study.

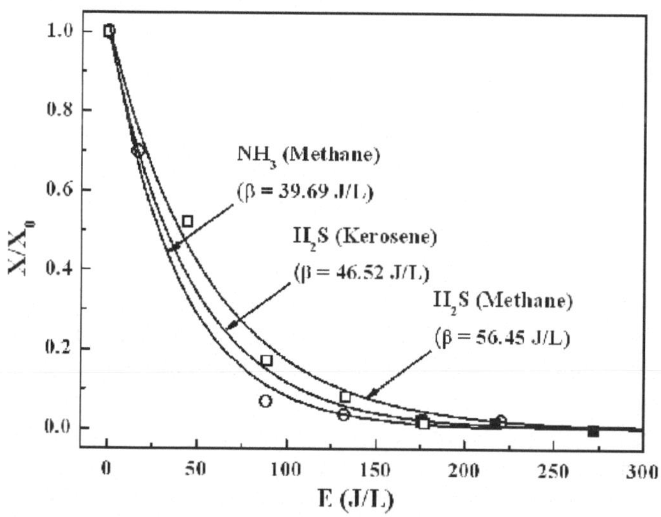

Fig. 15. Plots of leftover H_2S and NH_3 concentration in terms of the input energy density E. The closed and open square dots represent the data points of H_2S concentrations for kerosene and methane injection, respectively, and the open circle dots are NH_3 concentration data for methane injection. Each data point indicates the average value of eight repeated measurements (Hong et al., 2007).

4.3 Destruction of fluorinated compound gases

In NF_3 abatement, NF_3, N_2, O_2 and CH_4 were premixed in the gas-mixing vessel and injected from the side of the microwave plasma torch through FC and CH_4 gas injector. NF_3 can be directly ionized, attached, or dissociated to NF_x (x=0, 1, 2) radicals by electron impact processes, and the reaction is expressed as

$$NF_3 \rightarrow NF_x + F_y \quad (k = 6.6 \times 10^{-8} e^{(-24160/T)}\ cm^3/s), \tag{5}$$

where y is 1 or 2, and T is gas temperature. The microwave plasma burner produces high-temperature, large-volume plasma flame (Hong et al., 2006). NF_3 gas is easily decomposed in high-temperature environment. For example, the reaction rate in Eq. (5) is 4.2×10^{-12} cm^3/s at 2500 K. In fact, average temperature of the methane plasma burner from the CH_4 injector to 20 cm away is approximately 2500 K (Bang et al., 2006). Therefore, abatement of FC gases using the methane microwave plasma burner is accomplished by both plasma and thermal decomposition. The chemical reactions described below are considered in a standpoint of the additive gas used for effective abatement, although there are many other possible reactions. When O_2 as an additive gas is used to abate NF_3, the desired reaction pathway of O_2 is to oxidize the nitrogen in NF_3 to N_xO_y. Whenever diatomic oxygen molecules meet electrons, they undergo dissociative attachment, which produces an O radical and O^- ion (Hong et al., 2003) for the electron temperature in the range of the present experiment. These oxygen atoms react with the NF_x radicals. The chemical reaction equations are

$$O + NF_2 \rightarrow NF + OF \quad (k = 10^{-12}\ cm^3/s), \tag{6}$$
$$O + NF \rightarrow NO + F \quad (k = 10^{-12}\ cm^3/s), \tag{7}$$
$$O + OF \rightarrow O_2 + F \quad (k = 5 \times 10^{-11}\ cm^3/s). \tag{8}$$

Based on Eqs. (6)–(8), the final byproducts are nitrogen monoxide and fluorine at the downstream of the reactor. Also CH_4 electron impact dissociation produces H radicals. H radicals are precursors for FC remediation. As an example, the chemical reactions of NF_x by H radicals (Chang et al., 2000) are presented:

$$H + NF \rightarrow N + HF \quad (k = 1.5 \times 10^{-13}\ cm^3/s), \tag{9}$$
$$H + F \rightarrow HF \quad (k = 1.6 \times 10^{-9}\ cm^3/s). \tag{10}$$

As shown in Eqs. (9) and (10), the stable byproduct HF is formed by CH_4. It is well known that HF is water soluble and is easily captured by passing through a commercial wet scrubber. Although the plasma temperature decreases with the radius and the length of the plasma torch flame, NF_3 is easily decomposed to NF_x radicals with a high reaction rate at given temperatures, as shown Eq. (5). Therefore, we expect that all reactions presented in Eqs. (5) - (10) occur in the core of the plasma burner flame and HF contents increases in the afterglow.

In addition to NF_3 abatement, oxygen atoms also react with SF_x (x=1–5) radicals produced by electron impact processes, creating additional SO_2 or SO molecules and forming SOF_2 and SO_2F_2 molecules by F_2 reactions downstream of the plasma. The chemical reaction equations (Plumb & Ryan, 1988) are

$$O + SF_5 \rightarrow SOF_4 + F \quad (k = 2 \times 10^{-11}\ cm^3/s), \tag{7}$$
$$O + SF_4 \rightarrow SOF_4 \quad (k = 2 \times 10^{-14}\ cm^3/s), \tag{8}$$

$$O + SF_2 \rightarrow SOF_2 \quad (k = 1.08 \times 10^{-10}\ cm^3/s), \tag{9}$$
$$O + SF_4 \rightarrow SOF + F \quad (k = 7.63 \times 10^{-11}\ cm^3/s), \tag{10}$$
$$O + SF \rightarrow SOF \quad (k = 1.7 \times 10^{-10}\ cm^3/s), \tag{11}$$
$$O + SOF \rightarrow Products \quad (k = 7.9 \times 10^{-11}\ cm^3/s). \tag{12}$$

In the CF_4 abatement, the desired reaction pathway of O_2 is to oxidize the carbon in CF_4 to CO_2. When diatomic oxygen molecules meet electrons, they undergo dissociateve attachment that producing O radical and O^- ion at the electron temperature in this range of presented experiment. These oxygen atoms react with the CF_x radicals. The chemical reaction equations are (Hong et al., 2003)

$$O + CF_3 \rightarrow COF_2 + F \quad (k = 3.1 \times 10^{-11}\ cm^3/s), \tag{13}$$
$$O + CF_2 \rightarrow COF + F \quad (k = 1.4 \times 10^{-11}\ cm^3/s), \tag{14}$$
$$O + CF_2 \rightarrow CO + 2F \quad (k = 4.0 \times 10^{-11}\ cm^3/s), \tag{15}$$
$$O + CF \rightarrow CO + F \quad (k = 2.4 \times 10^{-11}\ cm^3/s). \tag{16}$$

The hydrogen radicals produced from the decomposition of CH_4 react with fluorine species and form simple, stable byproduct HF, as shown in Eq. (10).

As previously mentioned, FTIR was employed to identify the concentration changes of NF_3, SF_6, CF_4 and the plasma byproduct before and after the plasma burner treatment. The performance of the microwave plasma abatement device was described in terms of DRE. The DRE represents the percentage of FC gas that has been destroyed. In other words, the definition of DRE is

$$DRE\ (\%) = (S_{before} - S_{after}) / S_{before} \times 100, \tag{17}$$

where S_{before} and S_{after} are the main peak area of the FC gases before and after the plasma burner treatment, respectively.

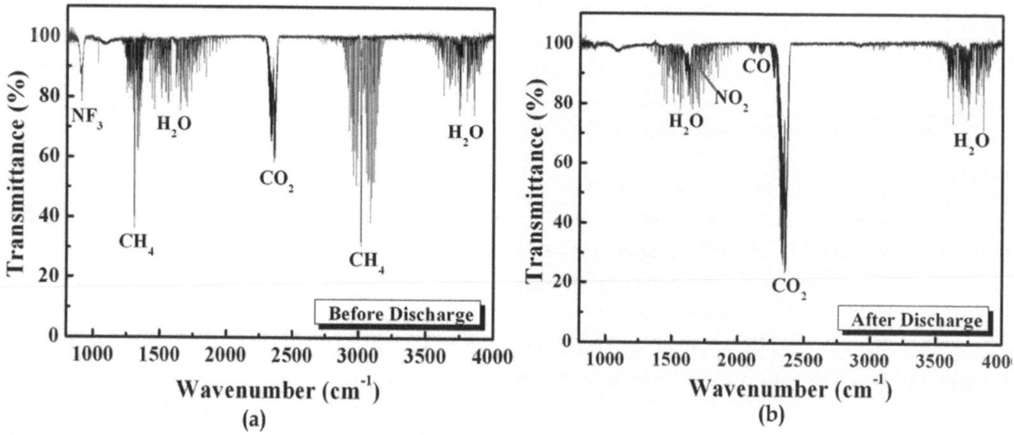

Fig. 16. FTIR spectra (a) before and (b) after the microwave plasma burner abatement of NF_3 with the components of 40 lpm compressed air as swirl gas, 250 lpm N_2, 30 lpm O_2, 15 lpm CH_4 and 0.6 lpm NF_3 at the applied plasma power of 1.2 kW (Hong et al., 2010).

Figures 16(a) and (b) show infrared transmitted spectra, demonstrating before and after the application of the microwave plasma burner to the components of 40 lpm swirl air, a mixture of 30 lpm O_2, 250 lpm N_2, 15 lpm CH_4 and 0.6 lpm NF_3 at the applied microwave power of 1.2 kW, respectively. The energy contained in the CH_4 is about 9.52×10^6 cal/m^3. The CH_4 injection rate can be translated into watt. For example, 15 lpm injection rate of methane is 9.98 kW. Therefore, total applied power is approximately 11.2 kW. As mentioned earlier, the microwave plasma torch showed a treatment limitation at high flow rates with low removal efficiency due to the drastic decrease of plasma volume and the short residence time of contaminants. The wavenumbers 910 and 1032 cm^{-1} on the horizontal line represent the signature of the NF_3 species. After the microwave plasma burner was turned on, the gas stream was analyzed. As expected, the CH_4 plasma burner destroyed almost all of the NF_3 contaminants with CH_4 decomposition rate close to 100%. The NO_2 curve in Fig. 16(b) represents the increases of the NO_2 molecules during the microwave discharge, because nitrogen in NF_3 is oxidized by O_2. Moreover, in air plasma, N_xO_y neutral molecules, $N_xO_y^-$ and $N_xO_y^+$ ions, will be formed, where NO_2 and N_2O, $N_2O_2^-$ and N_3O^+ are expected to be dominant neutral species, negative and positive ions, respectively. Therefore, NO_2 is formed. Generally, NO in the FITR spectrum is identified at 1950–1810 cm^{-1}. As shown in Fig. 16(b), the NO peak may be screened by strong H_2O peaks or the formation of NO may be minimal if there is any. HF can be formed through the reaction of F_2 with H_2O which may be in downstream of the plasma burner.

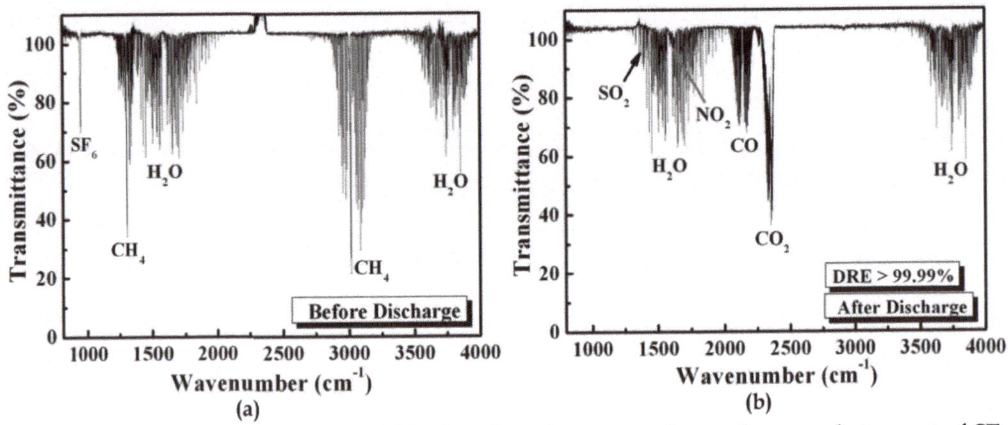

Fig. 17. FTIR spectra (a) before and (b) after the microwave plasma burner abatement of SF_6 with the components of 40 lpm compressed air as swirl gas, 120 lpm N_2, 30 lpm O_2, 15 lpm CH_4 and 0.1 lpm SF_6 at the microwave plasma power of 1.4 kW (Hong et al., 2010).

Figure 17 displays FTIR spectra before and after the application of the microwave plasma burner to the components of 40 lpm swirl air, a mixture of 30 lpm O_2, 120 lpm N_2, 15 lpm CH_4 and 0.1 lpm SF_6 at the applied microwave power of 1.4 kW. The translated total power for 15 lpm CH_4 is approximately 11.2 kW. The wavenumber 948 cm^{-1} on the horizontal line represents the signature of the SF_6 concentration, as shown Fig. 17(a). After the plasma burner treatment in Fig. 17(b), SF_6 concentration was significantly reduced, showing high DRE of 99.92%. SO_2 molecules at 1361 cm^{-1} as an important byproduct for decomposition of SF_6 were detected. The microwave plasma burner simultaneously destructed and burned out the SF_6 contaminants by exothermic reactions with O_2 in compressed air injected as a swirl gas, forming HF molecules.

Figure 18 shows infrared transmitted spectra, demonstrating the microwave plasma burner abatement for CF_4. In the same experimental condition with the case of SF_6, a mixture of 40 lpm N_2 and 0.05 lpm CF_4 as contaminant was injected. In Fig. 18(a), the wavenumber 1283 cm^{-1} on the horizontal line represents the signature of the CF_4 concentration. In addition to NF_3 and SF_6, CF_4 was also abated in the experimental condition, showing 98.1% DRE in Fig. 18(b). Although the trace of CO and CO_2 as resultant products of CF_4 oxidation was not identified due to kerosene combustion, main byproducts are CO and CO_2.

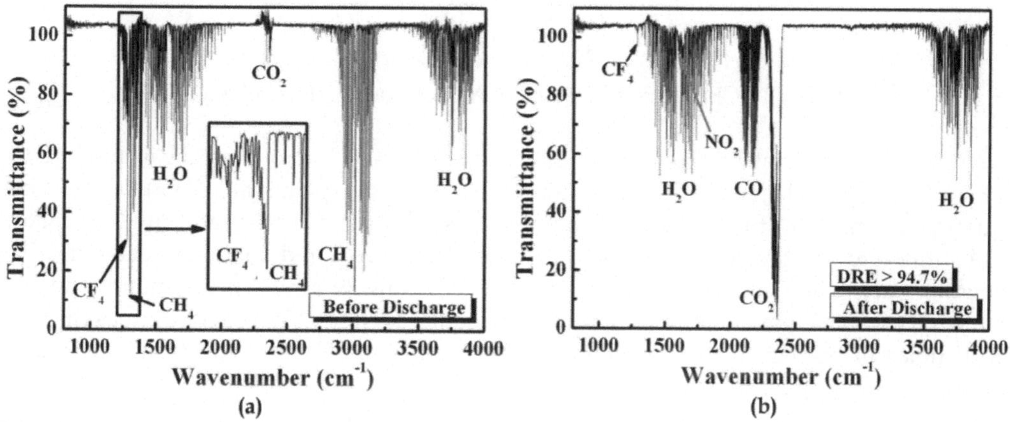

Fig. 18. FTIR spectra (a) before and (b) after the microwave plasma burner abatement of CF_4 (Hong et al., 2010).

Figure 19 shows the plot of the DREs of SF_6 and CF_4 versus nitrogen flow rates, corresponding to Figs. 17 and 18, respectively. The components of 40 lpm swirl air, a mixture of FC gas which is 0.1 lpm SF_6 or 0.05 lpm CF_4, 30 lpm O_2, 15 lpm CH_4 and various N_2 gas flow rates at the applied microwave power of 1.4 kW. DREs were measured in terms of N_2 flow rates. The N_2 flow rate of SF_6 and CF_4 abatement were from 40 lpm to 160 lpm and 20 lpm to 120 lpm, respectively. In case of SF_6 abatement, DREs were more than 99.9% until 120 lpm N_2, however, DREs of over 140 lpm N_2 were decreased drastically. At the N_2 flow rate over 140 lpm, plasma burner flame temperature, volume and residence time were decreased drastically. In this context, DRE of CF_4 was decreased drastically in terms of over 60 lpm N_2 flow. It is apparent that the abatement of CF_4 is much more difficult than that of SF_6. For instance, the DRE of SF_6 is more than 99.99% at 120 lpm N_2, while that of the CF_4 gas is only 46.2%. In particular, CF_4 has proven to be difficult to destroy and remove because of its chemical thermal stability due to the strong covalent nature of its bonding. Dissociation energies in SF_6 and CF_4 are represented by $e + SF_6 \rightarrow SF_5 + F + e$ (4.01 eV) and $e + CF_4 \rightarrow CF_3 + F + e$ (12.5 eV), respectively. The CF_4 dissociation energy is about three times of the SF_6 dissociation energy, which explains the observations in Fig. 19.

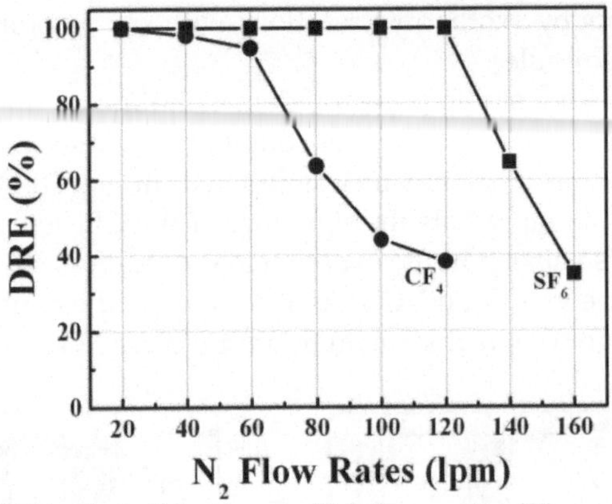

Fig. 19. DREs of SF_6 and CF_4 plotted in terms of N_2 flow rates (Hong et al., 2010).

Greenhouse gases that are subject to reduction under the Kyoto Protocol include HFCs, PFCs, SF_6, N_2), CH_4, and CO_2. Since the global warming potential (GWP) of the three fluorinated gas substitutes is several hundred to several tens of thousand times of that of CO_2, utmost efforts towards regulating their emission are required. In the present work, CO_2 emission from the combustion of 15 lpm CH_4 is about 29 grams per minute. And NF_3 emission from 0.5 lpm is about 1.5 grams per minute. GWP of NF_3 is very high as many as 9720 times in respect to CO_2. Therefore, we expect that CO_2 emission from the present plasma system is ignorable (Hong et al., 2010).

5. Conclusion

In order to significantly increase the volume and temperature of the microwave plasma torch, we have developed a microwave plasma-burner by injecting hydrocarbon fuel in liquid or gaseous state into the microwave plasma torch generated by air and a mixture of air and oxygen. The microwave plasma-burner implies that a plasma flame of high-power level can be obtained by only injecting small quantity of a hydrocarbon fuel. For example, the energy contained in diesel and in methane are approximately 10^7cal/kg and 9.52×10^6 cal/m^3, respectively. The fuel injection rate can be translated into watts. Eventually, a 0.019 lpm ($\approx$ 1.15 kg/hr) injection rate of diesel is 13.3 kW and a 10 lpm CH_4 corresponds to approximately 6.6 kW. A significant temperature increase of the CH_4 microwave plasma burner was observed in relative to those of the microwave plasma torch and CH_4 fuel-only flame, showing the influence of the microwave plasma on the combustion flame.

In decontamination test of chemical agents, the plasma flame could be a very effective mean for decontaminating the airborne chemical and biological warfare agents. The plasma flame also provided a unique opportunity of a simultaneous elimination and burnout of chemical and biological warfare agents diluted in air. However, it is essential for the nations and their military, worldwide, to carry out the decontamination experiment of the real chemical and biological warfare agents for the experimental data, with which the plasma flame system makes the warfare agents obsolete. We also reported the experimental results for removing FC gases by using the methane microwave plasma burner and described the simple model for effective abatement of FC gases. In NF_3 abatement, we achieved a DRE of more than 99.9999% in the mixture of 0.6 lpm NF_3, 400 lpm N_2, 100 lpm O_2 and 15 lpm CH_4 by applying the microwave power of 1.2 kW. In case of SF_6, we achieved a DRE of more than 99.9% in the mixture of 0.1 lpm SF_6, 120 lpm N_2, 30 lpm O_2 and 15 lpm CH_4 by at the applied microwave power of 1.4 kW. Also, DRE of CF_4 is 94.72% in the mixture of 0.05 lpm CF_4, 60 lpm N_2, 30 lpm O_2 and 15 lpm CH_4 at the microwave power of 1.4 kW. N_2 gas is required in post-pump systems. The aforementioned characteristics make the microwave plasma burner attractive for the destruction and removal of chemically stable gases emitted by the semiconductor industry.

6. References

Bang, C.U., Hong, Y.C., Cho, S.C., Uhm, H.S. & Yi, W.J. (2006): Methane-Augmented Microwave Plasma Burner, *IEEE Trans. Plasma Sci.* Vol. 34, No. 5, pp 1751-1756, ISSN 0093-3813.

Chang, J.-S., Urashima, K.G., Yamamoto, K., Okayasu, T., Kato, Y., Iwaizumi, T. & Yoshimura, K. (2000): Removal of NF_3 from semiconductor-process flue gases by tandem packed-bed plasma and adsorbent hybrid systems, *IEEE Trans. Ind. Appl.* Vol. 36, No. 5, pp 1251-1259, ISSN 0093-9994.

Czernichowski, A. (1994): Gliding arc: applications to engineering and environment control, *Pure Appl. Chem.* Vol. 66, No. 6, pp 1301-1310, ISSN 1365-3075.

Delaine, V., Cormier, J.M, Pellerin, S. and Lefaucheux, P. (1998): H_2S destruction in 50 hz and 25 kHz gliding arc reactor, *J. App. Phys.* Vol. 84, No. 3, pp 1215-1221, ISSN .

de Izarra, C. (2000): Computer simulation of the UV OH band spectrum, *Int. J. Mod. Phys. C* Vol. 11, No. 5, pp 987-998, ISSN 1793-6586.

Hartz, C.L., Bevan, J.W., Jackson, M.W. & Wofford, B.A. (1998): Innovative surface wave plasma reactor technique for PFC abatement, *Environ. Sci. Technol.* Vol. 32, No. 5, pp 682-687, ISSN 0013-936X.

Fauchais, P. & Vardelle, A. (1997): Thermal plasmas, *IEEE Trans. Plasma Sci.* Vol. 25, No. 6, pp 1258-1280, ISSN 0093-3813.

Green, K.M., Borass, M.C., Woskov, P.P., Flores, G.J., Haddi, K. & Thomas, P. (2001): Electronic excitation temperature profiles in an air microwave plasma torch, *IEEE Trans. Plasma Sci.* Vol. 29, No. 2, pp 399-406, ISSN 0093-3813.

Gutsol, A. & Bakken, J.A. (1998): A new vortex method of plasma insulation and explanation of the Ranque effect, *J. Phys. D: Appl. Phys.* Vol. 31, No. 6, pp 704-708, ISSN 1361-6463.

Hong, Y.C., Kim, J.H. & Uhm, H.S. (2003): Abatement of CF_4 by atmospheric pressure microwave plasma torch, *Phys. Plasmas* Vol. 10, No. 8, pp 3410-3414, ISSN 1089-7646.

Hong, Y.C., Kim, J.H. & Uhm, H.S. (2004): Simulated experiment for elimination of chemical and biological warfare agents by making use of microwave plasma torch, *Phys. Plasmas* Vol. 11, No. 2, pp 830-835, ISSN 1089-7646.

Hong, Y.C., Uhm, H.S., Kim, H.S., Kim, M.J., Han, S.H., Ko, S.C. & Park, S.K. (2005): Decomposition of phosgene by microwave plasma-torch generated at atmospheric pressure, *IEEE Trans. Plasma Sci.* Vol. 33, No. 2, pp 958-963, ISSN 0093-3813.

Hong, Y.C., Cho, S.C., Bang, C.U., Shin, D.H., Kim, J.H., Uhm, H.S. & Yi, W.J. (2006): Microwave plasma burner and temperature measurements in its flame, *Appl. Phys. Lett.* Vol. 88, No. 20, pp 201502-201504, ISSN 0003-6951.

Hong, Y.C., Shin, D.H. & Uhm, H.S. (2007): Simulated experiment for elimination of air contaminated with odorous chemical agents by microwave plasma burner, *Appl. Phys. Lett.* Vol. 91, No. 16, pp 161502-161504, ISSN 0003-6951.

Hong, Y.C., Lho, T., Shin, D.H. & Uhm, H.S. (2010): Removal of fluorinated compound gases by an enhanced methane microwave plasma burner, *Jpn. J. Appl. Phys.* Vol. 49, No. 1, pp 017101-017106, ISSN 1347-4065.

Huang, L., Nakajo, K., Ozawa, S. & Matsuda, H. (2001): Decomposition of dichloromethane in a wire-in-tube pulsed corona reactor, *Environ. Sci. Technol.* Vol. 35, No. 6, pp 1276-1281, ISSN 0013-936X.

Kim, J.H., Hong, Y.C., Kim, H.S. & Uhm, H.S. (2003): Simple microwave plasma source at atmospheric pressure, *J. Korean Phys. Soc.* Vol. 42, No. 92, pp S876-S879, ISSN 1976-8524.

Kim, J.H., Hong, Y.C. & Uhm, H.S. (2007): Binary oxide material made from a mixture of Zn and Cd in a microwave plasma, *Chem. Phys. Lett.* Vol. 443, No. 1, pp 122-126, ISSN 0009-2614.

Lai, W., Lai, H., Kuo, S.P., Tarasenko, O. & Levon, K. (2005): Decontamination of biological warfare agents by a microwave plasma torch, *Phys. Plasmas* Vol. 12, No. 5, pp 023501-023506, ISSN 1089-7646.

Margot, J. (2001): Studies of emission spectra in helium plasmas at atmospheric pressure and local thermodynamical equilibrium, *Phys. Plasmas* Vol. 8, No. 5, pp 2525-2532, ISSN 1089-7646.

Masuya, G., Takita, K., Takahashi, K., Takatori, F. & Ohzeki, H. (2002): Effects of airstream mach number on H/N plasma igniter, *J. Propul. Power* Vol. 18, No. 3, pp 679-685, ISSN 0748-4658.

Moisan, M., Zakrzewski, Z. & Rostaing, J.C. (2001): Waveguide-based single and multiple nozzle plasma torches: the TIAGO concept, *Plasma Sources Sci. Technol.* Vol. 10, No. 3, pp 387-394, ISSN 1361-6595.

Moon, S.Y., Choe, W., Uhm, H.S., Hwang, Y.S. & Choi, J.J. (2002): Characteristics of an atmospheric microwave-induced plasma generated in ambient air by an argon discharge excited in an open-ended dielectric discharge tube, *Phys. Plasmas* Vol. 9, No. 9, pp 4045-4051, ISSN 1089-7646.

Ogawa, S., Sakai, Y., Sato, K. & Sega, S. (1998): Influence of microwave on methane-air laminar flames, *Jpn. J. Appl. Phys.* Vol. 37, No. 1, pp 179-185, ISSN 1347-4065.

Penetrante, B.M., Hsiao, M.C., Merritt, B.T., Vogtlin, G.E., Kuthi, A., Burkhart, C.P. & Bayless, J.R. (1997): Identification of mechanisms for decomposition of air pollutants by non-thermal plasma processing, *Plasma Sources Sci. Technol.* Vol. 6, No. 3, pp 251-2598, ISSN 1361-6595.

Plumb, I.C. & Ryan, K.R. (1998): Gas-phase reactions in plasmas of SF_6 with O_2: Reactions of F with SOF_2 and SO_2 and rections of O with SOF_2, *Plasma Chem. Plasma Prccess.* Vol. 9, No. 3, pp 409-420, ISSN 0272-4324.

Shi, Y., Ruan, J., Wang, X., Li, W. & Tan, T. (2005): Decomposition of mixed maldorants in a wire-plate pulse corona reactor, *Environ. Sci. Technol.* Vol. 39, No. 17, pp 6786-6791, ISSN 0013-936X.

Takita, K., Masuya G., Sato T. & Ju, Y. (2001): Effect of addition of radicals on burning velocity, *AIAA J.* Vol. 39, No. 4, pp 742-744, ISSN 0001-1452.

Uhm, H.S. (1999): Properties of plasma flames generated by electrical breakdown in flames, *Phys. Plasmas* Vol. 6, No. 11, pp 4366-4674, ISSN 1089-7646.

Uhm, H.S., Hong, Y.C. & Shin, D.H. (2006): Plasma flame for mass purification of contaminated air with chemical and biological warfare agents, *Appl. Phys. Lett.* Vol. 89, No. 12, pp 121504-121506, ISSN 0003-6951.

Weinberg, F.J., Hom, K., Oppenheim, A.K. & Teichman, K. (1978): Ignition by plasma jet, *Nature* Vol. 272, No. 5651, pp 341-343, ISSN 0028-0836.

Woskov, P.P. & Haddi, K. (2002): Large electrodeless plasmas at atmospheric pressure sustained by a microwave guide, *IEEE Trans Plasma Sci.* Vol. 30, No. 1, pp 156-157, ISSN 0093-3813.

11

The blast furnace trazability by helium

1. Introduction

Blast furnaces (BF) may be considered to be among the oldest type of industrial equipment that is still used regularly today. Blast furnaces existed in China from about the 5th century BC, and in the West from the High Middle Ages. They spread from the region around Namur in Wallonia (Belgium) in the late 15th century, being introduced to England in 1491. With a few variations, they have been used for over 600 years.

The steel industry uses blast furnaces to chemically reduce iron ore (mainly oxides), removing oxygen and enriching the ore into metallic iron with a high degree of metallization (total iron content higher than 95-96%), from which is obtained pig iron (Aranguren, & Mallol, 1962) which is the raw material used in the integrated steel industry to manufacture steel (UNESID, 1998).

A blast furnace is a vertical reactor comprising an assembly of cylindrical or conical elements (Figure 1). Among its most important features are the internal volume, which is determined by the diameter of the crucible, the furnace bottom, defined by the desired results, which collects both molten metal and slag. Thus a furnace crucible which is 12 meters in diameter can produce more than three million tons of pig iron annually.

The blast furnace is a countercurrent reactor in which the reducing gas is produced by coke gasification with the oxygen blowing via tuyeres. The reducing gas flows upwards reducing the iron ores charged at the top of the furnace. The blast furnace process is very complex with many influencing and correlating factors (Steiler, 1998; Kundrat, 1989).

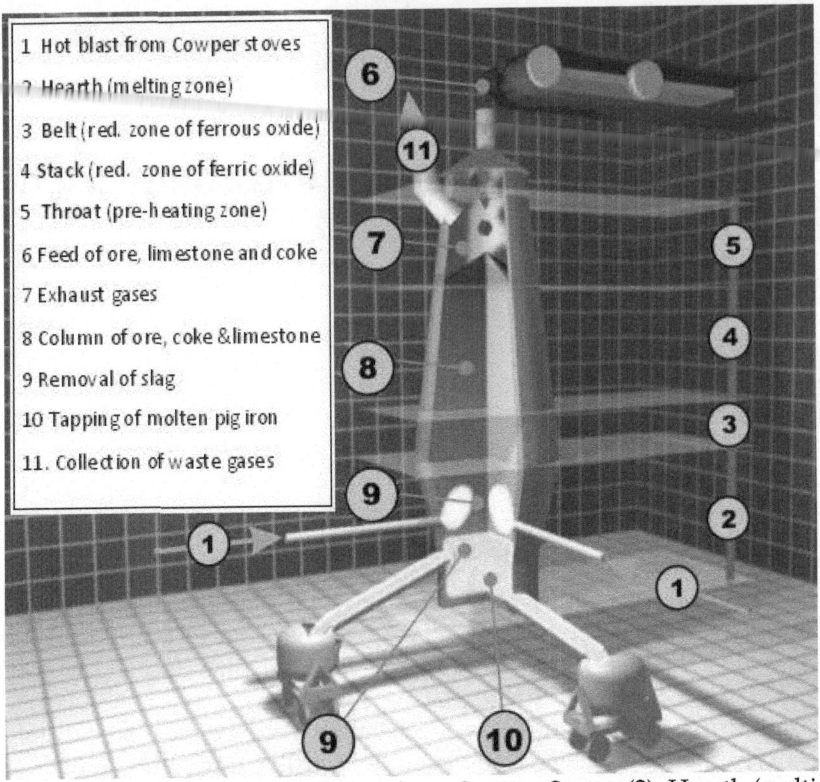

Fig. 1. A Blast furnace diagram:(1) Hot blast from Cowper Stones (2). Hearth (melting zone). (3). Belt (reduction zone of ferrous oxide) (4). Stack (reduction zone of ferric oxide (5). Throat (pre-heating zone) (6). Feed of ore, limestone and coke (7). Exhaust gases (8) Column of ore, coke and limestone (9). Removal of slag (10). Tapping of molten pig iron (11). Collection of waste gases

2. The Blast Furnace

2.1. The main parts of blast furnaces

The primary systems and most important components are:

(a) Gas Evacuation System: Consists of drainage pipes for the gas produced and comes equipped with relief valves and controls for pressure, temperature, gas composition, etc. (items 7 and 11 of the Figure 1).

(b) Oven Body: The oven itself, built with refractory materials with different characteristics, depending on the needs of each area and the furnace construction. Its design provides for the operations required, and the chemical exchange between solids and gases, which coexist in various proportions in each area. The design of the kiln is variable, along with its height, distinguishing different areas, which are described later.

(c) Cooling System: Varies according to each zone in order to provide the most intense, efficient performance to maximize system reliability, refractory lining durability, and, consequently, optimize the duration of the furnace campaign. It normally requires several closed cooling circuits, in addition to control systems for temperature, pressure, flow, heat loss calculations, detection of leaks or failures of refrigerated units, etc.

(d) Control System: To manage the operation of the furnace equipment, this system is made up of a large number of sensors (temperature controls, pressure, level sensors, gas analysis sensors, etc.) which, together with its ancillary facilities, form a global control system.

Inside the blast furnace there are different chemical zones:

(a) Throat (see Figure 2 and Item 5 in the Figure 1): The top of the oven where raw materials such as iron ore, coke and flux are charged. Formed by straight walls, loading should ensure a controlled distribution of each material inside the oven. Currently the most common system for controlling load distribution, designed by Paul Wurth, is a system using a gear box, allowing pellets to be downloaded inside the oven in a homogeneous way.

(b) Stack: (see Figure 2 and Item 4 in the Figure 1) located between the end of the hopper and the belly of the furnace, it contains most of the load and has a truncated cone shape. This part of the blast furnace has the job of encouraging the exchange of heat between solids and gases.

(c) Belt : (see Figure 2 and Item 3 in the Figure 1) This is the wider part of the furnace, and it's cylindrical in shape with straight walls. As the charge descends and the temperature ascends, more space is needed as volume increases. This area of the kiln is the link between the stack and the hearth.

(d) Hearth : (see Figure 2 and Item 2 in the Figure 1):This is the furnace area where air is injected into the hot blast from the stoves. The warm air reaches a circular tube that is inserted into the furnace through holes in nozzles. Its mission is to enable the combustion of coal. Auxiliary fuels are also injected through nozzles. The number of nozzles, which are always located equidistantly around the furnace, varies from one oven to another depending on the oven size.

(e) Crucible: (see Figure 2) This cylindrical shaped part is the bottom of the furnace, where the products obtained in the process, pig iron and slag, are collected through a trough. The crucible is the most critical part of the plan for the selection and installation of refractories. The material used is carbon (2 m thick in the walls and 2-3 m for the bottom). In all other areas of the BF, the refractories are generally of an aluminous type and, in some specific areas, more durable materials (SiC or Sial) are used.

2.2. Blast furnace control

Control of the blast furnace process requires significant measures to ensure adequate progress and manufacturing facilities related to the making and sintering of coke. This is to maintain a stable oven operation and to control and protect the functioning of the equipment installed. With this in mind, sensors and measuring equipment are used to control the movement of materials and fluids in and out of the oven.

Specific controls include:

(a) Input: raw materials, main and auxiliary fuels, hot blast, oxygen injected into the blast, steam injected with the hot blast (control of humidity, etc.).

(b) Inside the furnace itself: Screening and loading of raw materials, hopper and system of distributing charge, pressures and temperatures in different parts of the oven, charge levels, charge samples in different areas, etc.

(c) Output: iron and slag, gases, dust and sludge recovered, and so on. Of particular importance are controls for the distribution of the hopper charge and the metallurgical variables that define the process, drawing on metallurgical models of balancing heat and matter, in addition to statistical quality control, artificial intelligence, etc..

All of these factors combine to achieve stable operation of the BF in order to increase productivity, improve product quality (pig iron, slag), increase the duration of blast furnace functionality and reduce costs (minimum consumption of coke and auxiliary fuel, etc.).

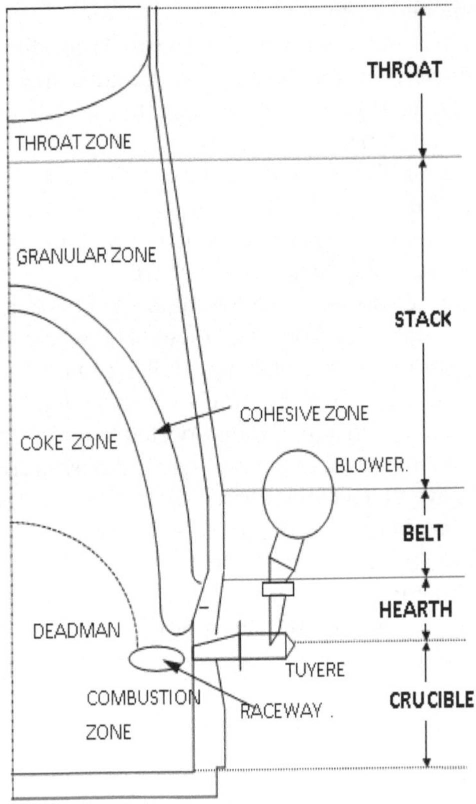

Fig. 2. The main parts of blast furnaces

One of the key aspects that influence productivity (t/m3 useful volume / day), production (t / year) and duration of a blast furnace campaign, measured in total output in relation to the size of the oven, is optimal control of the process that results in steady oven production.

Other aspects that influence production, productivity, and campaign duration are:

- The BF design (profile, crucible diameter, etc.)
- The strategy chosen to obtain a particular product (iron quality, etc.)
- Coke quality and the use of auxiliary fuels such as coal
- The ancillary equipment which facilitates peak charge distribution (hopper), high temperature blower (blast), optimal oxygen injection, work counter, level and regularity of the furnace functionality, and so on
- The original design quality of the equipment, such as the binomial refractory / cooling, equipment reliability, applied technologies, etc.
- The quality of operation and maintenance processes
- The quality and type of raw materials used (percentages and types of sinter, pellets, etc.).

2.3. Blast furnace functions

The functions to be performed by a BF are quite complex, if the process is to result in the required production volume in an economical fashion. These functions include a number of mechanical operations, since the furnace needs to constantly and steadily bring the charge of solid materials (coke, ore and flux) being placed in the hopper in rapid and evenly distributed contact with the ascending gas stream.

For this to happen on a regular basis, the charge must be highly porous and very well distributed so that its interaction with the hot gas flow is homogeneous. The greater or lesser porosity of the charge depends on the shape and dimensions of the pieces being used, with greater porosity being associated with spheres of equal size, tangent and centers at the vertices of a cube. In this case, the empty spaces left between the spheres represent 52.3% of volume. If the spheres were of equal tangent, but with centers at the vertices of a rhomboid, the empty spaces would be 47.7% at most.

In practice, the pieces of ore are not spherical and do not have the same dimensions, so fine pieces caught in larger pieces can significantly lower the porosity of the load. It can be seen, therefore, that preparing and classifying the charge is important, sintering the fine pieces while leaving the larger relatively homogenous ones untreated, as is the trend in BF operations. The type of hopper, Wurth or with bells, and the loading pattern are also essential for optimal distribution.

In the case of using a hopper with bells, IHI or MOHR, the inclination of the closing bell, the relative diameter of the bell and the hopper and the length of the bell, all influence the form of distribution of solid materials in the oven hopper. The specific weight of the materials in the charge and its angle of repose are also relevant.

In addition, the BF must distribute the load so as to permit a regular descent without forming particular passageways, which would give rise to problems. Nor should the charge preferentially descend by the furnace walls or through the center of the hot gases, which means that the distribution of both fine and larger pieces must be carefully controlled throughout the area of descent.

Besides all this, as the descending charge undergoes changes in composition, temperature and volume, the design of the BF must promote these changes and the rapid, steady and consistent descent of the charge, slag and iron formation and casting at the bottom, as well as the release of gases at the top.

2.4. The deadman

The part of the furnace called the deadman, between the belt and the crucible, greatly impacts the entire process. It plays a key role in hot metal quality. The condition of the deadman has a strong influence on hot metal temperature and composition, and flow conditions. When hot metal can flow freely towards the tap hole and the deadman is a porous coke bed, then conditions are good for the desulphurization and carbonization of hot metal. The deadman also has a significant effect on lining wear and campaign length as it controls hot metal flow in the hearth.

2.5 Improving the Productivity

The blast furnace productivity is the quotient between possible gas throughput per unit of time and required specific gas generation for one tonne of hot metal obtained (Harting et al., 2000). Consequently, a productivity increase requires an increase in the gas throughput, which implies, on the one hand, improvement in furnace permeability and, on the other hand, a reduction in the specific gas requirements so then finally means a reduction in the specific consumption of reducing agents. Permeability is a measure of the gas ability to pass through the bed of solid materials; if the permeability in the furnace is higher then the furnace burden movement and the reducing gas flow through the furnace are better (Pandey et al., 1996).

The search for improvement of overall blast furnace permeability results thus in further improvements in the following fields:
- burden composition and quality,
- behaviour of burden during reduction,
- the cohesive zone shape and position control
- liquid evacuation from the hearth (slag quantity, coke size and liquid flow conditions in the hearth and tapping practices).

In all these fields, the actions aiming at permeability improvements are generally beneficial for the consumption of reducing agents, making it possible to decrease the specific gas consumption, and thus to obtain high productivity easier. Beyond that, the oxygen enrichment of the blast is the major way to decrease the specific gas volume. However, this thing will only be applicable in combination with tuyere injection in order to maintain a sufficient gas quantity to overheat the burden (top gas temperature higher than 100°C); in addition, the oxygen enrichment of the blast will be adjusted to maintain the RAFT (Raceway Adiabatic Flame Temperature) in a sufficiently high value to ensure a good gasification of the reducing agent injected, but not too high to avoid mechanical problems on tuyeres (Formoso et al. 1999). The objective "high levels of injection of pulverized coal" (and thus low coke consumption) is not only compatible with productivity, but also even necessary to increase the blast furnaces productivity (Babich et al. 1996) (Babich et al. 1999).

Permeability of the ferrous burden and coke column for the gas flow is linked together with the increases of gas throughput. A burden column (structure of charged materials, ores and coke, with properties that extensively assure the necessary void for an adequate permeability, in spite of mechanical, thermal and melting area) are down to the lower furnace. The main function of coke is guaranteeing the permeability for the gas in the dry region above the cohesive area, in the cohesive area itself and in the hearth. In the cohesive zone the coke has an important role because the softening and melting iron-bearing materials can form an impermeable layer (Busby et al.1994)(Sert et al, 2004). Therefore, many quality criteria for ferrous burden materials and coke have been defined with ever increasing demands. In all cases, the general practice has shown that it is advisable to prepare an homogeneous mix of all iron bearing components before charging to the blast furnace, to achieve excellent permeability and suitable melting behaviour.

Then the gas distribution control is the results of a compromise among different requirements (Steiler, 1998):
To achieve a control gas flow in order to maintain the shaft permeability in spite of higher levels of injection of pulverized coal. .
To adjust the gas flow along the wall between two limits, in order to guarantee low heat losses and the absence of scaffold simultaneously
To achieve efficient gas solid flow in order to promote efficient reducing conditions and low reductant rate.

The blast furnace optimization implies that the process monitoring has reached a high standard to control the process in its inner part. Special importance has the optimisation and control of composition and gas distribution and the pressure losses inside the furnace.

Several indirect measurements (thermal losses, gas analysis and temperature recorded by burden probes) help to capture information over the gas distribution in the blast furnace (Nikus & Saxen, 1996)(Nicole et a. 2000). With the objective to develop new tools to improve the gas distribution monitoring the helium tracing techniques has been tested (Havelange, 2000). This method is employed to monitor the gas transfer time from tuyere to the burden probe. The analysis of this transfer time from the tuyere to the top and its relationship with the process would be explained.

3. Experimental method

3.1. Theory

Helium tracing technique consists in to inject helium in the blast furnace at the tuyeres level and its arrival at the blast furnace top is detected by a mass spectrometer. The spectrometer indicates the helium content in outboard gas and it is possible to define the transfer time as the delay between the injection moment and the time when helium concentration reaches 10 percent of the maximum detected level.

Inside the furnace the gas composition has an evolution from the tuyeres (where the gas is mainly composed for O_2 and N_2) to rotating chute (where the gas is mainly composed for CO_2, CO, H_2 and N_2). The initial oxygen, the oxygen produced in the decompositions of oxides and water, participates in the production of CO, CO_2 and H_2O. The detected He ppm would be lower when more mass of gases (CO_2, H_2 and CO) is produced. Using this data, it's possible to obtain an indirect measure of the oxidant reactions that has been produced inside the furnace. The air is more weightable after oxidant reaction then the measured of He ppm is lower. As gas composition introduced in tuyeres is similar and the He content is the same, the maximum He ppm content measured in top of the furnace can be related with the reactivity inside the furnace.

In this work the charge in the blast furnace will be treated like ceramic foams. In all cases perpendicular section to the flow and the average velocity may be based on the entire cross sectional area.

Using the Blake-Kozeny equation or Kozeny-Carman equation, when there is a laminar flow in a porous medium, and the Burke-Plummer equation, when the flow is turbulent, then the generalised Ergum`s equation (Ergum, 1957) is obtained for all-laminar to all turbulent flow.

$$\nabla P = -\left(\frac{150(1-\phi)^2 \eta}{\psi^2 D_p^2 \phi^3} + \frac{1,75(1-\phi)\rho}{\psi D_p \phi^3} \left| \vec{u} \right| \right) \cdot \vec{u} \tag{1}$$

where,
P = pressure (Pa)
u = vector of velocity (m/s)
ψ = shape factor (-)
η = viscosity of the fluid (Pa s)
ρ = fluid density (Kg/m^3)
Φ = porosity (-)
Dp = particle diameter of the granular medium (m)

In other way Darcy`s law (Gisonni 2003)(Hager 1994) was based on experiments with the pressure drop Δp measured over a finite length, L, in a sand pack of permeability tensor, k, and cross-sectional area, A. For linear flow of an incompressible fluid of viscosity η through this sand pack, the flow rate is related to other factor.

$$\vec{u} = -\frac{\bar{\bar{k}} \cdot \nabla P}{\eta} \tag{2}$$

The Darcy´s equation has been used to establish a relationship between the fluid speed and the permeability as function of the difference of pressures and gas viscosity. The gas speed is obtained dividing the distance travelled by the transfer time. Making the assumption that the instant permeability is constant in the furnace (an averaged permeability) then the permeability tensor is simplified to a constant. In this work the distance travelled by the gases and the gas permeability will be supposed constants for all the measures. The distance travelled by the gases and the permeability of the furnace k, can be calculated from the equation (3)

$$k = \frac{C}{t \cdot \nabla P} \quad (3)$$

Where C is a constant and t is the transfer time. It is possible to obtain transfer times in six points using the in-burden probe, so it is possible to establish a mean permeability burden distribution from the tuyeres up to the burden probe.
The sampling of gas for its analysis through the in-burden probe what allows comparing directly the results of the helium tracing technique with the gas composition and temperature obtained by means of the probe.
For the burden probe configuration, the journey carried out by the gas from the injection until the detector, does not vary with the measurement point, because the probe is a burden tube in whose end is the detector and filter. Its flow rate is measured to detect clogging of the sampling line. The helium concentration is measured by a mass spectrometer.
To clean the pipes before the injection of helium, a three ways electrovalve is attached to the system to let the injection of N_2. This will ensure not only the cleaning of the pipes but also will avoid any false measurement of helium concentration and time transfer due to the presence of residual helium in the pipes.

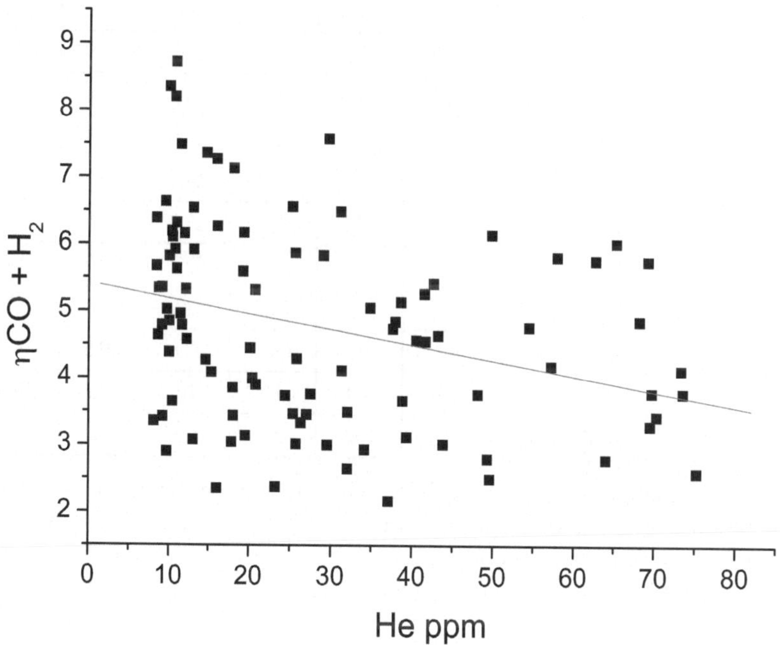

Fig. 3. It can see that higher maximum He ppm are detected, lower gases area reacting inner the furnace

2.2. Measurements

The probe measures in 6 points along a radius of the furnace from nearest area to the wall up to 1, 96 meters of the wall, near the furnace centre. When the temperature in the centre is very high (hotter than 950 °C) the probe is not introduced until the last point, and in those cases the last measures are lost.

t1 (s)	t2 (s)	t3 (s)	t4 (s)	t5 (s)	t6 (s)	Max1 (ppm)	Max2 (ppm)	Max3 (ppm)	Max4 (ppm)	Max5 (ppm)	Max6 (ppm)
99.3	204.7	260.7	417.9	338.9	517.5	18.0	17.9	10.8	14.6	11.4	10.7
31.3	167.7	334.2	388.9	364.7		32	25.4	15.3	12.2	10.9	
29.2	103.9	108.7	117.5	198.0	433.5	34.1	11.6	10	8.7	8.4	8.2
28.2	102.6	159.6	189.0	298.3	368.9	37.1	9.7	11.4	9.1	10.7	8
62.2	121.2	135.0	260.9	342.1	486.7	32	37.6	13.1	10.4	10.9	13
20.5	122.3	179.2	258.0	397.2		48.1	19.2	41.5	25.2	19.2	
123.0	181.7	299.3	288.8	303.2	337.2	31.3	37.9	62.7	42.6	31.1	34.3
99.0	117.7	123.9	186.4	283.2		25.7	73.3	43.2	25.6	10.3	
153.3	205.3	323.6	356.8	536.5		70.3	20.1	9.1	8.4	11.9	
136.3	217.8	300.4	328.0	478.7		64	27	9.7	15.9	29.6	
240.8	249.1	247.0	182.1	454.5		75.2	69.5	20.7	57.9	65.3	
84.4	99.2	147.8	275.5	144.5	517.1	23.2	19.5	8.1	27.4	54.5	17.5
176.4	268.2	262.7	438.2	212.8		16	10.4	25.8	20.4	34.8	
39.0	248.7	379.7	406.7	397.1		39.3	43.8	14.6	10.7	10.8	
119.6	116.8	181.1	259.3	478.8		17.8	41.6	19.3	18	10	
112.6	128.5	127.4	194.9	317.0		49.6	26.3	57.2	69.2	16	
135.6	233.2	369.9	433.8	496.1		29.5	73.5	10	29	13	
116.0	226.1	281.9	342.6	435.8	448.9	24.4	69.7	20.8	40.5	68.2	51.1
45.1	288.2	284.5	297.2	312.1	308.0	38.8	9.2	10	38.6	49.8	17.1
96.1	145.3	217.2	373.9	435.5	548.5	49.3	13	8.6	12.1	9.5	8.6

Table 1. Results of the measurements realised with He tracing technique. On the left of the table it can be observed the normalised transfer time in all points of measure from wall (point 1) up to centre of furnace (point 6). On the right side of the table we can read the He concentration at the top or furnace. When the furnace was hotter the last point (in the centre) did not measure.

It is necessary to correct the total transfer time previously obtained (ti) because the gases are picked up to different temperatures. When a furnace area is higher temperature, the gases go up quicker and the transfer time is reduced, but that does not indicate that the furnace has a better real permeability. As the hottest furnace is the central area if the influence of temperature is not corrected, seems as if a bigger permeability in the central area and a smaller flown in the wall exist. Then, it will be used a normalised transfer time (t_n) (equation) that will be independent of the temperature (T_b) inside the blast furnace. "t_n" enables more reliable comparisons than "t_i" when the gas flows faster because the gas temperature is higher although the permeability inside the furnace is not better.

$$t_n = \frac{T_b}{T_N} t_i \tag{4}$$

The helium concentration was measured by a spectrometer and its maximum was employed to predict the gases concentration in the furnace. Its concentration is a ratio between helium mass and air mass and although helium mass is constant, because is an inert gas, the air mass change with the reactions produced in the furnace.

$$ppm_{He\_measured} = \frac{m_{He}}{m_{Air}} \tag{5}$$

3. Discussion

The quantities of CO and CO_2 at the top gas are related to each other in the parameter etaCO ($\varsigma CO = \frac{CO_2}{CO + CO_2}$) calculated in each point of measure; however the lineal correlation among maximum He ppm in each point and measured etaCO is always negative and small (lower than 0.2) what would indicate that it is necessary to consider more influence factors. The detected fraction of H_2 in the top of furnace is higher than in the air due to the inner reaction in furnace. Consequently, the compounds form heavier gases; therefore, to more H_2 less He ppm should be detected. Whether etaCO and H_2 are added, a negative correlation with detected He ppm of the 0.3 is obtained (see Fig. 3). It is possible that it exist other variables that influence in this relation but we did not study in this paper. Therefore, higher maximum He ppm are detected, lower gases are reacting inner the furnace.

On the other hand, helium employs more time ascending when the burden is less permeable so that there is more time for the inner reactions takes place. It is possible to have an indicator about how the reaction in the burden evolves if it is computed the ratio between both variables (time in seconds/maximum He ppm). This ratio will name "Reaction" and when the time increases or the ppm diminishes, this parameter (Reaction) increases indicating that bigger reaction takes place in that area. The average measure of this variable in each point is show in Fig. 4:

The outliers can be used to detect anomalous situation. For this reason all measures evaluated as "normal state" were selected to calculate the mean value of the variable Reaction.

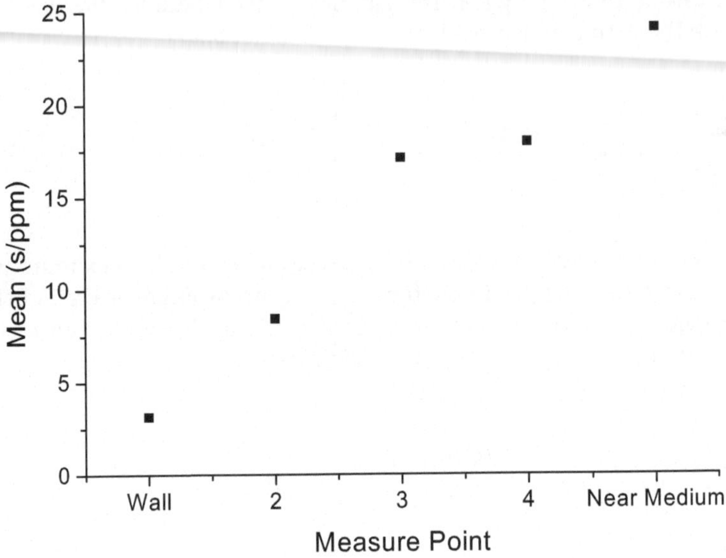

Fig. 4. In this figure, we can be see the mean value of the ratio between normalised transfer time and maximum He ppm. It can be observed that this ratio (it will name as Reaction) increase in the centre of furnace and its smaller in the wall.

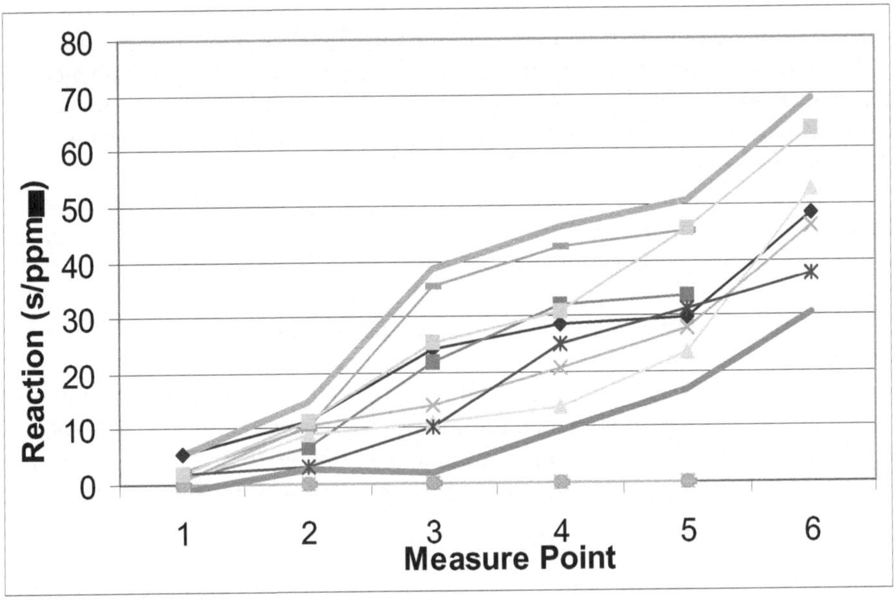

Fig. 5. In this figure has been represented the Reaction value obtained in all measures which technicians evaluated like "normal state" of furnace. In the figure have been included two gross lines that represent the mean value of Reaction plus/less the double of standard deviation for each point.

Additionally to normal data two calculated lines are drawing using the mean value plus/less the double of standard deviation for each point. These two lines will be considered as operation limits for a normal furnace state (see Fig. 5).
To represent all information in an optimum exposition, a ternary diagram representing the normalised variable named "Reaction" has been used. The six points of furnace define three specific working areas in the transversal section of the shaft: wall, medium and central. The wall area corresponds with the two first point's measure, the central area corresponds with the two last points and the two intermediate points correspond with medium area. Reactions profiles recorded during normal periods are shown in Fig. 6. The ranges of reaction in which furnace is correctly running are W (0-25%), M (40%-60%) and C (25%-50%).

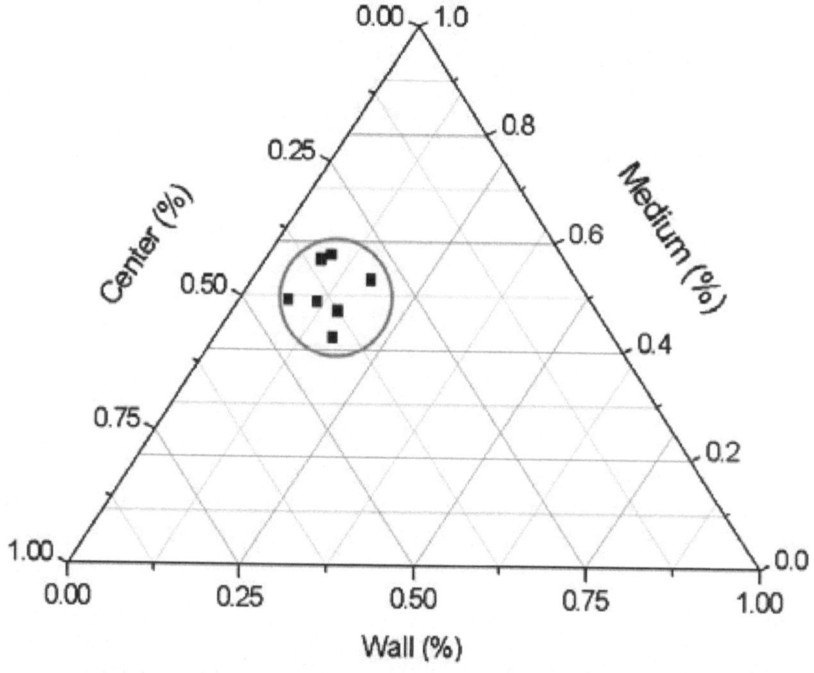

Fig. 6. In this figure it has been represented in a ternary diagram all measures evaluated by technicians as normal conditions. It can observed that all measures area grouped

It can be see that all points selected as "normal running" are grouped in the same area of the diagram. The points out this region represent 'non optimun' states of furnace, i.e, if the point is displaced at right then there are more reaction in the wall than in the centre and the furnace cooler can be observed.
The Fig. 7. represents all measures where the evaluation was certificated as "abnormal state". The points representing the measures are dispersed out of the area defined previously, except in two cases. It can be observed that almost always where the technicians indicated that the furnace was in abnormal state the measure realised by helium trazing

technique and evaluated with the variable Reaction showed an abnormal situation in the ternary diagram.
With this results, it is possible obtain only one parameter to characterise the evolution of blast furnace.

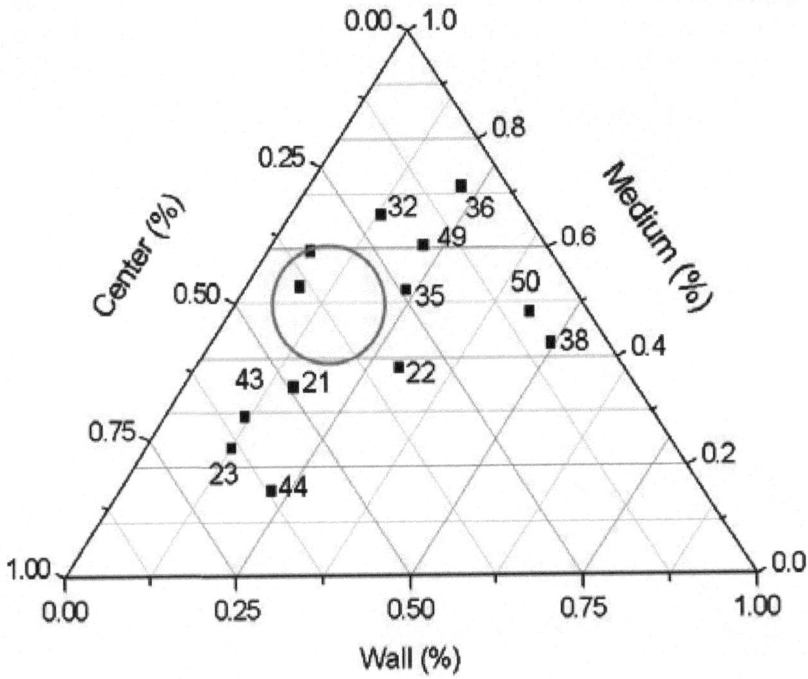

Fig. 7. In the figure it has been represented in a ternary diagram all measures evaluated by technicians as "abnormal state". It can observe that these measures are dispersed by the ternary. The origin of an abnormal state in each measure can be different

4. Conclusions

Calculated variables from the measurements allow a concise characterization of the blast furnace state. Gas transfer measurements can be considered as a new tool to evaluate the state of a furnace in a specific moment because although the results do not have an excessive repeatability they can be a help to the operator.
The main advantage will be that employed only one measure it will be possible to evaluate the furnace.

5. References

Aranguren, F. y Mallol, A.: (1962) *Siderurgia.* Editorial Dossat, S.A., 978-84-237-0049-3, Spain

Babich, A., Yarosevskii, S., Formoso, A., Isidro, A., Ferreira, S., Cores, A. and Garcia, L. (1996) *Increase of pulverized coal use efficiency in blast furnace* ISIJ Int., 36, 1250-1258, ISSN:0915-1559

Babich, A., Yarosevskii, S., Formoso, A., Cores, A., Garcia, L., and Nozdrachev, V. (1999), : *Co-injection of Noncoking Coal and Natural Gas in Blast Furnace* ISIJ Int., 39 229-238, ISSN:0915-1559

Busby, N.J., Fray, T.A.T., and Goldring, D.C. (1994), *Nature of cohesive zone in blast furnace* Ironmaking and Steelmaking, 21(3) 229-236

Ergum, S. (1952), *Fluid Flow Through Packed Columns, Chemical Engineering Progress,* Chem. Eng. Progress., 48 (2), 89-94

Formoso, A., Babich, A., Gudenau, H.W:, Garcia, L. and Cores, A (1999) *Heat Exchange in the Hearth of a Blast Furnace Operating with Combined Blast Parameters* ISIJ Int., 39 (11), 1134-1139, ISSN: 0915-1559

Gisonni, C. (2003). "*Henry Darcy and the pipe flow formula*", in Henry P. G. Darcy and Other Pioneers in Hydraulics: Contributions in Celebration of the 200th Birthday of Henry Philibert Gaspard Darcy, G.O. Brown, J.D. Garbrecht, and W.H. Hager (eds), ASCE, Reston, VA. pg. 24-36.

Hager, W. (1994). "*Die historische Entwicklung der Fliessformel*". Schwiezer Ingenieur und Architekt. 9(24), 123-133

Harting, W., Langer, K., Lüngen, H.B. and Stricker, K.P, (2000) *Proceding of Ironmaking Conf..,* Vol 59, Pittsburg, Pennsylvania, USA, : 9781886362437, 3

Havelange, O., Danloy, G., Franssen, R., and Bonte, L.: (2000), Rev. Metall.-CIT, 97 29

Havelange, O., Danloy, G., Franssen, R.(2004), *The dead man, floating or not?* Rev. Metall.-Pariz, 3, 195-201

Kundrat, D.M., *Conceptual model of the iron blast furnace considering thermodynamic and kinetic constraints on the process* Metall. Trans. B, 20B, 205-218, ISSN: 1073-5615

Nicole, R., Steiler, J.M., and Manin, M. (2000), : Rev. Metall.-CIT, 97, 61.

Nikus, M., and Saxen, H. (1996), : Ironmaking Conf. Proc., Vol. 55, 189.

Pandey, B.D., Sen, P.K. and Chattrjee, A (1996).: Ironmaking and Steelmaking, 23, 31.

Sert, D., Danloy, G., Havelange, O., and Saiz de Ayala, J. (2004) *New approach for the determination of the blast furnace cohesive zone",* ECSC, ISBN 92-894-7491-2. Luxembourg,

Steiler, J.M. (1998), *Proceding of Ironmaking Conf.,* Vol. 57, Toronto, Ontario, Canada,, 161

UNESID, (1998), *La fabricación del Acero,* Editorial Unión de Empresas Siderurgicas, 978-84-605-7290-9, Spain

12

Experimental Investigations into the Production Behavior of Methane Hydrate in Porous Sediment under Ethylene Glycol Injection and Hot Brine Stimulation

1. Introduction

Natural gas hydrates (NGH) are solid, non-stoichiometric compounds formed by host water molecules with small guest molecules, such as CH_4, C_2H_6, C_3H_8, CO_2, H_2S, etc (Sloan & Koh, 2008). Natural gas hydrate deposits involve mainly CH_4, and occur in the permafrost and in deep ocean sediment, where the necessary conditions of low temperature and high pressure exist for hydrate stability. Estimates of world hydrate reserves are very high, and vary from 0.2×10^{15} – 120×10^{15} m^3 of methane at STP (Standard Temperature and Pressure) (Sloan & Koh, 2008; Milkov, 2004; Klauda & Sandler, 2005). However, even with the most conservative estimates, it is clear that the energy in these hydrate deposits is likely to be significant compared to all other fossil fuel deposits, and was considered to be a potential strategic energy resource (Makogon et al., 2007; Collett, 2004; Moridis et al., 2009a).

Techniques for gas production from hydrate reservoir are based on three major dissociation principles, i.e.: 1) Depressurization (Yousif et al., 1991a; Moridis et al., 2004a; Moridis et al., 2007; Tang et al., 2007), to decrease the reservoir pressure below the hydrate dissociation pressure at a specified temperature; 2) Thermal stimulation (Kamath et al., 1991; Li et al., 2006, 2008a, 2008b; Tang et al., 2005a; Kawamura et al., 2007), to heat the reservoir above hydrate dissociation temperature with hot water, steam or hot brine injection; 3) Chemical inhibitor stimulation (Sira et al., 1990; Sung et al., 2002; Li et al., 2007a, 2007b; Kawamura et al., 2005a, 2005b), to inject chemicals, such as methanol or EG to shift the hydrate pressure-temperature equilibrium conditions. It is essential for safely and efficiently producing natural gas to characterize the hydrate dissociation mechanism and multiphase flowing mechanism of dissociated gas and water in hydrate-existing sediment.

Experimental investigations of hydrate dissociation behaviors under depressurization, thermal stimulation and chemical inhibitor stimulation in sediment have been reported. Yousif et al. (Yousif et al., 1990, 1991a, 1991b) developed a one-dimensional model to study depressurization-induced hydrate dissociation in berea sandstone cores suggested that a moving boundary model provide a satisfactory fit to hydrate dissociation measurements. Kono et al. (Kono et al., 2002) measured the dissociation rate of methane gas hydrate by depressurizing method and derived the kinetic dissociation rate equation and the order of reaction. Tang et al. (Tang et al., 2005a, 2007) experimentally investigated the temperature distribution, gas and water production rate, and the thermal efficiency during the hydrate dissociation process using depressurization method and the hot water injection using a one-dimensional physical model. Sung et al. (Sung et al., 2003) examined the flowing characteristics of the dissociated gas and water from hydrate in porous rock by the depressurization and methanol injection schemes, using electric resistance to distinguish the hydrate formation and dissociation. Kamath et al. (Kamath et al., 1991) investigated the dissociation characteristics of the methane hydrates during the brine injection with low salt concentrations and high temperatures, and measured the effect of the temperature, the salinity, and the injection rate of the brine on the hydrate dissociation rate. Kawamura et al. (Kawamura et al., 2005a, 2005b, 2006) analyzed the dissociation behavior of an artificial hydrate core sample in methanol aqueous solution and the experiment was carried out by varying the temperature and concentration of the methanol aqueous solution. The dissociation kinetics of mixed gas hydrates that contain propane as a guest molecule has been investigated as well. Sira et al. (Sira et al., 1990) reported the characteristics of hydrate dissociation process during methanol and EG injection. They concluded that the rate of hydrate dissociation is a function of chemical concentration, injection rate, pressure, temperature of chemical solution and hydrate-chemical interfacial area. Li et al. (Li et al., 2007a, 2007b) experimentally investigated the gas production behavior from methane hydrate in porous sediment by injecting ethylene glycol (EG) solution with different concentrations and different injection rates. The results showed that the production efficiency is affected by both the EG concentration and the EG injection rate. The above experimental work focused on the characterization of hydrate dissociation during methanol and ethylene glycol injection, while few reports are found about the investigation of the dissociation behavior methane hydrate in the porous media under hot brine injection.

Among these methods, the thermal stimulation is capable of producing substantial amounts of natural gas, and it will be more effective combining with the depressurization or the chemical injection methods. (Moridis et al., 2003, 2004b, 2009b) So far, the impacts of the higher concentrations (0~24 wt%) of the injected brine solution at a wide temperature ranges (-1~130 °C) are not well understood yet, especially at the medium temperature ranges. And ethylene glycol (EG) is widely known as a thermodynamic inhibitor of gas hydrate, studies of natural gas dissociation in the presence of EG are limited, especially the impacts of the EG concentration and injection rate on hydrate dissociation are not well understood yet (Sira et al., 1990; Li et al., 2007a; Kawamura et al., 2005a).

In this work, the production behaviors of MH in unconsolidated sediment under hot brine and EG injection were investigated in a developed one-dimensional experimental apparatus. In the experiments of the hot brine injection, NaCl aqueous solution with the concentration of 0~24 wt% and the temperature of -1~130 °C was injected into the vessel with the injection rate of 9.99ml/min. The dissociation kinetics of methane hydrate in porous sediment, the production behaviors of gas and water, and the thermal and the energy efficiencies of the hydrate recovery process under hot brine injection were investigated. The experiments of the EG injection were carried out by varying the concentration and injection rate of the EG solution. The relationship between these two parameters and the production behavior was obtained. In addition, the efficiency of the gas production process during MH dissociation was determined.

2. Experimental Apparatus

The schematic diagram for the one-dimensional experimental apparatus used in this work is shown in Figure 1. The pressure vessel was immersed in an air bath with the temperature range from -20 °C to 80 °C, ±0.5 °C to maintain a constant temperature. The vessel is made of quartz glass and has an internal diameter of 30mm and a length of 534mm, and it can be operated up to 10MPa. Four resistance thermometers and two pressure sensors with three differential pressure transducers were uniformly-spaced, shown in Figure 1, to measure the temperature and pressure profile along the vessel. The thermometers are Pt100 with the range of -20 °C to 200 °C, ±0.1 °C. The pressure transducers are KELLER PA-21S 80400, 0-20MPa, ±0.25%. Two gas flow meters, which were used to measure the gas injection or production rate and the cumulative gas produced from the vessel, are both of D07-11A/ZM, 0-1000 ml/min, ±1% from "seven star company". The pressure transducers, thermometers, gas flow meters, were calibrated using a pressure test gauge with the error of ±0.05%, a mercury thermometer with the tolerance of ±0.01 °C, and a wet gas meter with the accuracy of ±10 ml/min, respectively. The two balances, which were are used to measure the rates of liquid input and output, respectively, are Sartorius BS 2202S, 0-2200g, ±0.01g. In order to protect the metering pump from corrosion by the hot brine or chemical inhibitors, the middle containers have been used for the solution injection. The data acquisition system records pressure, differential pressure, temperature, gas/water injection rate and production rate. The vessel and the loading pipelines were covered from a heat-insulating material to avoid the heat loss during brine injection. The NaCl used was analytically pure supplied by Guangzhou Chemical Reagent Factory, China. The EG was analytically pure, and the methane gas was produced by Foshan Kody Gas Chemical Industry Co., Ltd, China with its purity of 99.99%.

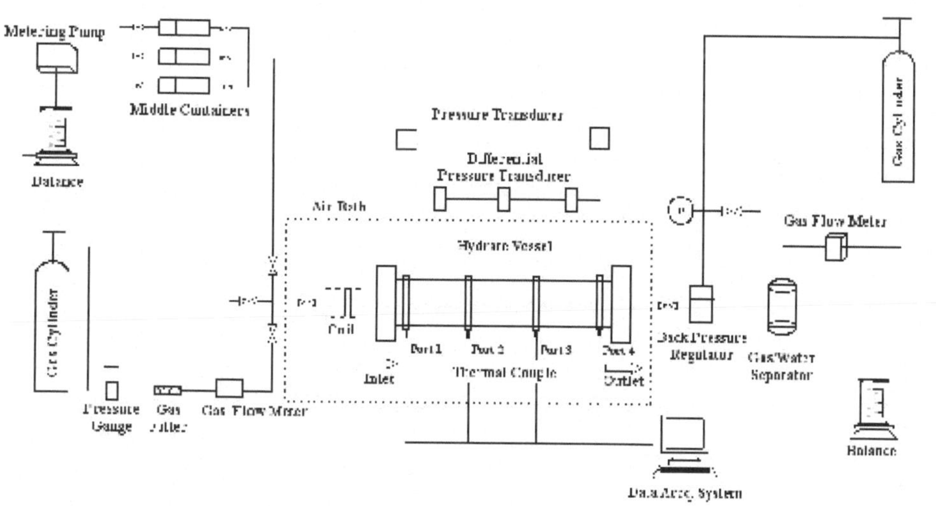

Fig. 1. The schematic plot of the experimental apparatus

3. Hot brine stimulation

3.1 Experimental Procedures

During the experiment, the raw dry quartz sand with the size range of 300-450 μm are tightly packed in the vessel, and then the vessel was evacuated twice to remove air in it with a vacuum pump. NaCl aqueous solution with the concentration of 2 wt% instead of distilled water was used for all experiments because of the higher formation rate of brine solution than that of the distilled water in sediments. (Tang et al., 2005b) Hence, in this work the dry quartz sand in the vessel was wetted at atmospheric pressure with 2 wt% NaCl aqueous solution, which was injected into the vessel from a the middle containers with a metering pump, when the amount of the liquid driven from the vessel was equal to those injected. Then the methane gas was injected into the vessel up to the pressure, which reached much higher than the equilibrium hydrate formation pressure at the working temperature. After that, the vessel was closed as an isochoric system, and the temperature in the vessel was gradually decreased to form the hydrate by changing the air bath temperature. In this work, the temperature of the air bath during the hdyrate formation is 0 °C. It was considered that the formation had not completed until there was no pressure decrease in the vessel.

Runs	T before formation °C	P before formation MPa	P before injection MPa	T of injected brine °C	Brine injected wt%
H1	25.7	5.898	3.041	130	0
H2	23.2	5.635	3.022		8
H3	20.3	5.617	3.095		16
H4	24.3	5.641	3.097		24
H5	25.2	5.641	2.973	90	0
H6	25.5	5.904	3.052		8
H7	28.7	5.849	3.095		16
H8	26.3	5.849	3.102		24
H9	24.0	5.922	2.888	50	0
H10	27.4	5.825	2.961		8
H11	28.8	5.934	2.955		16
H12	28.8	6.251	3.01		24
H13	27.5	6.013	2.967	-1	2
H14	27.8	6.007	3.047		8
H15	26.4	5.916	3.113		16
H16	26.2	5.507	2.845		24

Table 1. Experimental conditions of MH dissociation by hot Brine injection

Methane hydrate dissociation under hot brine injection was carried out in the following procedures. The brine with certain concentration was prepared in the middle container. The back pressure regulator was set to the certain required pressure value, which is the system pressure during the hydrate dissociation process under hot brine injection. The brine was firstly heated or cooled by the heater/cooler in the middle container. Then the brine was injected into the vessel. After a few minutes, gas and liquid began to release from the vessel through the outlet valve. During the entire formation and the dissociation runs of methane hydrate, the data thus as the temperature and the pressure in the vessel, the gas production, the amount of injected brine, and the liquid production were recorded at 6 second intervals. The brine injection rate is 9.99 ml/min. The detailed experimental arrangement is shown in Table 1.

3.2 Hydrate Dissociation

In this work, a total of 16 experimental runs were carried out to investigate into the dissociation behaviors of methane hydrate in the porous media by injecting the aqueous solutions of NaCl at -1, 50, 90 and 130 °C, respectively, and with the injection rate of 9.99 ml/min, as shown in Table 1.

In Run H1-Run H12, brine solutions were with 0, 8, 16 and 24 wt%, respectively. In Run H1, Run H5 and Run H9, the methane hydrate was dissociated with the effect of hot water (0 wt% brine solution) at 50, 90 and 130 °C, respectively. In Run H13- Run H16, the methane hydrate was stimulated with the brine solution with the concentrations of 2, 8, 16 and 24 wt%, respectively, and at -1 °C, lower than that of the hydrate system in the vessel (0 °C). Since the hydrate in the sediment in the vessel from 2 wt% NaCl solution was formed at 0 °C when, as did in Run H13, the brine was injected at 2 wt% and -1 °C into the vessel, the hydrate was not dissociated from the brine solution. Hence, the purpose of Run H13 is to determine the amount of free gas in the vessel. In addition, in Run H14-Run H16, .the hydrate was dissociated on account of the effect of the brine solution without thermal effect. In other experimental runs, the hydrate was dissociated from the synergistic effect of the heat and salt.

In this work, the following processes were involved, brine replacement of pore solution, heat transfer, and ion diffusion accompanying brine injection. Among these processes, the high salinity or high temperature can cause hydrate dissociate by shifting hydrate outside its stability regime. The high salinity can result either by the direct replacement of the original pore solution with high concentration of brine or by the diffusion of ions due to a concentration gradient, and the high temperature can be caused either by the direct replacement of the original pore solution with high temperature brine or by thermal transfer due to a temperature gradient.

3.2.1 Hot water stimulation

Figure 2 shows the curve of the temperature change with time at Port 2, Port 3 and Port 4 (the outlet) in the vessel with the injection of hot water at 50, 90 and 130 °C, respectively, as did in Run H9, Run H5 and Run H1. As shown in Figure 2, each temperature curve consists of the horizontal section and the ascending section. The horizontal section represents the hydrate non-dissociation processes and the isothermally endothermal dissociation processes resulting from the direct replacement of the original pore solution and heat transfer from hot water; the ascending section is the heterothermally endothermal process of the system in the porous media in the effect of hot water after the hydrate dissociation has fully completed. As shown in Figure 2, in the above three runs, the dissociation firstly happened in the inlet of the vessel, and then in Port 2, Port 3 and Port 4 with time in turn until the hydrate in the vessel was completely dissociated. Accordingly, it is considered that the dissociation process of the hydrate in the vessel is the moving-forward process of the hydrate dissociation boundary from the inlet to the outlet. In other words, the flowing of hot water injected in the vessel can be regarded as the moving of a piston from the inlet to the outlet.

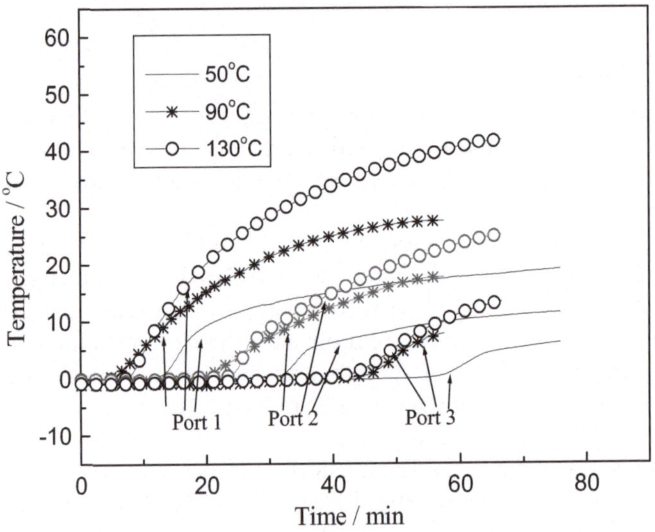

Fig. 1. The curve of the temperature change over time at Ports 1-3 in the vessel in Run H1, Run H5 and Run H9 with the effect of hot water at 50 °C, 90 °C and 130 °C

3.2.2 Brine stimulation

In Run H13-Run H16, the experiments of the brine stimulation was carried out. Figure 3 shows the curve of the temperature change over time at Port 4 in the vessel with the injection of the brine solution at -1 °C with 2, 8, 16 and 24 wt%, respectively. For Prot 1 and Port 3, the characteristics of the curve of the temperature change over time are similar. Since in the above experiments, the brine solution was injected into the vessel at -1 °C, lower than that of the hydrate system in the vessel (0 °C), the hydrate dissociation can be only caused from the inhibitors, not from the thermal effect.

As shown in Figure 3 and discussed above, under the injection of the brine at 2 wt% and -1 °C into the vessel, the hydrate was not dissociated. However, the hydrate dissociation can be caused by the effect of the brine solution with higher concentrations. As shown in Figure 3, the process of the hydrate dissociation is the process of the temperature decrease, which is the result of the presence of the brine solution. Since the temperature drop was caused by the heat balance between that needed for hydrate dissociation and that supplied from surrounding environment, the lowest point of temperature represents the occasion when hydrate dissociated most intensely. In addition, it was found that the time for the hydrate dissociation is shortened and the degree of depth (well depth) of the temperature drop increases with the increase of the concentration of the brine solution.

According to the calculation, about 16 minutes has been needed for brine to replace the pore water around the temperature sensors of Port 4 in Run H13-Run H16 with the effects of the different NaCl concentrations at -1 °C. However, the lowest points of temperature have occurred after lapse of time when the replacement had finished. This was caused by salinity change of pore water due to ion diffusion.

Figure 16 gives the curve of the temperature change with time at Ports 1-3 in the vessel in the presence of brine solution with 24 wt% and at -1 °C. As shown in Figure 16, there is a well depth of the temperature change in each temperature curve at Ports 1-3, and the wells appear with time in turn and the depths of the wells from Port 2 to Port 4 gradually increase. In the process of the hydrate dissociation, it might be caused by the direct replacement of pore water with brine at ports 1 and 2, resulting in the thermal homogenization, while the temperature change at Port 4 was caused by salinity change of pore water due to ion diffusion.

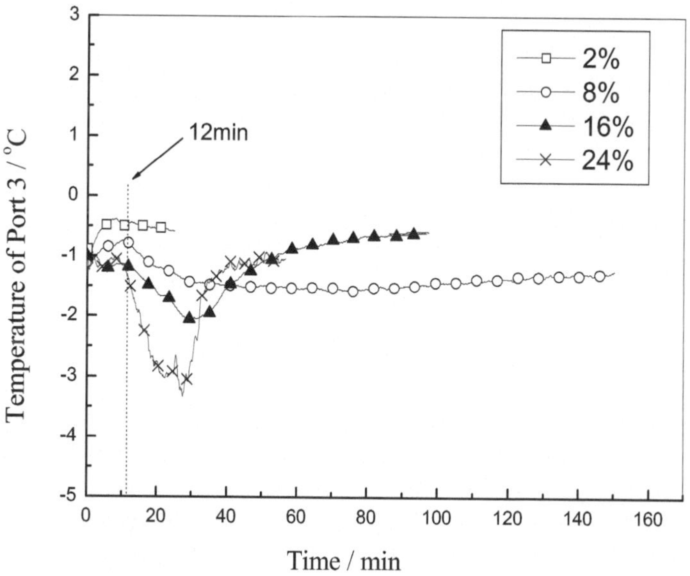

Fig. 3. The curve of the temperature change over time at Port 4 in the vessel in Run H13-Run H16 with the effects of the different brine concentrations at -1 °C

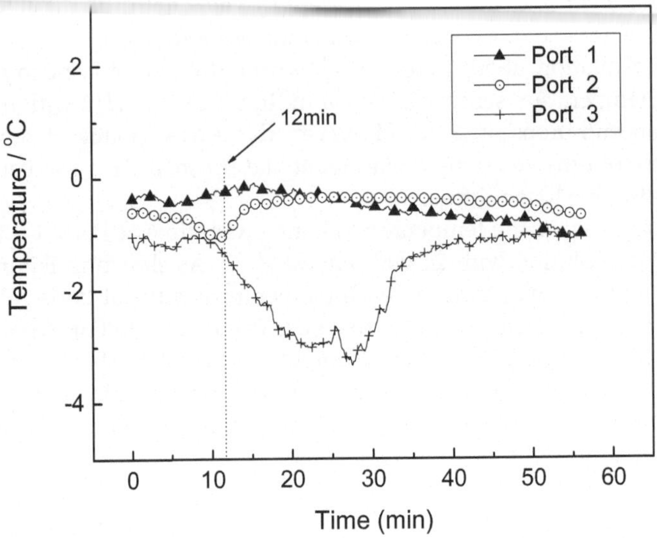

Fig. 4. The curve of the temperature change over time at Ports 1-3 in the vessel in Run H16 with the injection of 24 wt% brine solution at -1 °C

3.2.3 Hot Brine stimulation

Figure 5 gives the typical curve of the temperature change with time at Ports 1-3 in the vessel in the presence of hot brine solution with 24 wt% and at 90 °C. It is shown from the figure that at Port 4, the curve can be divided into three sections: the horizontal section, the downward section and the upward section. The horizontal section represents the non-dissociation and the isothermally endothermal dissociation (phase transformation) processes of the hydrate still without the effect of the inhibitor. The downward section is the cooling endothermal dissociation process of the hydrate on the effects of the hot water and brine solution. In this section, with the increase of concentration of brine solution with time, which acts on the surface of the hydrate, the temperature of the hydrate gradually decreases and the hydrate gradually dissociates until the dissociation is completed while the concentration of brine solution reaches the maximum value. The upward section is only the heterothermally endothermal process of the system in the porous media in the effect of heat after the hydrate dissociation has fully completed. In the section, there are no the phase transformation. As shown in Figure 5 that the characteristics of the temperature changes with Ports 1 and 2 are similar with Port 4. For other salt concentrations and other temperatures of the injected hot solutions, the characteristics of the temperature change are also similar with the above. In addition, as shown in the figure, the flowing of hot brine water injected in the vessel can be also regarded as the moving of a piston from inlet to outlet, as analyzed in Figure 2.

Temperature changes in Port 4 in Run H4, Run H8, Run H12 and Run H16 over time with the injection of the brine of 24 wt% at -1, 50, 90, 130 °C, respectively, have been shown in

Figure 6. The experimental results illustrate that with the brine injected at the same concentrations the same lowest value of temperature decrease of the hydrate system at the same port has been produced and it is independent of the initial temperatures of the injected solutions. The temperature changes over time with the brine injected at the other same concentrations at -1, 50, 90, 130 °C show the similar characteristics.

Figure 7 gives a typical curve of the temperature change over time at Port 4 with Run H1-Run H4 through injecting brine solution with the concentrations of 0, 8, 16, and 24 wt%, respectively, at 130 °C. As shown in Figure 7, it is noted that the time for the hydrate dissociation shortened and the degree of the depth (well depth) of the temperature drop increases with the increase of the concentration of brine solution. For other certain temperatures with the different injections of brine solution of 0, 8, 16 and 24 wt%, respectively, the similar characteristics can be obtained.

The dissociation processes of hydrate have been displayed through temperature curves at various ports changing over time. However, for 2 wt% and 8 wt% salinity curves in Figure 3, temperature shows an increase about 0.2- 0.3 °C during about 2 or 3 minutes early. This is due to heat transfer from the air bath after the air bath had been opened partially to turn on input valve and output valve on the purpose of the injection of liquid as shown in Figure 1. Heat transfer to or from the air bath affected all the temperature measurements during about 2 or 3 minutes early. In spite of that, this increase or drop does not demolish the data explain above because it was much lower than the well depth of the temperature change in the temperature curves occurring later.

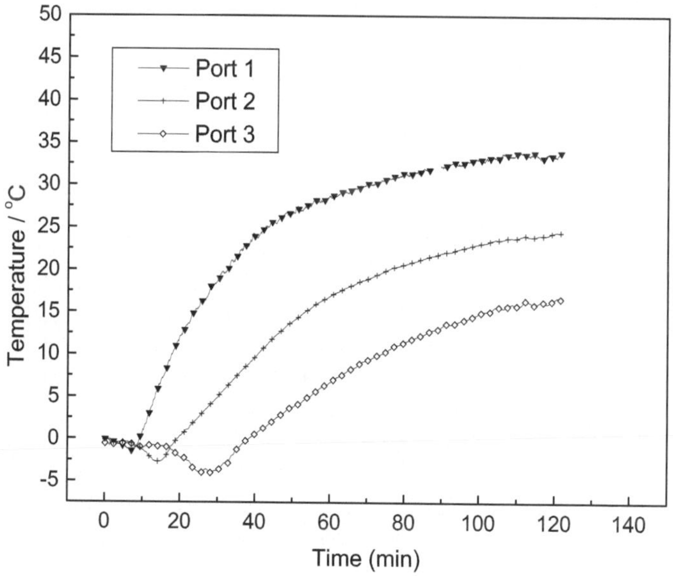

Fig. 5. The curve of the temperature change over time at Ports 1-3 in the vessel in Run H8 with the injection of 24 wt% brine solution at 90 °C

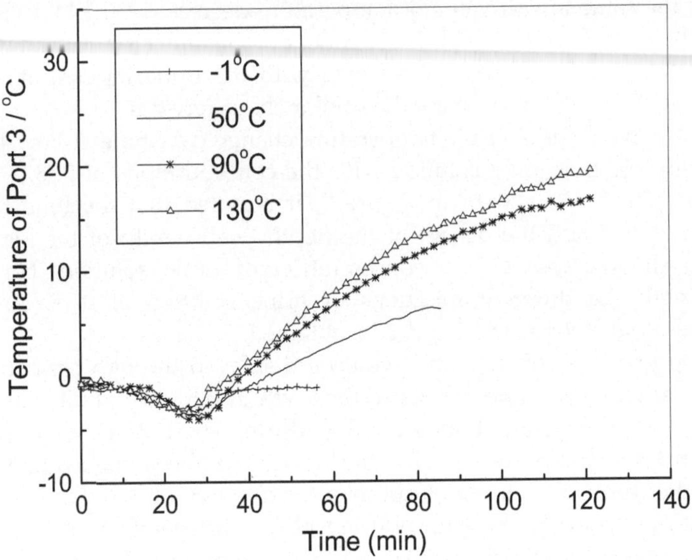

Fig. 6. The curve of the temperature change over time at Port 4 in the vessel in Run H4, Run H8, Run H12 and Run H16 with the injection of 24 wt% brine solution at the different temperatures

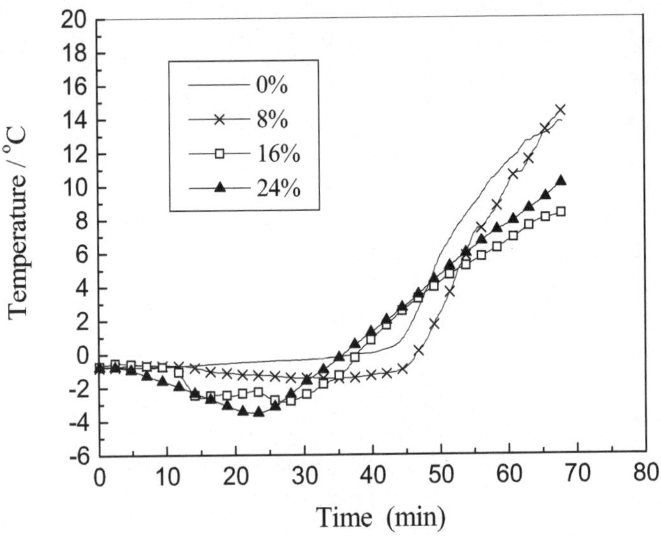

Fig. 7. The curve of the temperature change over time at Port 4 in the vessel in Run H1-Run H4 with the effects of the different brine concentrations at 130 °C.

3.3 Gas production

A typical curve of the accumulative gas production for the whole gas production process in Run H9 is given in Figure 8. As shown in Figure 8, the gas production process with the hot brine or hot water injection in the vessel can be divided into three sections. In Section I, the free methane gas in the vessel is released, and instantaneously gas production rate increases rapidly. The gas production rate could be expressed by the slope of the curve of the accumulative gas production. After the free gas released, the gas production rate decreases remarkably. This section is the hydrate dissociation and gas production process and considered to be Section II. Afterwards in Section III, the hydrate dissociation process has finished, and there is only the residual gas release from the vessel. (Sloan & Koh, 2008) As shown in Figure 8, there are two inflexion points on the curve of the accumulative gas production with time. The left point indicates the end of free gas release process (Section I) and the beginning of the hydrate dissociation process (Section II). The right one means the end of hydrate dissociation process and the beginning of production process of the residual gas (Section III).

Figure 9 gives the accumulative gas production over time with the 2 wt% brine solution injection at -1 °C, which is a typical case of the gas production without the effects of thermal and brine. It can be seen from the figure that there is only the free gas production without the dissociated gas from the hydrate in this case.

Figure 10 shows the accumulative gas production in Section II with the hot water injection at 50, 90 and 130 °C, respectively, as did in Run H9, Run H5 and Run H1. The hydrate dissociation rate increases with the increase of the temperature of the injected hot water during the hydrate dissociation process (Goel et al., 2001).

Figure 11 gives the accumulative gas production in Section II at 50 °C with the injections of the brine solution in the concentration range of 0~24 wt%. The hydrate instantaneous dissociation rate could be increased by injecting brine solution other than water, and it is related to the concentration of injected brine solution. When the brine concentration is less than 16 wt%, the dissociation rate increases with the brine concentration. It is noted that the hydrate instantaneous dissociation rate is approximately the same with the injection of brine solution of 16 wt% and 24 wt% at 50 °C. In other words, if the brine concentration continues rising after reaching certain value, the concentration has little effect on the hydrate instantaneous dissociation rate. Hence, in the process of hydrate dissociation with the injection of hot brine, it is not necessary to use the brine solution with very high concentrations. The accumulative gas production and the hydrate instantaneous dissociation rate at other certain temperature such as -1, 90, and 130 °C, with the injections of the brine solution in the concentration range of 0~24 wt% show the similar behavior.

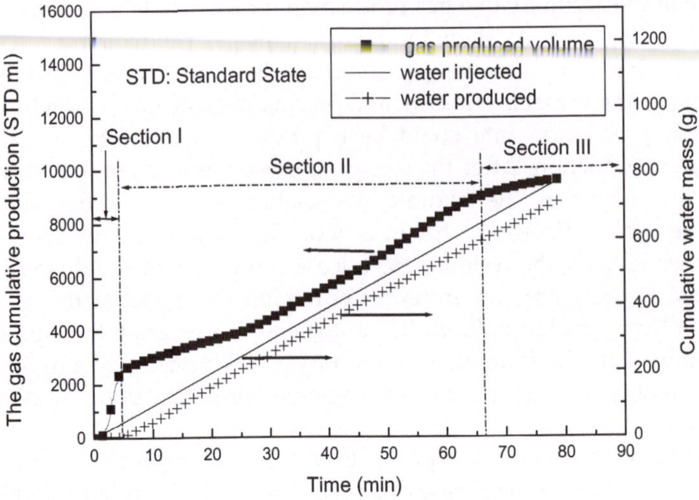

Fig. 8. The accumulative gas production and the accumulative mass of water injected and produced over time in Run H9 with the injection of hot water at 50 °C

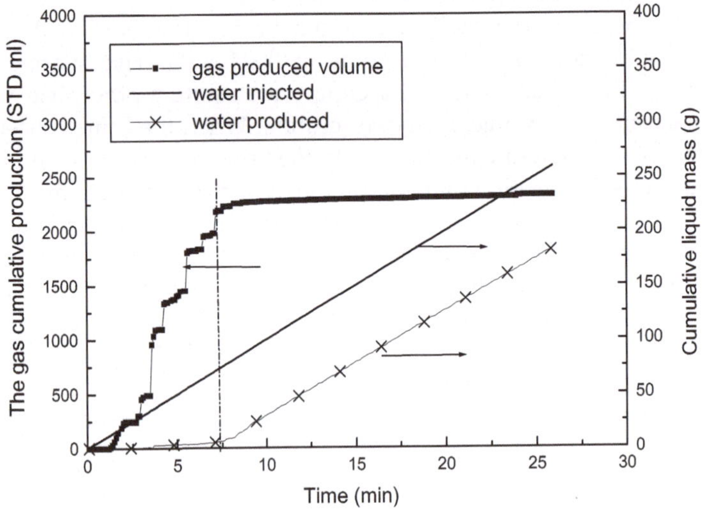

Fig. 9. The accumulative gas production and the accumulative mass of brine injected and produced in Run H13 with the injection of 2 wt% brine solution at -1 °C

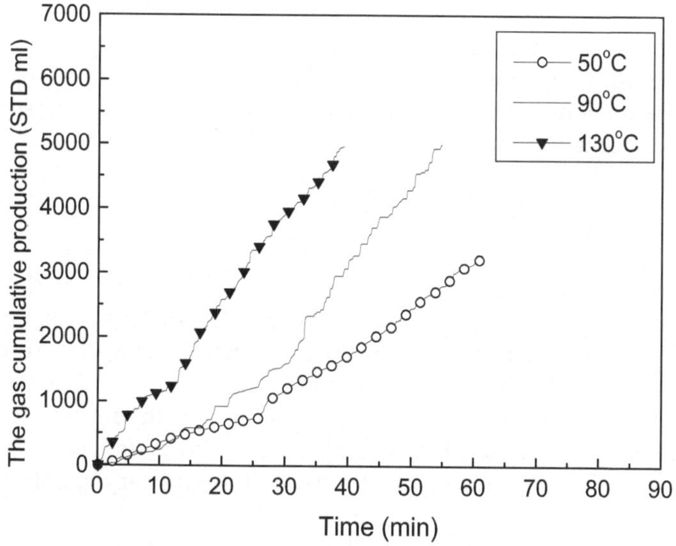

Fig. 10. The accumulative gas production at section II in Run H1, Run H5 and Run H9 with the effects of hot water at 50 °C, 90 °C and 130 °C

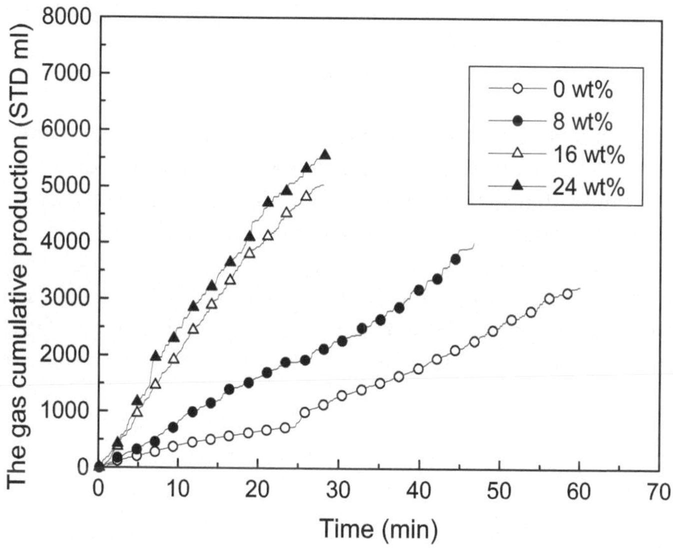

Fig. 11. The accumulative gas production at section II in Run H9-Run H12 with the effects of the different brine concentrations at 50 °C

3.4 Liquid production

As shown in Figure 8, during free gas production, with hot water or hot brine injection there is little liquid production. This stage is one process that free gas in the vessel is drived out, and in this stage, the injected liquid solution stays in the vessel. During the hydrate dissociation, the liquid production rate is slightly higher than the solution injection rate, due to the water produced from the hydrate dissociation. After the hydrate dissociation process finished, the liquid production rate is equal to the solution injection rate.

3.5 Production efficiency analysis

In this work, to determine the efficiency of gas production from the hydrate by hot brine injection, the thermal efficiency and the energy ratio are investigated. The thermal efficiency is defined as the ratio of the heat quantity for hydrate dissociation to the total heat input, which is defined as the amount of heat needed to raise the temperature of the hydrate system in the vessel up to the injection temperature. Thus, when the fluid is injected at 0 °C or less than 0 °C, the thermal efficiency is zero, and there is no thermal effect on the hydrate system in the vessel by the fluid injected. The energy ratio is defined as the ratio of the combustion heat quantity of produced gas to the total input heat quantity (Li et al., 2006, 2008b).

Thermal efficiencies and energy ratios for the hydrate production in the above various experimental runs under hot water and hot brine injections are shown in Figures 12 and 13, respectively. As shown in Figures 12 and 13, the thermal efficiency and the energy ratio decrease with the increase of the temperature of injected hot water at the 0 wt% salinity. For the case of the injection of hot brine solution, the thermal efficiency and the energy ratio increase with the increase of the concentration of injected hot brine with the certain temperature. For hydrate dissociation, more powerful temperature-driving force comes forth resulting from increasing salinity and thus hydrate dissociates more rapidly resulting in smaller the total heat input. Then, increasing thermal efficiency and energy ratio have been obtained.

However, with the differences of the temperatures of the injected hot brine, the degrees of the increases of the thermal efficiency and the energy ratio are different. As shown in Figures 12, 13, it is noted that at low temperature, 50 °C, the increase effectiveness of the thermal efficiency and the energy ratio is apparent with the increase of the concentration of hot brine. Whereas, at high temperature thus as 130 °C, there are only a little increase for them. Hence, it is suggested that in the gas hydrate production by the hot brine injection, the appropriate temperature in conjunction with the high concentration of brine solution brings relative high recovery efficiency. The injection with too high temperature results in the energy loss.

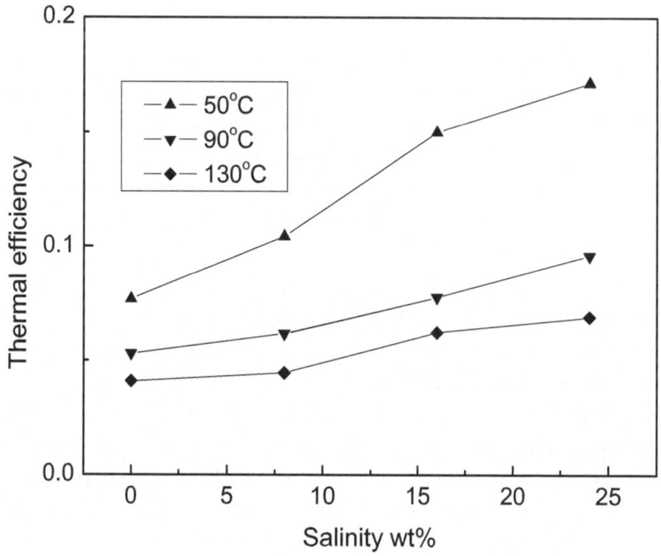

Fig. 12. Thermal efficiencies of gas production with the salinity at 50 °C, 90 °C and 130 °C

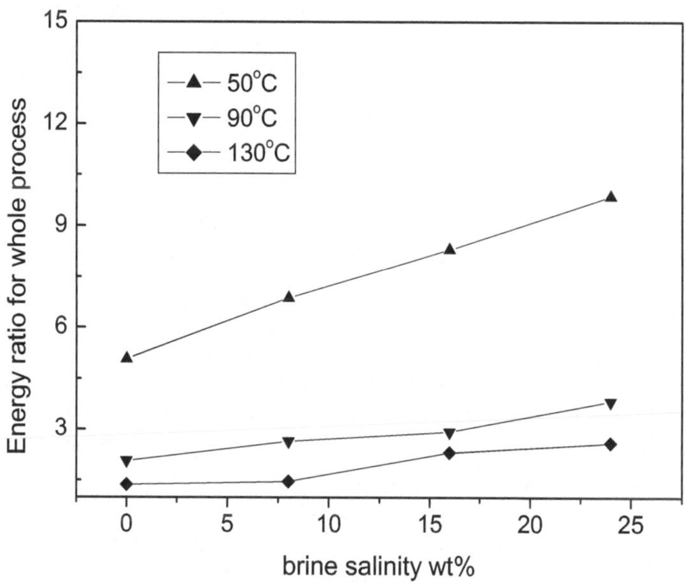

Fig. 13. Energy ratios of gas production with the salinity at 50 °C, 90 °C and 130 °C

4. EG stimulation

4.1 Experimental Procedures

During the experiment, the raw dry quartz sand with the size range of 300-450 μm are tightly packed in the vessel, and then the vessel was evacuated twice to remove air in it with a vacuum pump. The quartz sand in the vessel was wetted to saturation with distilled water using a metering pump. The sand sediment was saturated when the amount of water produced from the vessel was equal to the amount of water injected. It was assumed that the volume of water injected in the vessel was the total volume available in the vessel. Then the methane gas was injected into the vessel until the pressure in the vessel reaches much higher than the equilibrium hydrate formation pressure at the working temperature. After that, the vessel was closed as an isochoric system. The temperature was gradually decreased to form the hydrate by changing the air bath temperature. The hydrate formation was considered to be completed until there was no pressure decrease in the system. The hydrate formation process in general lasts for 2 to 5 days.

The hydrate dissociation by EG injection was carried out in the following procedures. Firstly, the EG solution with the desired concentration was prepared in the middle containers. The back pressure regulator was set to 3.8MPa, which is the system pressure during the hydrate dissociation process under EG injection. Then the dissociation run was started by injecting the EG solution from the middle containers into the vessel. The EG solution was cooled down to the temperature in the air bath before injected into the vessel. After injecting the EG solution for approximately 5 mins, hydrate began to dissociate and gas and water solution were observed to release from the vessel through the outlet valve. The gas production process lasted for 30-100 min, depending on the EG concentrations and injection rates. When there was no significant gas released, the EG injection was finished and the system pressure was released to 1 atm. gradually. During the entire dissociation run, the temperature and pressure in the vessel, the gas production, the amount of EG solution injected and the water production were recorded at 2 seconds intervals.

4.2 Hydrate Formation

Table 2 provides the hydrate formation conditions. The volume of the water and gas before hydrate formation is equal to the total volume of water, gas and hydrate after hydrate formation:

$$V_{w1}+V_{g1} = V_{w2}+V_{g2}+V_{h2} \tag{1}$$

It was assumed that there is 5.75 mol water in 1mol methane hydrate, and the density of methane hydrate is 0.94 g/cm^3 and water in the vessel is incompressible. The volume of the gas in the vessel after hydrate formation was calculated by the pressure and temperature conditions in the vessel using the Peng-Robinson equation. The inlet and outlet pressures of the vessel change simultaneously due to the high porosity and permeability of the sediment, so the pressure in the vessel in this work takes the average of the inlet and outlet pressures.

Figure 14 shows a typical experimental result of the pressure and temperature profiles with time during MH formation in the sediment. It can be seen from Figure 14 that the pressure profile during MH formation could be divided into four sections. In section I (0 min-175 min), the temperature decreased from 17.0 °C to 2.0 °C in isochoric condition, and the pressure decreases from 5.4 MPa to 5.1 MPa due to the gas adsorption on porous the quartz sand and the gas contraction in the vessel. After section I, the closed system was maintained at a constant temperature (2.0 °C) until the end of the experiment. In section II (175 min-280 min), the pressure of the closed system was above 5.0 MPa, which was much higher than the pure hydrate equilibrium pressure of 3.5 MPa at 2.0 °C. (Sloan & Koh, 2008) This section was considered to be the hydrate nucleation process, and in this period of time there was no hydrate formed in the vessel. (Fan et al., 2006) The section III is the hydrate formation process. In this section, the pressure gradually decreased due to the gas consumption during the hydrate formation, and this section takes much longer time than section I and II. In the last section (section IV), no further pressure decrease was observed, and the system was maintained at a constant temperature. Hence, the system reached the thermodynamic stable state.

Total 7 experimental runs of hydrate dissociation by EG injection have been carried out. Run E0 as the blank experiment, which injected the distilled water instead of EG solution, was used to eliminate the influence of the gas production by the liquid injection. Table 3 provides the experimental conditions during hydrate dissociation by EG injection, including the EG injection rate, the EG concentration and the average pressure and temperature during MH dissociation. The hydrate dissociation runs in Table 3 were related to the formation runs in Table 2.

	experimental runs							
	E0	E1	E2	E3	E4	E5	E6	E7
Initial Pressure (MPa)	5.403	5.519	5.488	5.476	5.306	5.311	5.416	5.409
Initial temperature (°C)	17.83	17.89	18.01	17.71	17.83	17.46	17.77	17.95
Final Pressure (MPa)	3.556	3.502	3.467	3.480	3.557	3.566	3.516	3.486
Final temperature (°C)	1.97	1.92	1.81	1.92	2.00	2.07	1.81	1.73
Final amount of water (ml)	43.73	47.53	46.22	45.53	42.18	41.95	42.92	43.26
Conversion of gas to hydrate (%)	33.03	36.77	36.82	36.22	31.44	31.49	33.83	34.52
Hydrate content (vol, %)	7.33	8.16	8.17	8.04	6.98	6.99	7.51	7.66

Table 2. Formation conditions of hydrate related to hydrate dissociation by EG injection

	experimental runs							
	E0	E1	E2	E3	E4	E5	E6	E7
EG injection rate (ml/min)	8.8	4.9	6.8	8.8	8.8	8.8	8.8	8.8
EG concentration (wt %)	0	30	30	30	40	50	60	70
Pressure (MPa)	3.889	3.862	3.926	3.862	3.864	3.85	3.901	3.825
Temperature (°C)	2.043	1.645	2.015	1.985	2.061	1.901	2.010	1.846

Table 3. Experimental conditions during Hydrate dissociation by EG injection

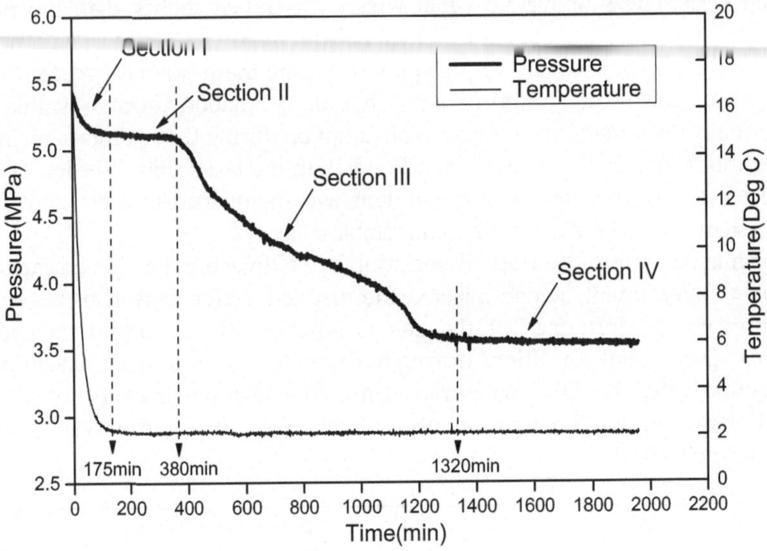

Fig. 14. The pressure and temperature profiles during hydrate formation in isochoric experiment

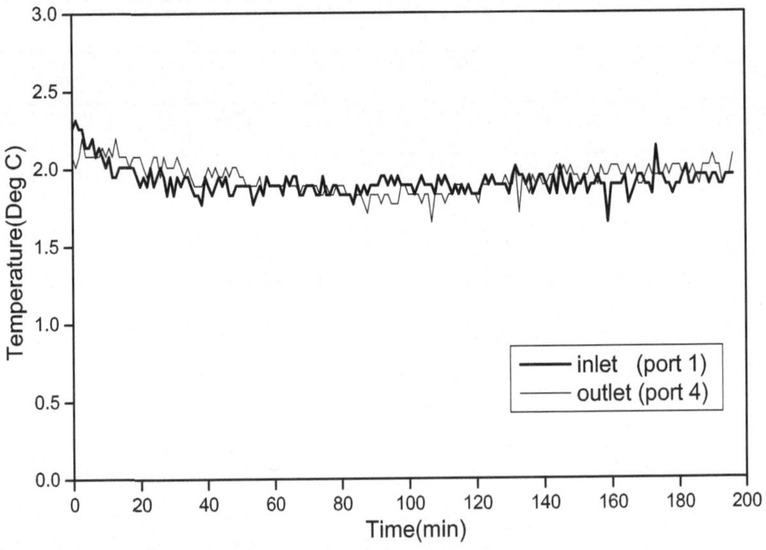

Fig. 15. The inlet and outlet temperature during the EG injection for Run E3

4.2 Hydrate Dissociation

4.3.1 Temperature distribution

Before injected into the vessel, the EG solution was cooled by the coil in the air bath. In an unstirred system, such as the vessel used in the experiment, it is difficult for the methane gas and water to form methane hydrate completely. Hence, only a little hydrate was formed and most part of water or gas was remained in the vessel. Besides, the existing of quartz sand disperses the formed hydrate. Thus, the temperature neither sharply increases in the process of methane hydrate formation shown in Figure 14 nor sharply decreases in the process of methane hydrate dissociation. So the temperature in the vessel maintained constant during EG solution injection.

The inlet (port 1) and outlet (port 4) temperature in the vessel during the EG injection for Run E3 is shown in Figure 15, and the temperature profiles for all other runs show a similar trend with Run3.

4.3.2 Gas and liquid production rate

The gas production rate for whole production process for Run E5, which is a typical one, is shown in Figure 4. The gas production rates for other runs show the similar characteristics. As shown in Figure 4, the process of the hydrate dissociation with the EG injection in one-dimensional vessel can be divided into four main sections. In section I, the free methane gas in the vessel was released. This caused the sudden increase of instantaneous gas release rate, up to 800 ml/min. After the free gas released, the gas production rate decreased and maintained about 30 ml/min. This section is considered to be section II. In section I and II, the EG was diluted by the remained water in the vessel after hydrate formation, and there were little hydrate dissociated due to the low concentration of EG solutions. With further injection of the EG, the concentration of the EG solution in the vessel increased gradually. The EG is a hydrophilic chemical that lowers the activity of water and destroys the structure of the hydrate. When the EG concentration increased high enough to make the hydrate dissociate, the gas production rate suddenly increased to about 100 ml/min, which indicated that the hydrate in the vessel began to dissociate. This is the hydrate dissociation section (section III). The dissociation section lasted approximately 25 min, which is the longest time among the four sections, as shown in Figure 16. Section IV was the last section of the experiment, with remain gas released.

Table 4 provides the Run Etime and gas produced from hydrate dissociation by EG injection for all runs. The EG injection time is from the beginning of EG injection to the end of hydrate dissociation. Onset time for hydrate dissociation is the starting point of section III, and the duration of hydrate dissociation is the length of time of section III. For example, in Run E5 in Figure 16, the onset time for the hydrate dissociation section is 8 min and the end of this section is 32 min, resulting in the duration of hydrate dissociation of approximately 24 min. The gas production ratio is defined as the ratio of the amount of gas generated from hydrate in the hydrate dissociation section and the initial amount of gas contained in all hydrate excluding the free gas in the vessel. Total gas produced after EG injection is also given in Table 4.

The rate of hydrate dissociation by EG injection is a function of EG concentration, injection rate of EG solution, pressure, temperature of the system and hydrate-EG interfacial area. (Sira et al., 1990) In this work, the pressure, temperature and the EG injection rate maintain constant after the EG injection. The instantaneous gas production rates during the whole process were unsteady as shown in Figure 16, while the hydrate dissociation rate decreased continuously with time as illustrated by a typical run (Run E5) in Figure 17. The hydrate dissociation rate was calculated by the gas production rate of section III in Figure 16, in which the gas production was caused by the hydrate dissociation at the certain pressure and temperature.

Figure 18 shows the effect of the EG injection rate on the cumulative gas produced from hydrate dissociation as a function of time for Runs 1-3. The cumulative gas produced from the vessel was measured by the gas flow meter in Figure 1. In Runs 1-3, the EG concentration was kept the same at 30 wt% and the injection rate was varied from 4.9 to 8.8 ml/min. As shown in Figure 18, in general, with the increase of the EG injection rate, the cumulative gas produced increased. As the EG injection rate increase, there were more EG injected into the vessel at the same time, which increased the hydrate-EG interfacial area and stimulated more hydrate dissociate at the same time. The general trend for gas production rate profile is similar in Runs 1-3, but the onset time and duration of hydrate dissociation section are all different with different EG injection rate. As shown in Table 4, from Run E1 to Run E3, the duration of hydrate dissociation section decrease from 73 min to 35 min, while the gas production ratio increased from 38.9% to 50.6%.

Figure 19 shows the effect of the EG concentration on the cumulative gas produced from hydrate dissociation as a function of time for fixed injection rate (Runs 3-7). From Runs 3 to 7, the EG injection rate was maintained same at 8.8 ml/min and the EG concentration was varied from 30 to 70 wt%. Run E0 was the blank experiment, which injected the distilled water instead of EG solution, with the same injection rate as Runs 3-7. Although the general trend for gas production rate profile is similar in Runs 3-7 with the same EG injection rate, the duration of hydrate dissociation decrease as the EG concentration increased from 30 wt% to 70 wt%. As shown in Table 4, from Run E3 to Run E7, the gas production ratio increased from 50.6% to 96.2%. The gas production ratio is larger than 90% while the EG concentration is over 60 wt% during hydrate dissociation. On the other hand, the EG injection time for all runs are different, which decreases with the increase of injection rate and concentration of the injected EG solution in general.

The EG injection and the solution production rate profiles are much simpler than that of the gas production, and Figure 20 gives a typical profile (Run E5). The solution produced from the outlet of the vessel was composed of the EG solution, water in the vessel before EG injection, and water produced from the hydrate dissociation. From Figure 20, the EG injection rate kept nearly constant for the whole production process. While there was fluctuation for the solution production rate, due to the unsteady state during hydrate dissociation process under the chemical stimulation.

	experimental runs							
	E0	E1	E2	E3	E4	E5	E6	E7
EG injection time (min)	-	107	71	43	33	32	29	24
Onset time for hydrate dissociation (min)	-	34	21	8	6	8	7	4
Duration of hydrate dissociation (min)	-	73	50	35	27	24	22	20
Gas produced from hydrate (ml)	-	977	1088	1252	1547	1800	2194	2268
Gas production ratio (%)	-	38.9	43.3	50.6	72.1	83.7	95.0	96.2
Gas produced after EG injection (ml)	-	3496	3334	4025	3210	3933	4180	3368

Table 4. Run time and gas produced from hydrate dissociation by EG injection

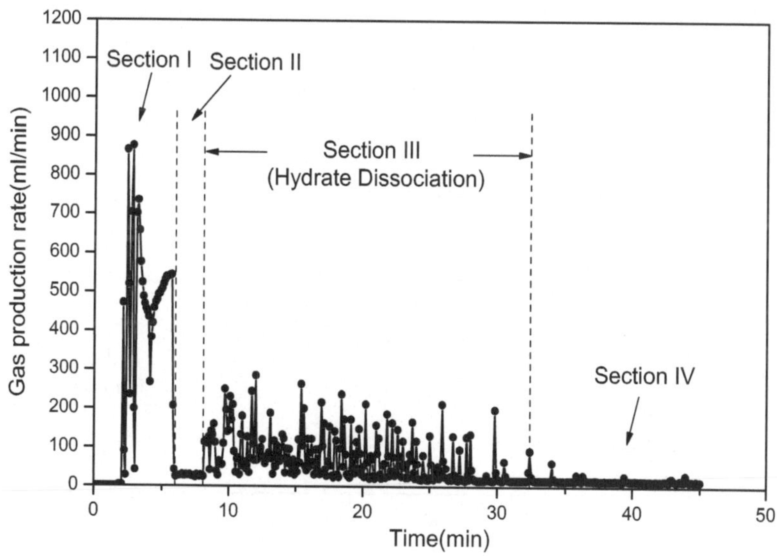

Fig. 16. The gas production rate for Run E5

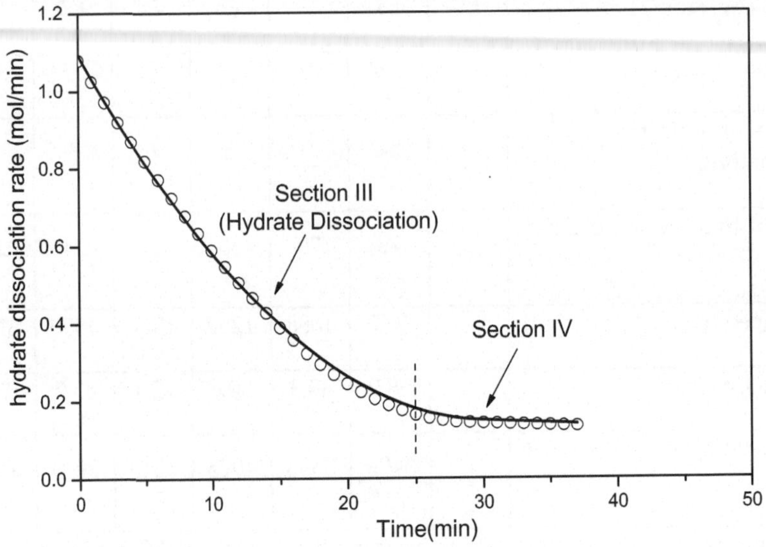

Fig. 17. The hydrate dissociation rate for Run E5

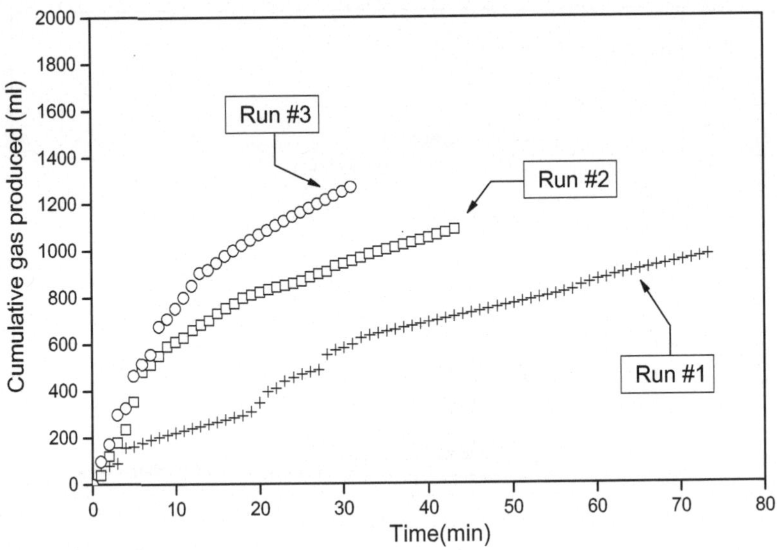

Fig. 18. The cumulative gas produced during the hydrate dissociation for Runs 1-3

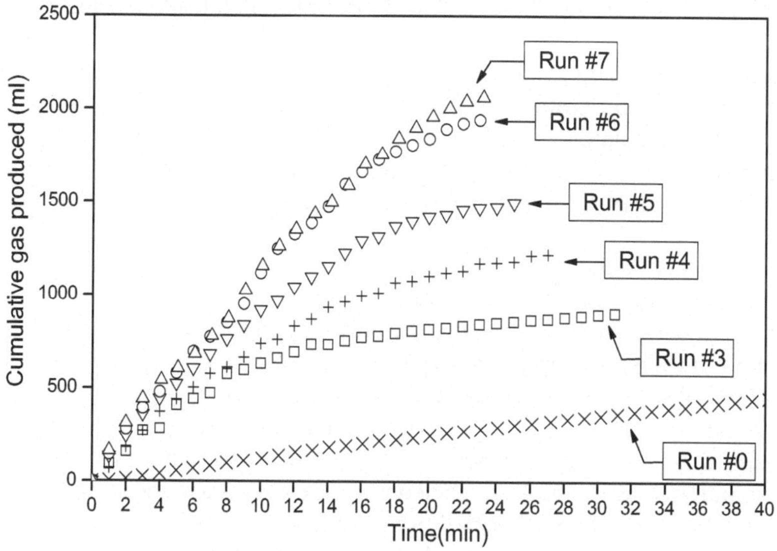

Fig. 19. The cumulative gas produced during the hydrate dissociation for Runs 3-7

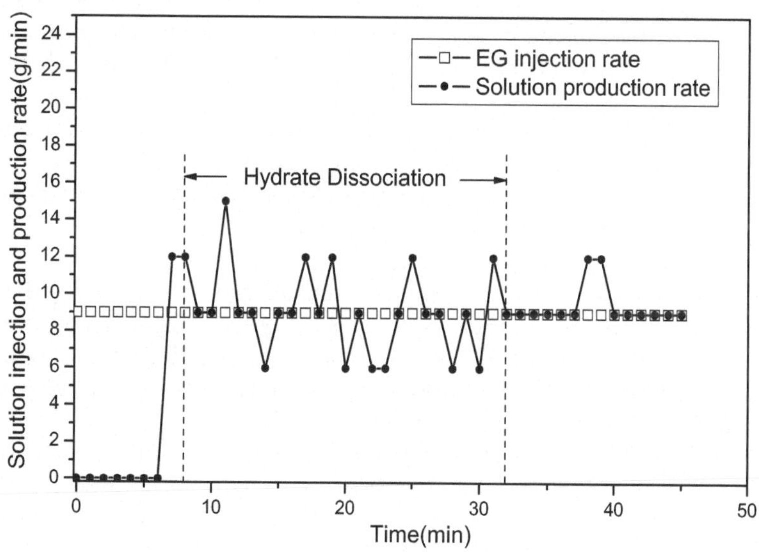

Fig. 20. Solution injection and production rate profile for Run E5

4.3.3 Production Efficiency Analysis

The efficiency of producing gas from hydrate by EG injection is investigated here. In order to compare the efficiency of different runs, the production efficiency has been defined as the ratio of the volume of produced gas to the mass of EG injected in unit time.

Under the EG stimulation, the hydrate dissociated only on the hydrate dissociation section (Section III shown in Figure 16). Table 5 shows three production efficiencies:
1) Production efficiency for hydrate dissociation section (section III) (ml/g/min)
2) Production efficiency at 50% hydrate dissociation (section III) (ml/g/min)
3) Production efficiency for whole injection process (section I to III) (ml/g/min)
The first one was calculated with the volume of gas, the mass of EG injected and the duration time of the hydrate dissociation section (section III in Figure 16). It can be used to measure the gas production efficiency of hydrate dissociation process by the effect of EG in section III.
The second one was calculated the same way with the first one, while it used the values of 50% hydrate dissociation point. In this work, the hydrate dissociation runs were divided into 4 sections, and the hydrate dissociation only happened in section III. The amount of dissociated hydrate was measured by the gas volume released from the vessel in section III. So 100% hydrate dissociation point was the end of section III, when the gas produced from hydrate dissociation all released from the vessel. In the same way, 50% hydrate dissociation point was some time in section III, when 50% gas produced from hydrate dissociation.
Both the first and second production efficiencies were calculated based on the experimental result of section III (the hydrate dissociation section).
The third one was used to measure the whole experimental runs, from the beginning of EG injection to the end of the hydrate dissociation process, which include section I to section III.
The formation conditions of hydrate used in the work was same for all runs, including pressure, temperature, amount of water and the hydrate content in the vessel. For all EG injection runs, the impact of the hydrate content, the amount of water and free gas in the vessel, and the difference of the operating conditions (including the pressure and temperature) during EG injection were all eliminated, while the impact of the EG concentration and injection rate on the production efficiency reflected in Table 5. As shown in Table 5, the production efficiency for the whole EG injection process was between 0.20 and 0.88 ml/g/min, while the efficiency for hydrate dissociation section was between 0.12 and 0.80 ml/g/min. But the production efficiency at 50% hydrate dissociation was much higher, with the maximum of 2.03 ml/g/min.
From the efficiency analysis, the following conclusions can be drawn: (1) the production efficiency of the hydrate dissociation section, 50% hydrate dissociation and the whole injection process varied with the EG concentration and injection rate, and the variation presented the same trend. (2) The hydrate dissociation rate decreased as the experiments go on, as shown in Figure 17. This can explain why the efficiencies at 50% hydrate dissociation were all higher than that of the whole injection process. (3) With the increase of the EG injection rate, the production efficiency increases, as Runs 1-3. The production efficiencies for hydrate dissociation section of Runs 1, 2 and 3 increase from 0.12 to 0.50 when the injection rate increase from 4.9 to 8.8 ml/min. (4) The EG concentration also influence the production efficiency, as Runs 3-7. From the result shown in Table 5, there was a maximum point for the EG concentration on the production efficiency with the same injection rate (9 ml/min), as Run E6 with 60 wt% EG solution. With the increase of the EG concentration, the gas production rate increases, as the result shown in Figure 19. While the mass of EG injected into the vessel increase with the increase of the concentration, the production efficiency reaches the maximum of 0.80 in Run E6. The same result was also concluded on

the production efficiency at 50% hydrate dissociation and the whole production process including the free gas release section.

	experimental runs							
	E0	E1	E2	E3	E4	E5	E6	E7
Production efficiency for hydrate dissociation section (section III) (ml/g/min)	-	0.12	0.28	0.50	0.57	0.60	0.80	0.76
Production efficiency at 50% hydrate dissociation (section III) (ml/g/min)	-	0.44	1.32	1.60	1.61	1.75	2.03	1.53
Production efficiency for whole injection process (section I to III) (ml/g/min)	-	0.20	0.31	0.79	0.79	0.81	0.88	0.86

Table 5. Production efficiency analysis for the hydrate dissociation by EG injection

5. Conclusions

1 The flowing of hot water or hot brine injected in the vessel can be regarded as the moving of a piston from the inlet to the outlet.

2 The hydrate dissociation process is divided into three stages: free gas production, hydrate dissociation and residual gas production.

3 The process of the hydrate dissociation is a process of the temperature decrease in the presence of the brine solution. The duration of the hydrate dissociation is shortened and the degree of the depth of the temperature drop increases with the increase of the brine concentration.

4 The rate of instantaneous hydrate dissociation increases with the increase of brine concentration with the injection of hot brine solution. However, when the brine concentration is higher than the certain degree, the rate of instantaneous hydrate dissociation no longer continues to increases.

5 During the hydrate dissociation, the rate of the liquid production is slightly higher than the rate of the solution injection, due to the water produced from the hydrate dissociation.

6 Thermal efficiency and energy ratio for the hydrate production can be enhanced by injecting hot brine solution, and the enhance effectiveness is apparent with the injection of high brine concentration at the relative low temperature.

7 After the EG injection, the hydrate dissociation in the vessel can be divided into four sections, that is the free gas producing section, EG diluting section, the hydrate dissociating section and the remnant gas producing section.

8 The gas and water production rate were both unsteady during hydrate dissociation rate decrease continuously with time under the EG stimulation, while the EG injection rate kept nearly constant for the whole production process.

9 Under the experiment conditions, with the EG injection rate increasing, the gas production ratio increased, the duration of hydrate dissociation shortened and the production efficiency increased.

10 Under the experiment conditions, with the EG concentration increasing, the gas production ratio increased, the duration of hydrate dissociation process shortened. And the EG concentration also affects the production efficiency. The production efficiency for the whole EG injection process increased with the EG concentration increasing from 0 to 60wt%, and after that the production efficiency began to decrease.

References

Collett, T. S. (2004). Gas hydrates as a future energy resource, *Geotimes*, 49, 11, (2004) pp 24-27, ISSN 0016-8556

Fan, S. S.; Zhang, Y. Z.; Tian, G. L.; Liang, D. Q.; Li, D. L. (2006). Natural gas hydrate dissociation by presence of ethylene glycol. *Energy & Fuels*, 20, 1, (2006) pp 324-326, ISSN 0887-0624

Goel, N.; Wiggins, M.; Shah, S., Analytical modeling of gas recovery from in situ hydrates dissociation. *Journal of Petroleum Science and Engineering* 2001, 29, (2), 115-127, ISSN 0920-4105

Kamath, V. A. M., P.N.; Sira, J.H.; Patil, S.L. (1991). Experimental study of Brine injection and depressurization methods for dissociation of gas hydrate. *SPE Formation Evaluation*, 6, 4, (1991) pp 477-484, ISSN 0885-923X

Kawamura, T.; Yamamoto, Y.; Ohtake, M.; Sakamoto, Y.; Komai, T.; Haneda, H. (2005a). Dissociation experiment of hydrate core sample using thermodynamic inhibitors, *15th International Offshore and Polar Engineering Conference (ISOPE 2005)*, pp. 346-350, Seoul, SOUTH KOREA, Jun 19-24, 2005, ISBN 1-880653-64-8

Kawamura, T.; Yamamoto, Y.; Ohtake, M.; Sakamoto, Y.; Komai, T.; Haneda, H. (2005b). In Experimental study on dissociation of hydrate core sample accelerated by thermodynamic inhibitors for gas recovery from natural gas hydrate, *The 5th International Conference on Gas Hydrate*, pp 3023-3028, Trondheim, Norway, 12-16 June, 2005, ISBN 9781615670666

Kawamura, T.; Sakamoto, Y.; Ohtake, M.; Yamamoto, Y.; Haneda, H.; Yoon, J. H.; Komai, T. (2006). Dissociation behavior of hydrate core sample using thermodynamic inhibitor, *International Journal of Offshore and Polar Engineering*, 16, 1, (2006) pp 5-9, ISSN 1053-5381

Kawamura, T.; Ohtake, M.; Sakamoto, Y.; Yamamoto, Y.; Haneda, H.; Komai, T.; Higuchi, S. (2007). Experimental study on steam injection method using methane hydrate core samples, *Proceedings of The Seventh (2007) ISOPE Ocean Mining Symposium*, pp 83-86, Lisbon, PORTUGAL, Jul 01-06, 2007, ISBN 1-880653-57-X

Klauda, J. B.; Sandler, S. I. (2005). Global Distribution of Methane Hydrate in Ocean Sediment, *Energy & Fuels*, 19, 2, (2005) pp 459–470, ISSN 0887-0624

Kono, H. O.; Narasimhan, S.; Song, F.; Smith, D. H. (2002). Synthesis of methane gas hydrate in porous sediments and its dissociation by depressurizing, *Powder Technology*, 122, 2-3, (2002) pp 239-246, ISSN 0032-5910

Li, G.; Tang, L.; Huang, C.; Feng, Z.; Fan, S. (2006). Thermodynamic evaluation of hot brine stimulation for natural gas hydrate dissociation, *Huagong Xuebao/Journal of Chemical Industry and Engineering (China)*, 57, 9, (2006) pp 2033-2038, ISSN 0438-1157

Li, G.; Li, X.; Tang, L.; Zhang, Y. (2007a). Experimental investigation of production behavior of methane hydrate under ethylene glycol stimulation in unconsolidated sediment, *Energy & Fuels*, 21, 6, (2007) pp 3388-3393, ISSN 0887-0624

Li, G.; Li, X.; Tang, L.; Zhang, Y.; Feng, Z.; Fan, S. (2007b). Experimental Investigation of Production Behavior of Methane Hydrate under Ethylene Glycol Injection, *Huagong Xuebao/Journal of Chemical Industry and Engineering (China)*, 58, 8, (2007b) pp 2067-2074, ISSN 0438-1157

Li, X. S.; Wan, L. H.; Li, G.; Li, Q. P.; Chen, Z. Y.; Yan, K. F. (2008a). Experimental Investigation into the Production Behavior of Methane Hydrate in Porous Sediment with Hot Brine Stimulation, *Industrial & Engineering Chemistry Research*, 47, 23, (2008) pp 9696-9702, ISSN 0888-5885

Li, G.; Li, X.; Tang, L.-G.; Li, Q.-P. (2008b). Control Mechanisms for Methane Hydrate Production by Thermal Stimulation, *Proceedings of the 6th International Conference on Gas Hydrates (ICGH 2008)*, Vancouver, British Columbia, CANADA, July 6-10, 2008.

Makogon, Y. F.; Holditch, S. A.; Makogon, T. Y. (2007). Natural gas-hydrates - A potential energy source for the 21st Century, *Journal of Petroleum Science and Engineering*, 56, 1-3, (2007) pp 14-31, ISSN 0920-4105

Milkov, A. V. (2004). Global estimates of hydrate-bound gas in marine sediments: how much is really out there? , *Earth-Science Reviews*, 66, 3-4, (2004) pp 183-197, ISSN 0012-8252

Moridis, G. J. (2003). Numerical studies of gas production from methane hydrates, *SPE Journal*, 8, 4, (2003) pp 359-370, ISSN 0036-1844

Moridis, G. J.; Collett, T. S.; Dallimore, S. R.; Satoh, T.; Hancock, S.; Weatherill, B. (2004a). Numerical studies of gas production from several CH4 hydrate zones at the Mallik site, Mackenzie Delta, Canada. *Journal of Petroleum Science and Engineering*, 43, 3-4, (2004) pp 219-238, ISSN 0920-4105

Moridis, G. J. (2004b). Numerical Studies of Gas Production from Class 2 and Class 3 Hydrate Accumulations at the Mallik Site, Mackenzie Delta, Canada, *SPE Reservoir Evaluation and Engineering*, 7, 3, (2004) pp 175-183, ISSN 1094-6470

Moridis, G. J.; Kowalsky, M. B.; Pruess, K. (2007). Depressurization-induced gas production from class 1 hydrate deposits, *SPE Reservoir Evaluation & Engineering*, 10, 5, (2007) pp 458-481, ISSN 1094-6470

Moridis, G. J.; Collett, T. S.; Boswell, R.; Kurihara, M.; Reagan, M. T.; Koh, C.; Sloan, E. D. (2009a). Toward Production From Gas Hydrates: Current Status, Assessment of Resources, and Simulation-Based Evaluation of Technology and Potential, *SPE Reservoir Evaluation & Engineering*, 12, 5, (2009) pp 745-771, ISSN 1094-6470

Moridis, G. J.; Reagan, M. T.; Kim, S. J.; Seol, Y.; Zhang, K. (2009). Evaluation of the Gas Production Potential of Marine Hydrate Deposits in the Ulleung Basin of the Korean East Sea, *SPE Journal*, 14, 4, (2009) pp 759-781, ISSN 0036-1844

Sira, J. H. Patil, S. L.; Kamath, V. A. (1990). Study of hydrate dissociation by methanol and glycol injection, *Proceedings - SPE Annual Technical Conference and Exhibition*, pp 977-901, Publ by Soc of Petroleum Engineers of AIME, Richardson, TX, USA, 1990.

Sloan, E. D.; Koh, C. A. (2008). *Clathrate Hydrates of Natural Gases*. 3nd ed ed.; Boca Raton, FLorida : Taylor and Francis, Inc.: 2008, ISBN 0-8493-9078

Sung, W. M.; Lee, H.; Lee, C. (2002). Numerical study for production performances of a methane hydrate reservoir stimulated by inhibitor injection, *Energy Sources*, 24, 6, (2002) pp 499-512, ISSN 0887-0624

Sung, W. M.; Lee, H.; Kim, S.; Kang, H. (2003). Experimental investigation of production behaviors of methane hydrate saturated in porous rock, *Energy Sources*, 25, 8, (2003) pp 845-856, ISSN 0887-0624

Tang, L.; Xiao, R.; Huang, C.; Feng, Z.; Fan, S. (2005a). Experimental investigation of production behavior of gas hydrate under thermal stimulation in unconsolidated sediment, *Energy & Fuels*, 19, 6, (2005) pp 2402-2407, ISSN 0887-0624

Tang, L. G.; Li, G.; Hao, Y. M.; Fan, S. S.; Feng, Z. P. (2005b). Effects of salt on the formation of gas hydrate in porous media, *The 5th International Conference on Gas Hydrate*, pp 155-160, Trondheim, Norway, 2005, ISBN 9781615670666

Tang, L.; Li, X.; Feng, Z.; Li, G.; Fan, S. (2007). Control mechanisms for gas hydrate production by depressurization in different scale hydrate reservoirs, *Energy & Fuels*, 21, 1, (2007) pp 227-233, ISSN 0887-0624

Yousif, M. H.; Li, P. M.; Selim, M. S.; Sloan, E. D. (1990). Depressurization of natural gas hydrates in Berea sandstone cores, *J.Inclusion Phenom&Mol Recognition Chem*, 8, (1990) pp 71-88, ISSN 0923-0750

Yousif, M. H.; Abass, H. H.; Selim, M. S.; Sloan, E. D. (1991a). Experimental and theoretical investigation of methane-gas-hydrate dissociation in porous media, *SPE Reservoir Engineering*, 6, 4, (1991) pp 69-76, ISSN 0885-9248

Yousif, M. H.; Sloan, E. D. (1991b). Experimental Investigation of Hydrate Formation and Dissociation in Consolidated Porous Media, *SPE Reservoir Engineering*, 11, (1991b) pp 452-458, ISSN 0885-9248

INDEX

9 789392 969614

Printed at Repro India Ltd.